WITHDRAWN
UTSA Libraries

# CONTINUOUS TRANSCUTANEOUS MONITORING

# ADVANCES IN EXPERIMENTAL MEDICINE AND BIOLOGY

Editorial Board:

NATHAN BACK, *State University of New York at Buffalo*

EPHRAIM KATCHALSKI-KATZIR, *The Weizmann Institute of Science*

DAVID KRITCHEVSKY, *Wistar Institute*

ABEL LAJTHA, *N. S. Kline Institute for Psychiatric Research*

RODOLFO PAOLETTI, *University of Milan*

---

Recent Volumes in this Series

Volume 214
THE NEW DIMENSIONS OF WARFARIN PROPHYLAXIS
Edited by Stanford Wessler, Carl G. Becker, and Yale Nemerson

Volume 215
OXYGEN TRANSPORT TO TISSUE IX
Edited by I. A. Silver and A. Silver

Volume 216 A
RECENT ADVANCES IN MUCOSAL IMMUNOLOGY, Part A: Cellular Interactions
Edited by Jiri Mestecky, Jerry R. McGhee, John Bienenstock, and Pearay L. Ogra

Volume 216 B
RECENT ADVANCES IN MUCOSAL IMMUNOLOGY, Part B: Effector Functions
Edited by Jerry R. McGhee, Jiri Mestecky, Pearay L. Ogra, and John Bienenstock

Volume 217
THE BIOLOGY OF TAURINE: Methods and Mechanisms
Edited by Ryan J. Huxtable, Flavia Franconi, and Alberto Giotti

Volume 218
CORONAVIRUSES
Edited by Michael M. C. Lai and Stephen A. Stohlman

Volume 219
REGULATION OF OVARIAN AND TESTICULAR FUNCTION
Edited by Virendra B. Mahesh, Dharam S. Dhindsa, Everett Anderson, and Satya P. Kalra

Volume 220
CONTINUOUS TRANSCUTANEOUS MONITORING
Edited by Albert Huch, Renate Huch, and Gösta Rooth

Volume 221
MOLECULAR MECHANISMS OF NEURONAL RESPONSIVENESS
Edited by Yigal H. Ehrlich, Robert H. Lenox, Elizabeth Kornecki, and William O. Berry

---

A Continuation Order Plan is available for this series. A continuation order will bring delivery of each new volume immediately upon publication. Volumes are billed only upon actual shipment. For further information please contact the publisher.

# CONTINUOUS TRANSCUTANEOUS MONITORING

Edited by

Albert Huch
Renate Huch

and

Gösta Rooth

Department of Obstetrics
University Hospital
Zurich, Switzerland

PLENUM PRESS • NEW YORK AND LONDON

Library of Congress Cataloging in Publication Data

International Symposium on Continuous Transcutaneous Monitoring (3rd: 1986: Zurich, Switzerland)

Continuous transcutaneous monitoring.

(Advances in experimental medicine and biology, v. 220)

Bibliography: p.

Includes index.

1. Transcutaneous blood gas monitoring—Congresses. I. Huch, Albert, date. II. Huch, Renate. III. Rooth, Gösta. IV. Title. V. Series.

RB45.2.I58 1986 616.07'561 87-15358

ISBN 0-306-42661-7

Proceedings of the Third International Symposium on Continuous Transcutaneous Monitoring, held October 1–4, 1986, in Zurich, Switzerland

© 1987 Plenum Press, New York

A Division of Plenum Publishing Corporation
233 Spring Street, New York, N.Y. 10013

All rights reserved

No part of this book may be reproduced, stored in a retrieval system, or transmitted in any form or by any means, electronic, mechanical, photocopying, microfilming, recording, or otherwise, without written permission from the Publisher

Printed in the United States of America

LIBRARY
The University of Texas
at San Antonio

To Dietrich W. Lübbers on the
occasion of his 70th birthday

## PREFACE

The international symposia on transcutaneous monitoring have dealt with the interaction between ideas and research, the introduction of unconventional techniques into clinical practice, and the joint efforts of researchers, clinicians, and industry to design and manufacture practical equipment for noninvasive monitoring.

The First International Symposium on Continuous Transcutaneous Blood Gas Monitoring took place in Marburg, West Germany, from May 31 to June 2, 1978. This was the first major international meeting exclusively devoted to transcutaneous blood gas monitoring, and it was attended by the scientists who had developed this technique or had been working with it, by a large number of doctors, mainly neonatologists who had just begun to use the technique or hoped to do so, and, finally, a rather large number of representatives of industry.

The second symposium, with the same title, was held in Zurich, Switzerland, October 14-16, 1981. This time the focus was, to a large extent, on transcutaneous $P_{CO_2}$ monitoring, for which equipment had become commercially available only a short time before. Fetal monitoring was also discussed at length, as was the use of the transcutaneous techniques in other fields, such as vascular surgery and experimental animal research.

The third symposium, October 1-4, 1986, was again held in Zurich. It was entitled "Continuous Transcutaneous Monitoring," indicating that not only blood gases but also other parameters could be monitored transcutaneously.

Pulse oximetry, which came rapidly into widespread use in the USA and is a technique which has now also become available in Europe, was this time the center of interest. A special session was devoted to comparisons between the transcutaneous $P_{O_2}$ technique and oxygen saturation monitoring with pulse oximeters. An extensive discussion was held on the advantages and disadvantages of these two, partly competitive, techniques.

In order to properly evaluate transcutaneous monitoring, a knowledge of skin circulation is of paramount importance. Therefore, a joint session was held with the Swiss Society of Microcirculation (A. Bollinger, Chairman).

The proceedings of the first two international symposia were reported in extenso in two books:

Continuous Transcutaneous Blood Gas Monitoring, A. Huch, R. Huch, and J.F. Lucey (eds), The National Foundation--March of Dimes, Birth Defects: Original Article Series XV: 4, Alan R. Liss, Inc, New York, 1979.

Continuous Transcutaneous Blood Gas Monitoring, R. Huch and A. Huch (eds). Reproductive Medicine 5, Marcel Dekker, Inc, New York and Basel, 1983.

These two books have become standard references for transcutaneous research and clinical use. We hope that the third book will be as useful as the previous ones.

The present publication contains the most important contributions not published previously. Abstracts of the papers not included in this book can be obtained from the editors.

Albert Huch
Renate Huch
Gösta Rooth

Zurich
March, 1987

CONTENTS

## REVIEWS OF MAIN THEMES

History, Status and Future of Pulse Oximetry ........................... 3
J.W. Severinghaus

Possibilities and Limitations of the Transcutaneous Measuring Technique. A Theoretical Analysis ........................ 9
D.W. Lübbers

Growth and Development of Transcutaneous Monitoring in the U.S.A. - 1978-1986 ........................................................ 19
J.F. Lucey

Transcutaneous $Po_2$ and $Pco_2$ Monitoring at 37°C. Cutaneous $Po_2$ and $Pco_2$ .............................................. 23
G. Rooth, U. Ewald, and F. Caligara

## PERFORMANCE OF TRANSCUTANEOUS ELECTRODES

The Measurement of $tcPo_2$ and $tcPco_2$ in Newborn Infants at 44°C, 42°C and 37°C after Initial Heating to 44°C ..................... 35
B. Friis-Hansen, P. Voldsgaard, J. Witt, K.G. Pedersen, and P.S. Frederiksen

Drift in Vivo of Transcutaneous Dual Electrodes ...................... 41
C. Lanigan, J. Ponte, and J. Moxham

Evaluation of Single Sensor Transcutaneous Measurement of $Po_2$ and $Pco_2$ in the Neonate .................................. 45
J. Messer, A. Livolsi, and D. Willard

Current Correction Factors Inadequately Predict the Relationship between Transcutaneous (tc) and Arterial $Pco_2$ in Sick Neonates ................................................................... 51
R.J. Martin, A. Beoglos, M.J. Miller, J.M. DiFiore, and W.A. Carlo

Computing the Oxygen Status of the Blood from Heated Skin $Po_2$ ......... 55
I.H. Gøthgen and E. Jacobsen

## APPLICATION IN ADULTS

Transcutaneous Monitoring of $Po_2$ and $Pco_2$ during Running - A Non-Invasive Determination of Gas Transport .................. 61
J.M. Steinacker and K. Röcker

Transcutaneous Oxygen Tension during Exercise in Patients with Pulmonary Emphysema ........ 67
D.C.S. Hutchison, B.J. Gray, J.M. Callaghan, and R.W. Heaton

The Effect of Inhaled Bronchoconstrictors on Transcutaneous Gas Tensions in Normal Adult Subjects ........ 71
B.J. Gray and N. Barnes

In Vivo Calibration of a Transcutaneous Oxygen Electrode in Adult Patients ........ 75
B.J. Gray, R.W. Heaton, A. Henderson, and D.C.S. Hutchison

Inflammation and Transcutaneous Measurement of Oxygen Pressure in Dermatology ........ 79
A. Ott

Diagnostic Assessment of Diabetic Microangiopathy by $tcPo_2$ Stimulation Tests ........ 83
N. Weindorf, U. Schultz-Ehrenburg, and P. Altmeyer

APPLICATION IN PERINATOLOGY

Transcutaneous Blood Gases and Sleep Apnea Profile in Healthy Preterm Infants during Early Infancy ........ 89
K.H.P. Bentele, U. Ancker, and M. Albani

Transcutaneous Monitoring as Trigger for Therapy of Hypoxemia during Sleep ........ 95
M.E. Schaefke, T. Schaefer, H. Kronberg, G.J. Ullrich, and J. Hopmeier

Effectiveness of Combined Transcutaneous $Po_2/Pco_2$ Monitoring of Newborns ........ 101
H. Schachinger and C. Kolz

Incidence and Severity of Retinopathy in Low Birth Weight Infants Monitored by $tcPo_2$ ........ 105
I. Yamanouchi, I. Igarashi, and E. Ouchi

Transcutaneous Oxygen Monitoring and Retinopathy of Prematurity ........ 109
E. Bancalari, J. Flynn, R.N. Goldberg, R. Bawol, J. Cassady, J. Schiffman, W. Feuer, J. Roberts, D. Gillings, and E. Sim

Combined Transcutaneous Oxygen, Carbon Dioxide Tensions and End-expired $CO_2$ Levels in Severely Ill Newborns ........ 115
W.B. Geven, E. Nagler, Th. de Boo, and W. Lemmens

Transcutaneous $Po_2$ and $Pco_2$ during Surfactant Therapy in Newborn Infants with Idiopathic Respiratory Distress Syndrome ........ 121
S. Bambang Oetomo, B. Robertson, and A. Okken

Continuous Fetal Acid-Base Assessment during Labour by $tcPco_2$ Monitoring ........ 123
C. Nickelsen and T. Weber

Microvascular Dynamics during Acute Asphyxia in Chronically Prepared Fetal Sheep near Term ....127
A. Jensen, U. Lang, and W. Künzel

PULSE OXIMETRY

Pulse Oximetry: Physical Principles, Technical Realization and Present Limitations ....135
M.R. Neuman

Pulse Oximetry - an Alternative to Transcutaneous $Po_2$ in Sick Newborns ....145
J.L. Peabody, M.S. Jennis, and J.R. Emery

Application of the Ohmeda Biox 3700 Pulse Oximeter to Neonatal Oxygen Monitoring ....151
W.W. Hay Jr., J. Brockway, and M. Eyzaguirre

Pulse Oximetry and Transcutaneous Oxygen Tension for Detection of Hypoxemia in Critically Ill Infants and Children ....159
S. Fanconi

Is Pulse Oximetry Reliable in Detecting Hyperoxemia in the Neonate? ....165
P, Baeckert, H.U. Bucher, F. Fallenstein, S. Fanconi, R. Huch, and G. Duc

Comparison between Transcutaneous $Po_2$ and Pulse Oximetry for Monitoring $O_2$-Treatment in Newborns ....171
E. Bossi, B. Meister, and J. Pfenninger

The Accuracy of the Pulse Oximeter in Neonates ....177
A. Hodgson, J. Horbar, G. Sharp, R. Soll, and J. Lucey

Pulse Oximetry and Transcutaneous Oxygen Tension in Hypoxemic Neonates and Infants with Bronchopulmonary Dysplasia ....181
H.N. Lafeber, W.P.F. Fetter, A.R. v.d. Wiel, and T.C. Jansen

Control of Pedicle and Microvascular Tissue Transfer by Photometric Reflection Oximetry ....187
H.P. Keller and D.W. Lübbers

Comparison of in Vivo Response Times between Pulse Oximetry and Transcutaneous $Po_2$ Monitoring ....191
F. Fallenstein, P. Baeckert, and R. Huch

Errors in the Measurement of CO-Hemoglobin in Fetal Blood by Automated Multicomponent Analysis ....195
P. Tuchschmid

A Simple Method for HbF Analysis ....201
U. v. Mandach, P. Tuchschmid, A. Huch, and R. Huch

## SKIN BLOOD FLOW

Capillary Blood Pressure ....................................................209
J.E. Tooke and S.A. Williams

Microvasculatory Evaluation of Vasospastic Syndromes ..................215
H. Saner, H. Würbel, F. Mahler, J. Flammer, and P. Gasser

Infrared Fluorescence Videomicroscopy with Indocyanine Green (Cardiogreen ®)....................................................219
M. Brülisauer, G. Moneta, K. Jäger, and A. Bollinger

Measurement of Tissue Blood Flow by High Frequency Doppler Ultrasound ....................................................223
S. Basler, A. Vieli, and M. Anliker

Laser-Doppler Probes for the Evaluation of Arterial Ischemia ..........227
H. Seifert, K. Jäger, and A. Bollinger

Skin Reactive Hyperemia Recorded by a Combined $tcPo_2$ and Laser Doppler Sensor ....................................................231
U. Ewald, A. Huch, R. Huch, and G. Rooth

Comparison of Laser-Doppler-Flux and $tcPo_2$ in Healthy Probands and Patients with Arterial Ischemia.............................235
L. Caspary, A. Creutzig, and K. Alexander

The Use of the Hellige Oxymonitor to Study Skin Blood Flow Changes ....................................................241
P.M. Gaylarde and I. Sarkany

## CALCULATION OF SKIN BLOOD FLOW

Estimated Peripheral Blood FLow in Premature Newborn Infants ..........249
W.A. van Asselt, J.J. Geerdink, G. Simbruner, and A. Okken

Skin Blood Flow Calculations from Transcutaneous Gas Pressure Measurements ....................................................253
F. Caligara, G. Rooth, and U. Ewald

The Percentual Initial Slope Index of $tcPo_2$ as a Measure of the Peripheral Circulation and its Measurements by different $tcPo_2$ Electrodes ....................................................259
R. Lemke and D.W. Lübbers

Examinations on the Blood Flow Dependence of $tcPo_2$ Using the Model of the "Circulatory Hyperbola" ...........................263
J.M. Steinacker, W. Spittelmeister, and R. Wodick

Estimation of the Determinants of Transcutaneous Oxygen Tension Using a Dynamic Computer Model ....................................269
A. Talbot-Pedersen, M.R. Neuman, G.M. Saidel, and E. Jacobsen

TECHNICAL ASPECTS OF CONTINUOUS MONITORING

Quality and Safety Aspects in the Development and Fabrication of Transcutaneous Sensors ........ 277
E. Konecny and U. Hölscher

A New System for $tcPo_2$ Long-Term Monitoring Using a Two-Electrode Sensor with Alternating Heating ........ 285
F. Fallenstein, P. Ringer, R. Huch, and A. Huch

A Novel Approach for an ECG Electrode Integrated into a Transcutaneous Sensor ........ 291
U. Hölscher

Microelectronic Sensors for Simultaneous Measurement of $Po_2$ and pH ........ 295
C.C. Liu, M.R. Neuman, L.T. Romankiw, and E.B. Makovos

A Modified Electrode Ring for Use in Transcutaneous Measurement of $Po_2$ ........ 299
H. Schachinger and Th. Lauhoff

Multichannel Recording and Analysis of Physiological Data Using a Personal Computer ........ 305
A.R. van der Wiel, T.C. Jansen, H.N. Lafeber, and W.P.F. Fetter

Continuous Non-Invasive Beat-by-Beat Blood Pressure (B.P.) Measurement in the Newborn ........ 311
P.Rolfe, P.P. Kanjilal, C. Murphy, and P.J. Burton

The Relative Accuracy of Three Transcutaneous Dual Electrodes at 45° in Adults ........ 315
C. Lanigan, J. Ponté, and J. Moxham

Contributors ........ 321

Index ........ 325

# REVIEWS OF MAIN THEMES

# HISTORY, STATUS AND FUTURE OF PULSE OXIMETRY

John W. Severinghaus

Department of Anesthesia, University of California
San Francisco, CA 94143 USA

## Introduction

Within the past 4 years, pulse oximetry has become widely used in anesthesia and many associated critical care situations. At least 15 manufacturers are now marketing these instruments. They have had a major negative effect on the sales of transcutaneous oxygen electrodes. It is thus appropriate to discuss oximetry in this 3rd International Symposium on Transcutaneous Blood Gas Monitoring. Poul Astrup and I have recently published a book and papers on this history [1-2] and Petersen briefly reviewed pulse oximetry history [3].

## Origin of pulse oximetry

Oximetry, the measurement of hemoglobin oxygen saturation in either blood or tissue, depends upon changes of the optical density of hemoglobin with changes of its oxygen saturation, particularly in the red region. Millikan [4] coined the term oximeter in connection with his light weight ear hemoglobin saturation meter during World War II to train pilots and control $O_2$ delivery to them. In order to separate the absorption due to blood, the wanted signal, from that due to tissue, several British workers had tried the idea of compressing the tissue to expell its blood at the beginning of a test. Wood and Geraci [5] used this idea, adding a pneumatic cuff to Millikan's ear oximeter to obtain a bloodless zero, which in optimal conditions permitted an absolute reading oximeter. Brinkman and Zijlstra [6] in Groningen showed that red light reflected from the forehead could measure saturation. Zijlstra initiated cuvette and catheter reflection oximetry. Using multiple wavelengths, blood COHb and MetHb were also measured in cuvette oximeters, and Shaw devised the Hewlett Packard 8 wavelength pre-calibrated ear oximeter. Oximeters were applied widely in pulmonary medicine and physiology, but not in anesthesiology or critical care as monitors, and their use regressed in the 20 years after the introduction of Clark's $O_2$ electrode.

The idea of using pulsatile light variation to measure arterial oxygen saturation was conceived by the Japanese physiological bioengineer Takuo Aoyagi. In the early 1970's, Aoyagi was working at the Nihon Kohden Corporation, one of Japan's leading medical electronics firms. His research concerned using light transmission through the ear to measure cardiac output by dye dilution. The transmitted light signal had been noted to contain pulsatile variations which were considered undesirable noise in the dye curve. Aoyagi conceived the idea of measuring only the pulsatile changes in light transmission through living tissues to compute the arterial saturation, realizing that these changes of light transmission at all wavelengths would be due solely to pulsatile alteration of the intervening arterial blood volume. Thus the variable absorption of light by tissue, bone, skin, pigments etc. would be eliminated from analysis. It was this key idea which permitted development of oximeters which required no calibration after initial factory setting, since all human blood has essentially identical optical characteristics in the red and infra red bands chosen for oximetry.

In January, 1974 an abstract in Japanese entitled "Improvement of the earpiece oximeter" describing the invention was submitted to the Japanese Society of Medical Electronics and Biological Engineering, by Aoyagi and associates [7]. The first prototype pulse oximeter was made by Aoyagi between Sept., 1973 and March 1974. That instrument was first used by Nakajima and his associates at the Sapporo Minami National Sanatorium.

Because of the difficulty of finding and reading original Japanese bioengineering meeting abstracts, the origin of pulse oximetry has been incorrectly ascribed variously to surgeons Nakajima et al in 1975 [8], or to anesthesiologists Yoshiya et al in 1980 [9] and attributed to the wrong company as well.

Several groups began developmental work using Aoyagi's idea in the late 1970's. Initially the main problem was extreme sensitivity to motion. The Minolta-Mochida Oximet MET 1471 instrument was tested clinically by several Japanese groups [8-11] and by one American group [12] and found to provide reasonably accurate continuous convenient non-invasive monitoring of oxygen saturation. However, in most reports it overestimated saturation, averaging a reading of 70% at 50% actual saturation.

However, limited interest was generated by the development of pulse oximetry during the first 8-10 years. Almost no one foresaw its value in anesthesia, intensive care and other emergent situations, probably because conventional oximetry had never been convenient enough for these uses. The Hewlett Packard ear oximeter had been used almost entirely in pulmonary function laboratories and other physiologic environments. Indeed, the initial work on pulse oximetry in the USA by several firms (Minolta, Corning, Biox) considered such laboratories to be the probable market.

Credit for the present enormous interest belongs to anesthesiologist William New of Stanford University Medical School who with an engineer, Jack Lloyd, founded the Nellcor Company. New recognised the potential importance and market for a convenient accurate oximeter in the operating room, and all the other hospital and clinic sites where patients are sedated, anesthetized, unconscious, comatose, paralyzed or in some way limited in their ability to regulate their own oxygen supply. The Nellcor pulse oximeter was evaluated by Yelderman and New, the manufacturers, in 1983 [13].

In 1984, Shimada et al [14] showed that overestimation of saturation noted with the Minolta pulse oximeter was due to multiple internal scattering, and that the theoretic equations did not apply. All the current devices use empiric algorithms to make the reported saturation fit a set of data.

## Current status of pulse oximetry

Major advances in pulse oximetry have been reduction in size of the light source and detectors, use of analog to digital converters, the incorporation of digital computer chips with memory and with logic algorithms to reduce artifacts and compute saturation from empiric relationships [13-19].

Pulse oximeters may work on ear, bridge of nose, nasal septum, finger, or temple over the temporal artery. In infants the foot or palm and have now been found to work well.

Pulse oximeters estimate $SaO_2$ to within $\pm 5\%$ at the top of the curve (2 s.d.), but somewhat less accurately at low saturations. Their response time is typically 6 seconds plus transit time of the blood to the finger (about 20 sec) or ear (about 5 sec). Only the Nellcor and Biox (Ohmeda) instruments have been extensively tested by others than the manufacturers.

The interest can be characterized as greater than for any new instrument ever introduced for monitoring of patients in anesthesia and associated critical care situations. By the autumn of 1986, at least 15 manufacturers were known to be developing or marketing pulse oximeters.

## Methodology of pulse oximetry

Intense and efficient light emitting diodes with wavelengths of 660 (red) and 940 nm (infra red) are pulsed alternatively at high frequency, e.g. 480 Hz, with an off-period used to correct for ambient light. The single photo-diode detects both wavelengths, generating signals proportional to the transmission of light at these two wavelengths, which are separately analyzed by the known time of pulsing each source. The light intensity of both red and infrared varies at the much slower pulse frequency, and the instrument separates the slowly varying signal from the steady light transmission through the tissue. The ratio of the red to infra-red pulsatile signals at 0.5-3 Hz is used to compute saturation. Originally the logarithm of this ratio was thought to be a linear function of saturation, but inaccuracies due to scattering have required empiric solutions and inbuilt calibration algorithms. In some instruments only the peak and trough of the pulse-added signal is used to compute saturation, while in others, the changes of signal strength are measured 30 times per second and averaged 2-3 times per second, and then averaged again over the last 3-6 seconds to produce saturation data.

## Pulse oximetry versus transcutaneous $Po_2$ monitoring

It is appropriate to compare pulse oximetry with transcutaneous $Po_2$ determination at this conference. At high $Po_2$, oximetry only shows 100% saturation, where $PsO_2$ indicates changes in $PaO_2$. At normal levels, oximetry has a standard deviation of the order of 2.5% saturation, and skin $Po_2$ uncertainty is about $\pm 15$ mmHg (comparable changes at about 95% saturation). At lower saturation and $Po_2$ both are reasonably accurate estimators of arterial blood, until low $Po_2$, when the skin $Po_2$ falls too low, reaching zero at about 30 mmHg $PaO_2$.

Pulse oximetry has the following advantages:

1. No calibration is ever needed.
2. Accuracy and calibration do not drift.
3. No heating of skin is used.

4. No membrane changes are needed.
5. The probes are inherently inexpensive, and may be disposable (Nellcor).
6. The probes are not fragile.
7. The initial time from connection to the patient to obtaining a reading is about 15 seconds rather than 5-15 minutes.
8. The response time is less.
9. A pulse waveform is usually displayed or indicated, and heart rate is computed and displayed.
10. Hypotension does not induce a fall in the $SpO_2$ reading compared to $SaO_2$.

Transcutaneous blood gas analysis has the following advantages:

1. The combined electrodes measure $Pco_2$ as well as $Po_2$.
2. A falling $Po_2$ at a high level may signal undesirable changes long before arterial saturation begins to fall.
3. Pulse oximetry is somewhat more difficult in heavily pigmented individuals.
4. Pulse oximeters sometimes cease indicating when saturation falls, or when pulse amplitude is weak, or absent due to vasoconstriction, shock or hypotension, of for no obvious reasons, perhaps at critical times. Transcutaneous electrodes may be more dependable in such circumstances.
5. Transcutaneous $Po_2$ might be more successful at avoiding hyperoxia if $SaO_2$ were to be kept at the 95-97% level as we next discuss.

**Pulse oximetry in premature infants**

Manufacturers are only beginning to consider probe design for infants, and choice of a site may be difficult. Transillumination of the palm seems satisfactory.

The decision between pulse oximetry and transcutaneous $Po_2$ methods depends heavily upon the set-point of saturation (and $Po_2$) chosen by the neonatologists. If one wished to keep the premature infant at 95% saturation, there would be some danger in a few infants because of the 5% range of 95% confidence limits. If 90% is the chosen set point, even this range may be considered to provide safe limits of 85-95% $SaO_2$.

There is no concensus among physicians or physiologists whether $PaO_2$ or $SaO_2$ is the more important for proper delivery of $O_2$ to tissue, but there is agreement that high $Po_2$ inhibits retinal vascular development. The decision needed is whether a deliberately low $PaO_2$ and $SaO_2$ (by adult standards) is optimal for premature infants.

**Future directions**

Pulse oximetry has initiated a re-education of physicians toward thinking of oxygen in terms of saturation and content rather than $Po_2$, reversing the change that occurred after Clark's invention of the oxygen electrode tended to displace the ear oximeter. The lack of drift and dependability of pulse oximetry may make possible closed loop servo control of the oxygen saturation of premature infants, greatly reducing the tendency to over and under shoot when personnel over respond to transients and then are diverted to other tasks.

Pulse oximetry is arguably the most significant technological advance ever made in monitoring the well being and safety of patients during anesthesia, recovery and critical care. Its use in patients who are unconscious or unable to maintain adequate ventilation and gas exchange without assistance may become mandatory, with insurance companies providing the motivation.

**References:**


1. Astrup P, Severinghaus JW. The history of blood gases, acids and bases. Copenhagen: Munksgaard, 1986:332pp

2. Severinghaus J, Astrup P: History of blood gas analysis. VI. Oximetry. J Clin Monitoring 1986;2:270-288

3. Petersen J: The development of pulse oximetry. Science 1986;232:G135-136

4. Millikan GA: The Oximeter, an instrument for measuring continuously oxygen saturation of arterial blood in man. Rev Scient Instr 1942;13:434-444

5. Wood E, Geraci JE: Photoelectric determination of arterial oxygen saturation in man. J Lab Clin Med 1949;34:387-401

6. Zijlstra WG. A manual of reflection oximetry. Van Gorcum, Ed. Assen, Nederland, 1958.

7. Aoyagi T, Kishi M, Yamaguchi K, Watanabe S: Improvement of an earpiece oximeter. Abstracts of the 13th annual meeting of the Japanese Society for Medical Electronics and Biological Engineering, 1974, 90-91 (Jap)

8. Nakajima S, Hirai Y, Takase H, Kuse A, Aoyagi S, Kishe M, Yamaguchi K. Performances of new pulse wave earpiece oximeter. Respir Circ 1975;23: 41-45

9. Yoshiya I, Shimada Y, Tanaka K. Spectrophotometric monitoring of arterial oxygen saturation in the fingertip. Med Biol Eng Computing 1980;18: 27-32

10. Asari M, Kemmotsu O. Application of the pulse wave ear oximeter in anesthesiology. Jap J Anesthesiol 1976;26: 205-207(in Japanese)

11. Suzukawa M, Fujisawa M, Matsushita F, Suwa K, Yamamura H. Clinical Use of Pulse-type Fingeroximeter in Anesthesia. Jap. J Anesthesiol 1978; 27:600-605. (transl Mochida Pharmaceutical Co Ltd, Tokyo)

12. Sarnquist F, Todd C, Whitcher C. Accuracy of a new non-invasive oxygen saturation monitor. Anesthesiology 1980;53,S163

13. Yelderman M, New W, Jr. Evaluation of pulse oximetry. Anesthesiol 1983;59: 349-352

14. Shimada Y, Yoshiya I, Oka N, Hamaguri K. Effects of multiple scattering and peripheral circulation on arterial oxygen saturation measured with a pulse-type oximeter. Med Biol Eng Comput 1984;22:475-8

15. Shippy MB, Petterson MT, Whitman RA, Shivers CR. A clinical evaluation of the BTI Biox II ear oximeter. Resp Care 1984;29: 730-735

16. Tyler IL, Tantisira B, Winter PM, Motoyama EK. Continuous monitoring of arterial oxygen saturation with pulse oximetry during transfer to the recovery room. Anesth Analg 1985;64:1108-12

17. Fanconi S, Doherty P, Edmonds JF, Barker GA, Bohn DJ. Pulse oximetry in pediatric intensive care: comparison with measured saturations and transcutaneous oxygen tension. J Pediatr 1985;107:362-6

18. Mihm FG, Halperin BD. Noninvasive detection of profound arterial desaturations using a pulse oximetry device. Anesthesiology 1985;62:85-7

19. Chapman KR, Liu FLW, Watson RM, Rebuck AS. Range of accuracy of two wavelength oximetry. Chest 1986;89:540-542

POSSIBILITIES AND LIMITATIONS OF THE TRANSCUTANEOUS MEASURING TECHNIQUE

A THEORETICAL ANALYSIS

D.W. Lübbers

Max-Planck-Institut für Systemphysiologie

4600 Dortmund, FRG

It is well known that oxygen and carbon dioxide can pass the human epidermis and that this transport occurs by diffusion[1]. Therefore, it is possible to measure noninvasively $Po_2$ and $Pco_2$ values on the skin surface. The first direct $Po_2$ measurements on the skin surface by Evans and Naylor (1967)[2], however, were disappointing because the $Po_2$ value was very close to zero. First $Pco_2$ measurements by Huch et al. (1967)[3] revealed rather large $Pco_2$ values. However, a short report of Baumberger and Goodfriend (1951)[4], in which they demonstrated that the $Po_2$ of a phosphate buffer heated up to 45°C and brought in contact with the skin of a finger approached arterial $Po_2$, was a challenge to study the different parameters which influence skin surface $Po_2$ and $Pco_2$ in order to find out the kind of information which can be obtained from such noninvasive transcutaneous $Po_2$ and $Pco_2$ measurements.

Fig. 1 (left side) shows a schematic drawing of the upper part of the skin. The epidermis (E) is supplied by capillary loops (str. papillare), which receive their blood from arterial plexus (black) and the venous limb of which leads to subepidermal venous plexus. The oxygen exchange between the blood in the upper part of the capillary loop, the so-called capillary dome, and the skin surface occurs by diffusion and is determined by the $Po_2$ difference between blood and skin surface, the diffusion properties, and the $O_2$ consumption of the tissue. The rather regular structure of the vasculature and the skin allows to construct a microcirculatory unit (Fig. 1, left, mu), by which the $O_2$ supply of the skin can be simulated. The microcirculatory unit (Fig. 1 , right side) consists of three layers. 1) ed: dead part of the epidermis (str. corneum and part of the str. granulosum). 2) ev: viable part of the epidermis (part of the str. granulosum, str. spinosum, and str. germanitivum). The dimensions are taken from values published in the literature. The tissue block has a side length of 140 $\mu$m and a height of 240 $\mu$m. The capillary loop is situated in the diagonal. The length of the arterial (ar) limb is 190 $\mu$m, that of the capillary dome (cd) 30 $\mu$m. The epidermis consists of a 25 $\mu$m thick viable (ev) and a 15 $\mu$m thick dead (ed) layer. Blood is simulated by a homogeneous, hemoglobin-containing plasma solution. For describing the $O_2$ dissociation of hemoglobin the ADAIR equation is used[5,6]. Simpler skin models according to the diffusion model of Krogh have been published by Eberhard (1976)[7], Quinn (1978)[8], Lübbers (1979)[9], Thunstrom et al. (1979)[10].

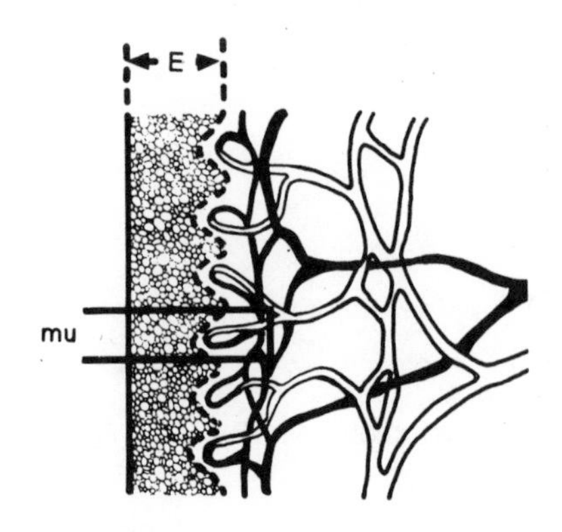

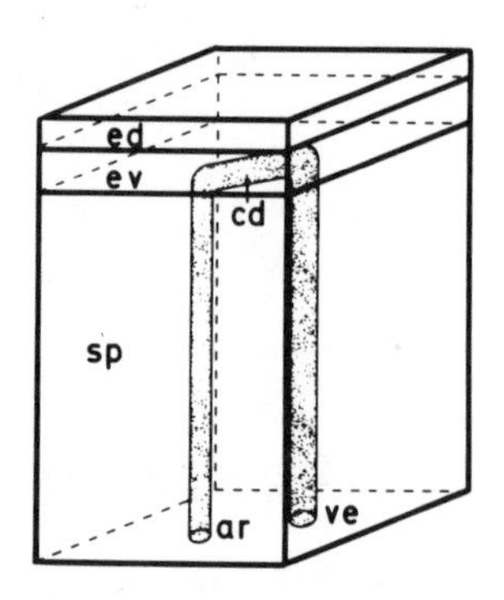


Fig. 1. Schematic drawing of the blood supply of the epidermis. Left side: epidermis (E), black: arterial vessels, mu: microcirculatory unit. Right side: capillary loop model. ed: dead , and ev: viable epidermis, ar: arterial, ven: venous capillary limb, cd: capillary dome, sp: stratum spinosum.

The anatomy of the adult skin demonstrates clearly that because of the perpendicular capillary loops there is no arterial blood in contact with the epidermis. Such a contact exists in newborns, since they have different capillary patterns, which slowly transform to the adult ones[11]. Therefore the question arises whether conditions can be found at which the blood in the capillary dome becomes sufficiently arterialized so that these values can be taken as representative for the arterial blood. It can be seen from a simple balance equation that this depends mainly on blood flow and $O_2$ content of arterial blood. Under steady state conditions the amount of oxygen delivery to the tissue (i.e. the difference between arterial and venous oxygen concentration, $(C_{a,O_2} - C_{v,O_2})$ times blood flow,($\dot{B}$) equals the tissue oxygen consumption, $\dot{V}_{O_2}$.

$$(C_{a,O_2} - C_{v,O_2}) \cdot \dot{B} = \dot{V}_{O_2} \quad (1)$$

or

$$C_{a,O_2} - C_{v,O_2} = \frac{\dot{V}_{O_2}}{\dot{B}} \quad (2)$$

$$C_{v,O_2} = C_{a,O_2} - \frac{\dot{V}_{O_2}}{\dot{B}} \quad (3)$$

$AVD_{O_2}$ (eq. 2) and $C_{v,O_2}$ (eq. 3) depend in a nonlinear hyperbolic way on the flow. From this hyperbolic relationship follows that dependent on the magnitude of the flow, similar flow changes have different effects on the $C_{v,O_2}$. Similar equations hold for $CO_2$, but with the opposite sign, since $CO_2$ is produced within the tissue.

$$C_{v,CO_2} = C_{a,CO_2} + \frac{\dot{V}_{CO_2}}{\dot{B}} \quad (4)$$

To investigate the relationship between arterial and venous concentration eq. (3) can be arranged in the following way:

$$C_{v,O_2} = C_{a,O_2}\left(1 - \frac{\dot{V}_{O_2}}{C_{a,O_2}\cdot\dot{B}}\right) = f_v\cdot C_{a,O_2} \tag{5}$$

$$\frac{C_{v,O_2}}{C_{aO_2}} = \left(1 - \frac{\dot{V}O_2}{C_{aO_2}\cdot\dot{B}}\right) = f_v \tag{6}$$

Since local peripheral blood flow is given by the product "flow conductivity, c, times effective blood pressure, $P_{eff}$,"

$$\dot{B} = c\cdot P_{eff} \tag{7}$$

we obtain with eqs. (6. and 7.)

$$\frac{C_{v,O_2}}{C_{a,O_2}} = \left(1 - \frac{\dot{V}_{O_2}}{C_{a,O_2}\cdot c\cdot P_{eff}}\right) = f_v \tag{8}$$

As eq. (5) shows, the venous value can be obtained by multiplying arterial $O_2$ concentration by factor $f_v$. Factor $f_v$ describes the effectiveness of the supply ($C_{a,O_2}\cdot\dot{B}$) in relation to consumption ($\dot{V}_{O_2}$). With increasing supply, $f_v$ approaches unity. Eqs. (6) and (7) show that in the oxygen offer, changes of $C_{a,O_2}$ and $\dot{B}$ are equivalent. It is important to realize that because of the hyperbolic course, three different flow regions exist in which changes of oxygen offer have different effects on $f_v$: At normal blood flow (region 1, $f_v < 0.7$) small changes of the oxygen offer cause large changes of $C_{v,O_2}$. At moderate hyperemia (region 2, $f_v > 0.7 < 0.9$) the effect becomes smaller, whereas at excessive hyperemia (region 3, $f_v > 0.9$) the effect becomes negligible.

Since our balance equations not only hold for the venous outflow, but also for other parts of the capillary, the equations describe also in which way oxygen consumption and oxygen offer influence the $O_2$ concentration in the capillary dome. To approach arterial blood values, one should increase the flow in such a way that the denominator in eq. (6) becomes large, compared to $\dot{V}_{O_2}$ (region 3). If one is interested in the behaviour of flow or peripheral blood pressure, one should decrease the denominator and preferably work in region 2. In region 1 it is difficult to draw any conclusions about the $O_2$ offer, since the conditions are rather unstable[9,12,13].

Since in the circulatory hyperbola the quotient $\dot{V}_{O_2}/\dot{B}$ is independent of the arterial oxygen content, there is a linear relationship between the $O_2$ content of the capillary dome and of the artery. For our measurements, however, the $O_2$ has to be transported from the capillary dome to the skin surface by diffusion, and for this transport the $O_2$ pressure and not the $O_2$ content is of importance. Because of the $O_2$ dissociation curve of hemoglobin, the $O_2$ content is related in a nonlinear way to the $O_2$ pressure. Therefore, inspite of constant $O_2$ consumption and constant flow, the skin surface $Po_2$ depends in a nonlinear way on the arterial $Po_2$. The nonlinearity is additionally enlarged by diffusional shunts between the adjacent arterial and venous limb of the capillary.

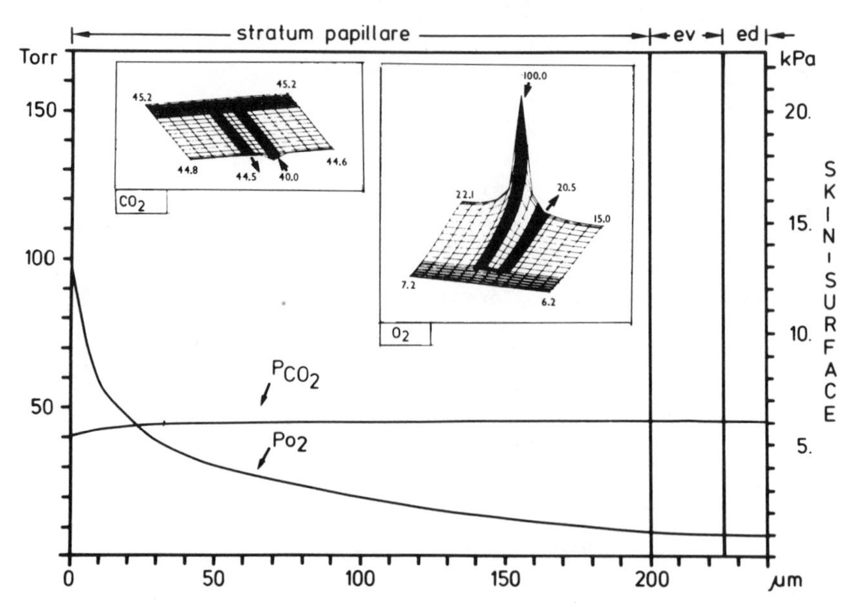


Fig. 2. $Po_2$ and $Pco_2$ profiles along the arterial capillary limb and across the epidermis. Inserts: the corresponding $Po_2$ and $Pco_2$ fields. Blood flow: 0.01 ml/(g·min), $O_2$ consumption: 0.003 ml/(g·min), hemoglobin concentration: 16 g/dl, temperature: 37°C, ev, ed: viable and dead part of the epidermis, skin surface impermeably covered[14].

We have calculated the $Po_2$ and $Pco_2$ distribution within the microcirculatory unit for normal skin, i.e. $P_{a,O_2}$ = 100 Torr, $P_{a,CO_2}$ = 40.0 Torr, pH = 7,4, Hb = 16 g/dl, $V_{O_2}$ = 0.003 ml $O_2$/(g·min), B = 0.01 ml/(g·min)[14]. Fig. 2 shows the $Po_2$ and $Pco_2$ profile along the arterial limb and across the epidermis, which is covered by a gas-impermeable layer. The $P_{a,O_2}$ decreases along the arterial limb first strongly, then levels off. The $Po_2$ decrease across the viable part of the epidermis is only small. There is no $Po_2$ gradient across the dead part of the epidermis. The $Po_2$ on the skin surface amounts to 7.0 Torr. Because of an $O_2$ shunt diffusion of 16 %, the venous $Po_2$ of 20.5 Torr is higher than the $Po_2$ in the capillary dome. The arterial $Pco_2$ increases from 40.0 Torr to 45.0 Torr at the skin surface. The venous $Pco_2$ amounts to 44.5 Torr and the shunt diffusion to 11 %. In the inserts the corresponding $Po_2$ and $Pco_2$ values of a cross section of the microcirculatory unit are shown as $Po_2$ and $Pco_2$ surfaces. It can be seen that there are distinct gradients between the arterial and venous capillary limb. The $Po_2$ and $Pco_2$ values at the skin surface are rather uniform. The gradients within the $Pco_2$ fields are much smaller than in the $Po_2$, because of the high $CO_2$ solubility (about 16:1).

If the skin is heated up to 43°C, oxygen and $CO_2$ dissociation curves are changed, producing a $P_{a,O_2}$ = 139 Torr and a $P_{a,CO_2}$ = 52.4 Torr. In Fig. 3 it is assumed that by heat-induced hyperemia, flow increased to 1.0 ml/(g·min). By the high flow the $Po_2$ decrease between arterial inflow and skin surface becomes smaller. The skin surface $Po_2$ amounts to 111 Torr, the venous $Po_2$ to 116.6 Torr. There is still $O_2$ shunt diffusion. The skin surface $Pco_2$ is 53.3 Torr and the venous $Pco_2$ 53.1. Whereas the $Pco_2$ measurement using a conventional $Pco_2$ electrode does not influence the $Pco_2$ field, the $Po_2$ field is changed by the $O_2$ consuming polarographic $Po_2$ measurement. This influence is shown in Fig. 4. The Pt wire is in direct contact with the skin surface and covers the microcirculatory unit. Since during the $Po_2$ measurement the $Po_2$ at the Pt surface is zero,

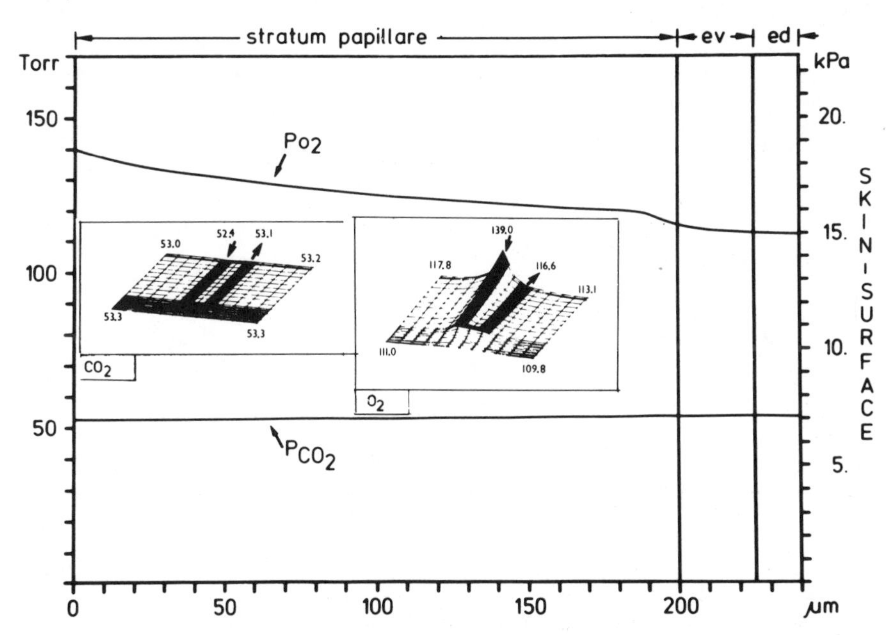


Fig. 3. $P_{O_2}$ and $P_{CO_2}$ profiles along the arterial capillary limb and across the epidermis. Inserts: the corresponding $P_{O_2}$ and $P_{CO_2}$ fields. Blood flow: 1.0 ml/(g·min), $O_2$ consumption: 0.004 ml/(g·min), hemoglobin concentration: 16 g/dl, blood temperature: 37°C, electrode temperature: 43°C, ev, ed: viable and dead part of the epidermis, skin impermeably covered.

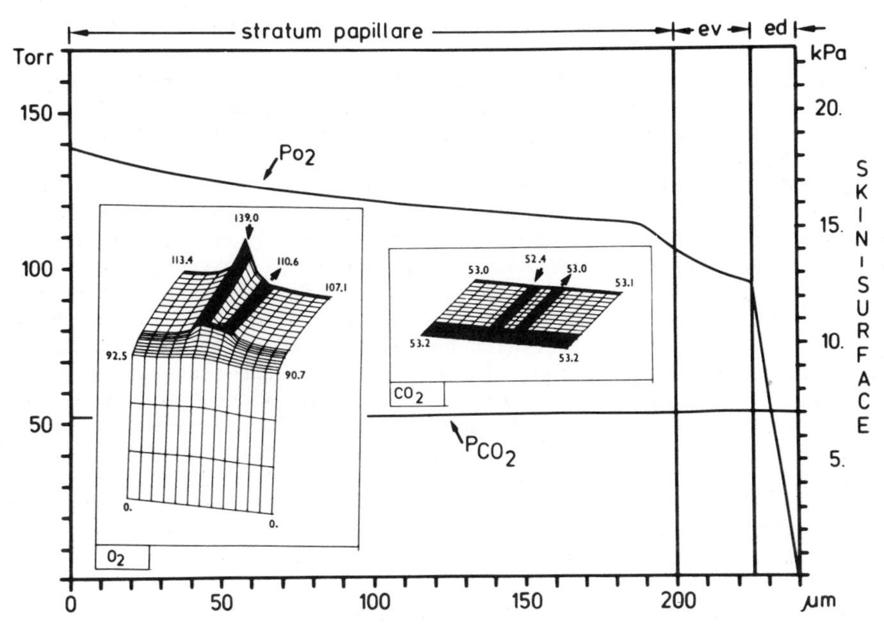


Fig. 4. $P_{O_2}$ and $P_{CO_2}$ profiles along the arterial capillary limb and across the epidermis. Same parameters as in Fig. 3, but with a bare platinum electrode on the skin surface.

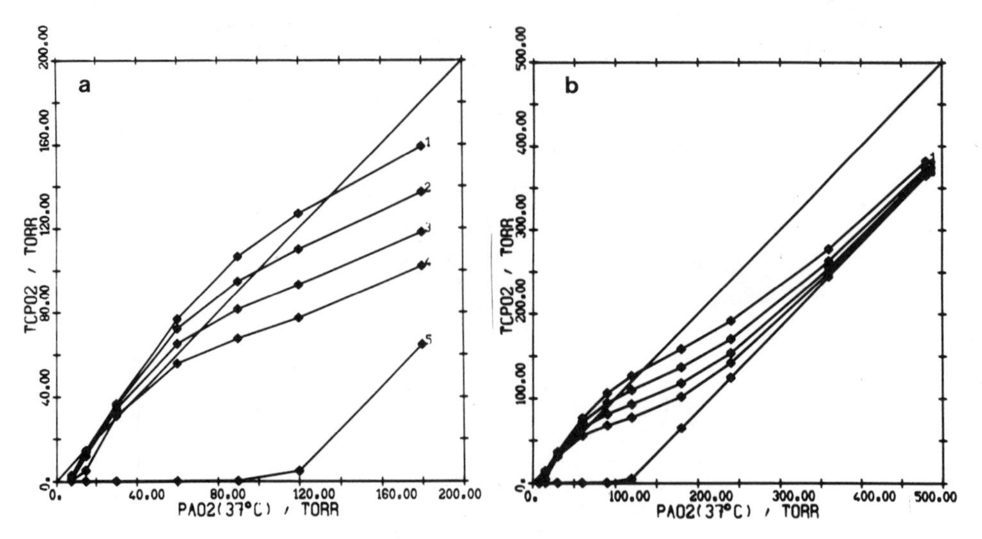


Fig. 5. Transcutaneous $Po_2$ "calibration curves" at different hemoglobin concentrations. $TcPo_2$ (43°C, ordinate) vs. $P_{a,O_2}$ (37°C, abscissa). Hemoglobin concentrations: 1) 16 g/dl, 2) 8 g/dl, 3) 4 g/dl, 4) 2 g/dl, 5) 0 g/dl. a) 0-200 Torr, b) 0-500 Torr. $O_2$ consumption: 0.004 ml/(g·min), blood flow: 1.0 ml/(g·min), skin impermeably covered.

the total $Po_2$ profile becomes much steeper. The insert demonstrates the strong influence of the $Po_2$ electrode. If we consider the dead epidermis as a membrane of the electrode, the figure shows that heating allows to obtain a $Po_2$ at the interface between membrane and skin which approximates arterial $Po_2$ at 37°C. With a normally membranized electrode the interface $Po_2$ would correspond to the $tcPo_2$ (membrane compensation[15]).

The model can also be used to calculate the relationship between arterial and transcutaneous $Po_2$ under different physiological conditions[16]. Fig. 5 demonstrates such a relationship, when hemoglobin is varied in 5 steps. 1) 16 g/dl, 2) 8 g/dl, 3) 4 g/dl, 4) 2 g/dl, 5) 0 g/dl. The flow is maintained at 1.0 ml/(g·min) and the $O_2$ consumption at 0.004 ml $O_2$/(g·min). The blood enters the microcirculatory unit with 37°C and is heated up to 43°C. Similar as in Fig. 4, the skin is covered by an impermeable layer; there is no $Po_2$ electrode. Without hemoglobin (Fig. 5,5) the $tcPo_2$ remains zero up to a $P_{a,O_2}$ of 90 Torr. From a $P_{a,O_2}$ of 120 Torr the "calibration curve" without hemoglobin runs in parallel to the identity line, but the $Po_2$ is about 115 Torr smaller. With hemoglobin the total curve becomes nonlinear. Dependent on the hemoglobin concentration, the temperature shift of the oxygen dissociation curve increases the $tcPo_2$ above the identity line, crosses this line and then slowly approaches the calibration curve without hemoglobin. $TcPo_2$ equals $P_{a,O_2}$ at a $Po_2$ of about 20 Torr and 135 Torr. Between a $P_{a,O_2}$ of 60 and 90 Torr, $tcPo_2$ is about 16 Torr larger, at a $P_{a,O_2}$ of 180 Torr, about 20 Torr smaller than the $P_{a,O_2}$. With smaller Hb concentrations, the increase above the identity line is smaller and also the range in which the $tcPo_2$ is larger than the $P_{a,O_2}$.

Table 1 shows the values when flow was varied with a constant hemoglobin content of 16 g/dl. As predicted from the circulatory hyperbola (see eq. (5)), the effect of flow changes are similar to the changes of hemoglobin concentration. Partly, values are higher, partly lower than

Table 1. Values, when flow was varied with a constant hemoglobin content of 16 g/dl.

| flow (ml/(g·min)) | 1.0 | 0.5 | 0.25 | 0.1 | 0.01 |
|---|---|---|---|---|---|
| $P_{a,O_2}$ (Torr) | | $tcPo_2$ (Torr) | | | |
| 30 | 37,7 | 37,7 | 36,0 | 29,1 | - |
| 60 | 75,4 | 72,0 | 65,1 | 51,4 | - |
| 80 | 106,3 | 96,0 | 80,6 | 61,7 | - |
| 120 | 126,9 | 109,7 | 92,6 | 68,6 | 12,0 |
| 480 | 383,6 | 340,5 | 306,0 | 275,9 | 250,0 |

the arterial $Po_2$. As predicted, the relationship between $tcPo_2$ and flow is a hyperbolic one, but the hyperbola is different for different $P_{a,O_2}$ values. The nonlinearities caused by hemoglobin were described also by Huch et al.[17] and Thunstrom et al.[10]. The existence of the circulatory hyperbola for the skin is experimentally verified by Wyss et al.[18] and Steinacker et al.[19].

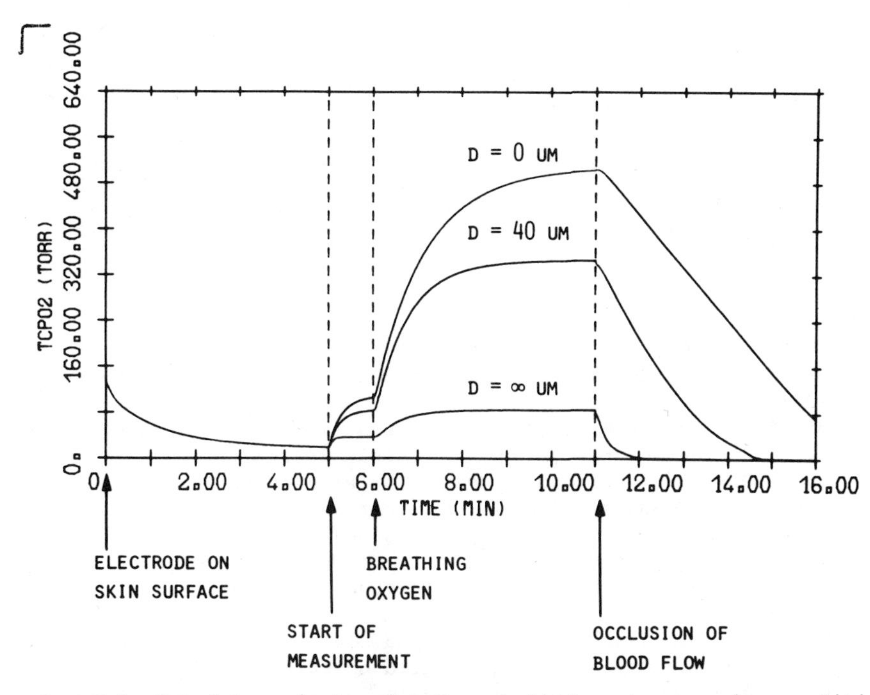


Fig. 6. Calculated transients of $tcPo_2$ at different measuring conditions. The skin surface is covered by an electrode membrane. D: diameter of the $O_2$ consuming electrode. Blood temperature: 37°C. Hemoglobin concentration: 16 g/dl.

Fig. 4 demonstrates that the actually measured $tcPo_2$ depends very much on the properties of the $Po_2$ electrode. This holds not only for the steady state value but also for $tcPo_2$ changes. In Fig. 6 the $O_2$ consumption of the electrode is varied by changing the diameter of the Pt wire: (D = 0 μm) is the trace of an idealized electrode without $O_2$ consumption, (D = 40 μm) is the trace of an electrode with a Pt wire of 40 μm diameter and (D = ∞ μm) is the trace of an electrode which covers the microcirculatory unit[20]. At time zero, the idealized electrode is brought in contact with the skin and covers the skin. Consequently, because of the $O_2$ consumption of the skin, $Po_2$ decreases slowly. At "start of the measurement", temperature is increased to 43°C and flow increases to 1.0 ml/(g·min). At an arterial $Po_2$ of 90 Torr (Hb = 16 g/dl) the $tcPo_2$ amounts to 104 Torr for (D = 0), to 80 Torr for (D = 40), and to 40 Torr for (D = ∞). Then, by breathing $O_2$, $P_{a,O_2}$ suddenly increases to 600 Torr. The quickest response shows the trace (D = ∞). (D = 0) reaches 504 Torr, (D = 40) 344 Torr, and (D = ∞) 88 Torr. This demonstrates that the $tcPo_2$ signal is strongly influenced by the construction of the electrode. After flow stop, the steepness of the $tcPo_2$ decrease is determined by tissue respiration, the $O_2$ dissociation curve of hemoglobin, and $O_2$ consumption of the electrode.

Comparing the results of practical measurements, they are in good agreement with the prediction of the capillary loop model. Only the absolute $tcPco_2$ values, after temperature correction, should be closer to the arterial $Pco_2$. The reason for this discrepancy must be clarified by further experiments. The analysis of the parameter which influences the $tcPo_2$ and $tcPco_2$ shows clearly that at a given hemoglobin content of the blood, the transcutaneous values always depend on both parameters flow and arterial concentration. Only if it is possible to increase flow to high values, the transcutaneously measured values become independent of flow and reflect the arterial values. The great advantage of such an arterial blood gas analysis is that it can be monitored non-invasively and continuously. To be on the safe side, it is recommendable to "calibrate" the electrode by an invasive blood gas analysis. The $tcPo_2$ index, $tcPo_2/P_{a,O_2}$ (comparable to $f_v$, eq. (6)) allows to clarify whether the transcutaneous values reflect arterial values or peripheral flow. I think it is a challenge to develop in future noninvasive methods for the assessment of local flow to facilitate and improve the application of transcutaneous measuring technique.

## REFERENCES


1. R. Huch, A. Huch, and D.W. Lübbers, "Transcutaneous $Po_2$," Thieme-Stratton Inc., New York, Georg Thieme Verlag, Stuttgart-New York (1981).
2. N.T.S. Evans, and P.F.D. Naylor, The systemic oxygen supply to the surface of human skin, Respir. Physiol. 3:21-27 (1967).
3. A. Huch, D.W. Lübbers, and R. Huch, Patientenüberwachung durch transcutane $Pco_2$-Messung bei gleichzeitiger Kontrolle der relativen lokalen Perfusion, Anaesthesist 22:379-380 (1973).
4. J.P. Baumberger, and R.B. Goodfriend, Determination of arterial oxygen tension in man by equilibration through intact skin, Fed. Proc. 10: 10 (1951).
5. U. Grossmann, J. Huber, K. Fricke, and D.W. Lübbers, A new method for simulating the oxygen pressure field of skin, in: "Oxygen Transport to Tissue", Adv. Physiol. Sci., Vol. 25, A.G.B. Kovach, E. Dóra, M. Kessler, I.A. Silver, eds., Pergamon Press, Akadémiai Kiadó, Budapest, pp. 319-320 (1980).

6. U. Grossmann, Simulation of combined transfer of oxygen and heat through the skin using a capillary loop model, Math. Biosci. 61: 205-236 (1982).
7. P. Eberhard, Continuous oxygen monitoring of newborns by skin sensors, Dissertation, Basel (1976).
8. J.A. Quinn, Gas transfer through the skin: A two layer model relating transcutaneous flux to arterial tension, in: "Oxygen Transport to Tissue III", I.A. Silver, M. Erecińska, H.I. Bicher, eds., Plenum Press, New York, pp. 175-181 (1978).
9. Lübbers, D.W., Cutaneous and transcutaneous $Po_2$ and $Pco_2$ and their measuring conditions, in: "Continuous Transcutaneous Blood Gas Monitoring", A. Huch, R. Huch, J.F. Lucey, eds., Birth Defects, Original Article Series, Vol. XV, 4, The National Foundation March of Dimes, A.R. Liss, New York, pp. 13-31 (1979).
10. A.M. Thunstrom, M.J. Stafford, and J.W. Severinghaus, A two temperature, two $Po_2$ method of estimating the determinations of $tcPo_2$, in: "Continuous Transcutaneous Blood Gas Monitoring", A. Huch, R. Huch, J.F. Lucey, eds., Birth Defects, Original Article Series, Vol. XV, 4, The National Foundation March of Dimes, A.R. Liss, New York, pp. 167-182 (1979).
11. T.J. Ryan, The blood vessels of the skin, in: "The Physiology and Pathophysiology of the Skin", A. Jarrett, ed., Academic Press, London, pp. 577-801 (1973).
12. D.W. Lübbers, Theoretical basis of the transcutaneous blood gas measurement, Crit. Care Med.: 9:721-733 (1981).
13. D.W. Lübbers, and U. Grossmann, Gas exchange through the human epidermis as a basis of $tcPo_2$ and $tcPco_2$ measurements, in: "Continuous Transcutaneous Blood Gas Monitoring", R. Huch, A. Huch, eds., M. Dekker, New York-Basel, pp. 1-34 (1983).
14. U. Grossmann, P. Winkler, and D.W. Lübbers, Coupled transport of $O_2$ and $CO_2$ within the upper skin simulated by the capillary loop model, in: "Oxygen Transport to Tissue V", Adv. Exp. Med. & Biol., Vol. 169, D.W. Lübbers, H. Acker, E. Leniger-Follert, T.K. Goldstick, eds., Plenum Press, New York-London, pp. 125-132 (1984).
15. R. Huch, A. Huch, and D.W. Lübbers, Transcutaneous measurement of blood $Po_2$ ($tcPo_2$) - Method and application in perinatal medicine, J. Perinat. Med. 1:183-191 (1973).
16. U. Grossmann, P. Winkler, and D.W. Lübbers, The effect of different parameters (temperature, $O_2$ consumption, blood flow, hemoglobin content) on the $tcPo_2$ calibration curves calculated by the capillary loop model, in: "Oxygen Transport to Tissue VI", Adv. Exp. Med. & Biol., Vol. 180, D. Bruley, H.I. Bicher, D. Reneau, eds., Plenum Press, New York-London, pp. 793-802 (1984).
17. R. Huch, A. Huch, K. Meinzer, and D.W. Lübbers, Eine schnelle beheizte Pt-Oberflächenelektrode zur kontinuierlichen Überwachung des $pO_2$ beim Menschen, Med. Tech., Stuttgart (1972).
18. C.R. Wyss, F.A. Matsen III, R.V. King, C.W. Simons, and E.M. Burgess, Dependence of transcutaneous oxygen tension on local pressure gradient in normal subjects, Clin. Sci. 60:499 (1981).
19. J.M. Steinacker, W. Spittelmeister, R.E. Wodick, F. Fallenstein, and D.W. Lübbers, Untersuchungen über die Blutflußabhängigkeit des transcutanen Sauerstoffpartialdruckes, Biomed. Tech. 30:92-93 (1985).
20. U. Grossmann, and P. Winkler, Transients of gas exchange processes in the upper skin calculated by the capillary loop model, in: "Oxygen Transport to Tissue VI", Adv. Exp. Med. & Biol., Vol. 180, D. Bruley, H.I. Bicher, D. Reneau, eds., Plenum Press, New York-London, pp. 35-41 (1984).

GROWTH AND DEVELOPMENT OF TRANSCUTANEOUS MONITORING IN THE U.S.A. - 1978-1986

Jerold F. Lucey

Professor of Pediatrics
University of Vermont College of Medicine
Burlington, Vermont 05401 USA

The exact gestational age of a new device is often difficult to establish. The European commercial birth date of the transcutaneous oxygen electrode is 1972. It did not appear in the United States, however, until 1978. The growth and development of this device in the United States has been a spectacular success. One index of growth has been the number of articles written on the subject which rose from none in 1970 to over 110 in English alone in 1986, with well over a total of 300 articles since 1972.

This technique is now in its fifteenth year and is entering its adolescent phase of life. As pediatricians we all know this can be a turbulant time.

Once electrode manufacturing left Marburg, over a dozen commercial companies began to produce electrodes. These were of several slightly different designs which could be mass produced at a lower cost. Sales in the USA of three companies are shown in Figure 1.

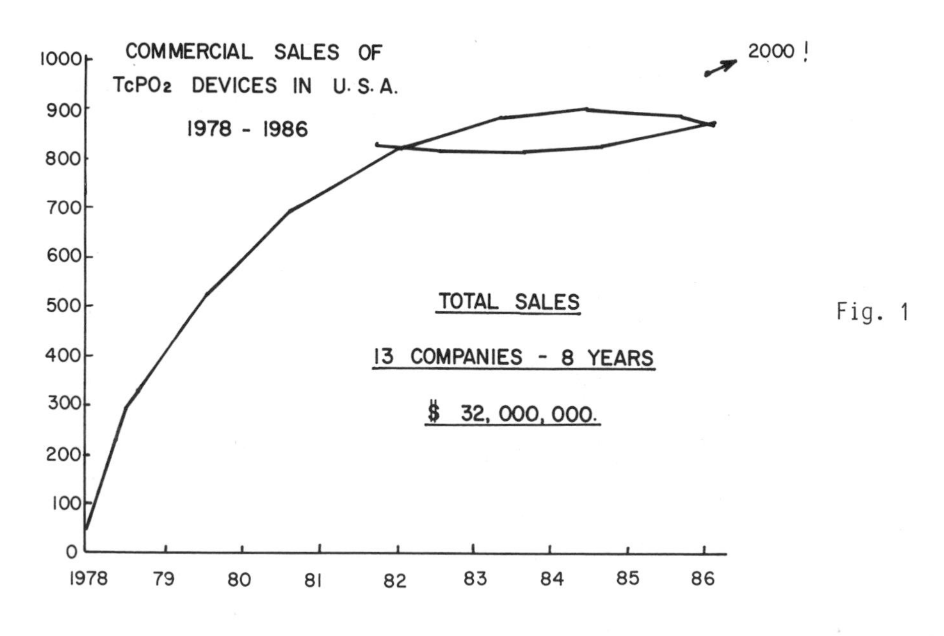


Fig. 1

The first company started in 1978 with a few monitors being sold and by 1985 was selling over 800 per year. Another company started in 1981 and sold 800 in its first year alone. In 1986 a new company entered the field and sold over 2,000 the very first year! The electrode is clearly a financial success story with estimated sales of over $32,000,000 in the first eight years. It is widely used in nearly all intensive care nurseries in the USA.

How is the electrode working? Well, I am concerned! I am not sure that some of the electrodes are working as well as the original electrodes! The reasons for this are complex. It's undoubtedly a multi-factor problem.

The companies are under constant competitive pressure to make a cheaper electrode. Electrodes can be tested in the laboratory for accuracy but the only meaningful test is in the field in clinical practice on a patient's skin. This testing has not been carried out. Some recent results are very disturbing. Dr. Jay of the Health Protection Branch of the Canadian Department of Health issued an Alert Letter Nr. 73 in 1985. The results are summarized in Table I.

Of the nine commercial $tcPo_2$ devices tested, <u>none</u> of them performed very well.

Our own experience with the Corometrics-Biochem electrode system was also very disappointing. We noted that it was erratic and particularly prone to underestimating arterial oxygen tension at high (>80 torr) arterial $Po_2$. We warned the company about this serious and potentially dangerous defect. We have not heard from them since that warning.

I have summarized nine significant sources of errors which can occur in Table II. This audience is familiar with these but unfortunately many clinicians are not. The majority of users are also not aware that many of the new electrodes have <u>not</u> been subjected to any careful or extensive clinical trials to prove they are accurate, especially at hyperoxemic levels!

Table I

**CANADIAN FIELD TEST $TcPO_2$ VS $PaO_2$**

A. JAY. et. al. 1985

| | NO. | CORR. |
|---|---|---|
| *KONTRON* | 103 | .68 |
| *LITTON* | 41 | .77 |
| *NARCO* | 56 | .58 |
| *RADIOMETER* | 44 | .16 |
| *CRITIKON* | 87 | .84 |
| *NOVAMETRIX* | 34 | .76 |
| *IL* | 23 | .77 |
| *MICROGAS* | 22 | .81 |
| *HP* | 31 | .84 |

Table II

**COMMON PROBLEMS AND UNCONTROLLED USE**

1. WRONG SITE - BONE, ? DUCTUS
2. LOW TEMPERATURE - CUSTOM, POOR ADVICE
3. NO TIME CORRECTION
4. PRESSURE ON ELECTRODE - TAPE, ON BACK
5. ? BLOOD GAS LAB
6. WRONG PATIENT - BPD
7. ELECTRODE DESIGN - CHANGES BY MFG ??
8. UNTRAINED PERSONNEL
9. NO CALIBRATION - HEEL STICKS, CRYING

On May 28, 1976 the Medical Device Amendment to the Food, Drug and Cosmetic Act was passed in the USA. The F.D.A. never approved $tcPo_2$ monitors for sale. It only granted permission to market the device if it was substantially equivalent to a device sold before May 28, 1976. One research model device had been sold before that date in the USA!

No clinical trials were required of the dozen companies who have since produced modified $tcPo_2$ electrodes "substantially equivalent" to the original Huch electrode. I should point out that 98% of 5,000 devices submitted to the F.D.A. have passed through this same "loophole" and that pulse oximeters have been approved under the same loophole!

## SUMMARY

Surface blood gas monitoring is an invaluable clinical technique. It is, unfortunately, subject to many errors unless carefully carried out. This can be done by clinicians. Clinicians, however, cannot be responsible for the cumulative effects of changes in the electrodes made by manufacturers. Before a modified electrode is approved for use its accuracy under clinical conditions and at high arterial $Po_2$ tensions has to be established. In America this has not and is not being done. The F.D.A. met in December, 1986, to consider this problem. Recommendations are expected to be published in 1987. In any new regulations similar but not identical requirements for accuracy should be demanded from pulse oximeters, as these devices have their own unique limitations under hyperoxic and hypoxic conditions.

TRANSCUTANEOUS $Po_2$ AND $Pco_2$ MONITORING AT 37°C

# CUTANEOUS $Po_2$ AND $Pco_2$

G. Rooth, U. Ewald and F. Caligara

Departments of Obstetrics, University of Zurich, Switzerland, Pediatrics, University of Uppsala, Sweden and the Joint Research Centre of the Commission of the European Communities, Karlsruhe Establishment, FRG

## INTRODUCTION

It was the aim of the transcutaneous blood gas technique, as developed into a practical, clinical instrument by Huch et al.[1], to monitor blood gas tensions continuously. In order to achieve this, particularly for $Po_2$ monitoring, the temperature has to be elevated to 44° or 45° C. Thereby the capillary blood in the skin becomes arterialized because of the maximal, or close to maximal, skin blood flow obtained.

Used at 37° C the transcutaneous electrodes both $Po_2$ and $Pco_2$ vary very much with skin capillary blood flow. Consequently the terms cutaneous $Po_2$($cPo_2$) and $Pco_2$ ($cPco_2$) are used.

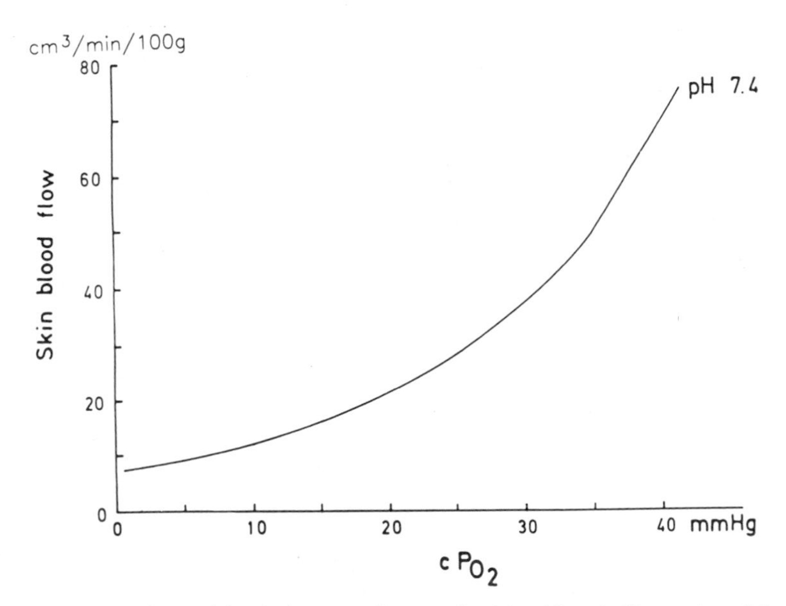


Fig.1. Relationship between $cPo_2$ and skin blood flow at a blood pH of 7.4 and a hemoglobin concentration of 160 gram/l.

## CUTANEOUS $Po_2$ AS INDICATOR OF SKIN BLOOD FLOW

The main factors which govern cutaneous $Po_2$ and standard, steady state values are given in (1).

$$cPo_2 = \text{Capillary } Po_2 - O_2 \text{ (skin } O_2 \text{ consumption)} \quad (1)$$
$$2 = 42 - 40 \text{ mmHg}$$

The main factor which causes changes in skin capillary $Po_2$ is variations in skin blood flow. As Luebbers[2] stressed, there is a parabolic relationship between skin capillary blood $Po_2$ and blood flow. Using an approach reported separately in this volume[3], the curve shown in Fig 1 was obtained assuming a pH of 7.40 and a hemoglobin concentration of 160 g/l. It will be seen that cutaneous $Po_2$ initially is very sensitive to increases in skin blood flow. If $cPO_2$ is higher than 40 mm Hg, its usefulness as an indicator of blood flow is limited.

It is not always necessary to express skin blood flow in ml/min/100 g. It often suffices to monitor cutaneous $Po_2$ changes in order to study skin blood flow variations, and the cutaneous $Po_2$ value itself may be used as an indicator of skin blood flow.

### Postischemic hyperemia

After standardizing the experimental conditions, including maintaining the stasis for four minutes, duplicate measurements performed within 15 minutes showed a coefficient of variation of 14 per cent[4]. Fig. 2 shows a typical curve and it suffices to measure the peak value after the stasis.

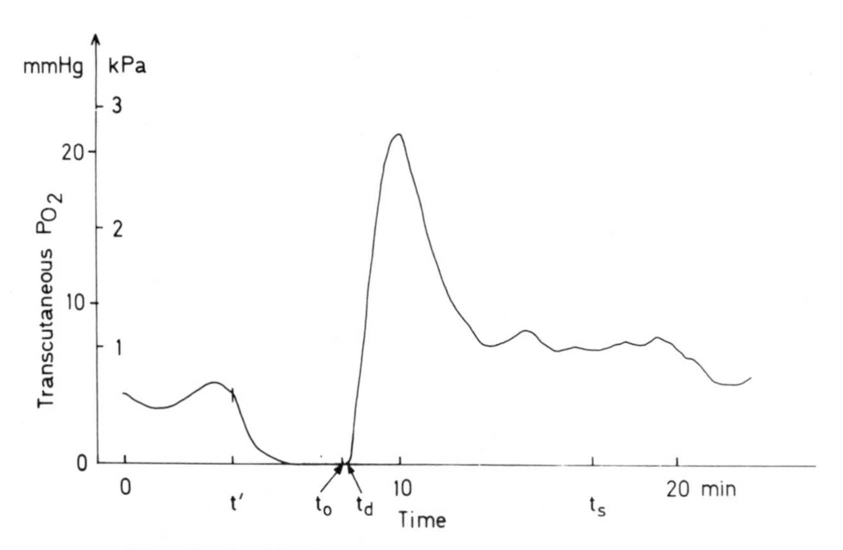


Fig. 2. Postischemic hyperemia as revealed by $cPo_2$. Stasis from t' to $t_o$.

Reactive hyperemia constitutes the physiologic vascular reaction to tissue hypoxia. Since the reaction occurs after denervation and atropinization of the tissue it is considered to be mediated locally. Prostaglandins, particularly prostacyclin, are locally released in response to hypoxia and are potent vasodilators. Indomethacin, a cyclooxygenase inhibitor, reduces reactive hyperemia in the forearm monitored with the $cPo_2$ technique as well as plethysmographically[5].

Smoking also impairs vascular reactivity in vivo. With the cutaneous $Po_2$ method, Ahlsten et al[6] demonstrated a 25 per cent reduction of the postocclusive peak in smokers. Even more interesting is that the newborn infants of smoking mothers exhibited a 30-40 per cent reduction in vascular reactivity on the second day after birth compared to infants of non-smoking mothers. Five days after birth the difference from the control group had disappeared. This indicates that some substance or substances transmitted from the smoking mother to her baby before birth impaired vascular function for at least 48 hours after birth.

Age also influences vascular function. We found a significant negative correlation between the postocclusive cutaneous $Po_2$ peak and log age. 34 children between 5 and 18 years of age had a $cPo_2$ peak of about 30 mm Hg (4.0 kPa) compared to 20 mm Hg (2.6 kPa) in 67 adults age range 22 to 67 years.[4]

The vascular reactivity is stronger in females than in males, but this only becomes manifest after puberty.

The postischemic reactivity is of particular interest in diabetic disease. We found a 50 per cent reduction in the postocclusive $cPo_2$ peak in 28 children with a mean duration of their disease of 5 years compared to healthy children. None of the diabetic children exhibited any signs of structural vascular disease[7].

In a longitudinal prospective study including more than 60 newly diagnosed type I diabetic children, Kobbah et al[8] found that the reactivity was already reduced when the diagnosis was made and before insulin was given. Repeated examinations during the first six months revealed a gradual normalization within this time period. Thereafter the reactivity again began to be reduced and two years after diagnosis the reactivity was significantly reduced compared to a healthy control group.

Indices of carbohydrate control were only weakly correlated to the impaired vascular reactivity indicating that other factors determining the development of diabetic angiopathy need to be found.

## CUTANEOUS $Pco_2$

As already mentioned, cutaneous $Po_2$ is mainly a function of skin blood flow. Cutaneous $Pco_2$ is also influenced by skin blood flow, but its greatest importance lies in the fact that it is also influenced by changes in the metabolic acidosis.

In analogy with eq (1) we have for cutaneous $Pco_2$:

$$\text{Cutaneous } Pco_2 = \text{Capillary } Pco_2 + Pco_2 \text{ (skin metabolism)} \quad (2)$$
$$48 = 46 + 2 \text{ mm Hg}$$

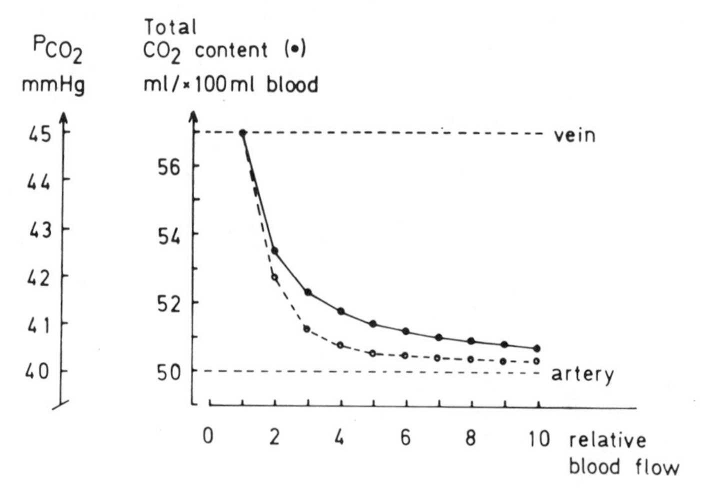


Fig. 3. Changes in total $CO_2$ and $Pco_2$ with increased skin blood flow. Relative blood flow at steady state = 1.

Should skin blood flow increase so much that the capillaries contain arterial blood, $Pco_2$ would decrease from 46 to 40 mm Hg assuming an arterial $Pco_2$ of 40 mm Hg.

The relation between the increase in skin blood flow and capillary $Pco_2$ is shown in Fig. 3. A relative blood flow of 1 indicates pure venous blood. A 2 means twice that flow, i.e. an equal mixture of venous and arterial blood, etc. The relative blood flow is obtained from the cutaneous $Po_2$ signal. The flow read off from the actual cutaneous $Po_2$ is divided by the basic flow to obtain the relative flow. For this reason it is convenient to use a combined $Po_2$ and $Pco_2$ sensor.

For practical purposes the correction of cutaneous $Pco_2$ for skin blood flow may be simplified by using the following table.

Table 1
Corrections for calculating skin capillary $Pco_2$ from cutaneous $Pco_2$ when skin blood flow and hence cutaneous $Po_2$ is increased.

| $cPo_2$ mm Hg | mm Hg to add to actual $cPco_2$ |
|---|---|
| 0 - 2 | 0 |
| 3 - 4 | 1 |
| 5 - 8 | 2 |
| 9 -15 | 3 |
| 16 -30 | 4 |
| 31 -49 | 5 |
| 50 | 6 |

## Cutaneous $Pco_2$ and changes in metabolic acidosis

That transcutaneous $Pco_2$ increases when there is a hypoxic metabolic acidosis in the skin itself was indicated already in the first publication on the $tcPco_2$ electrode by Huch, Luebbers and Huch[9]

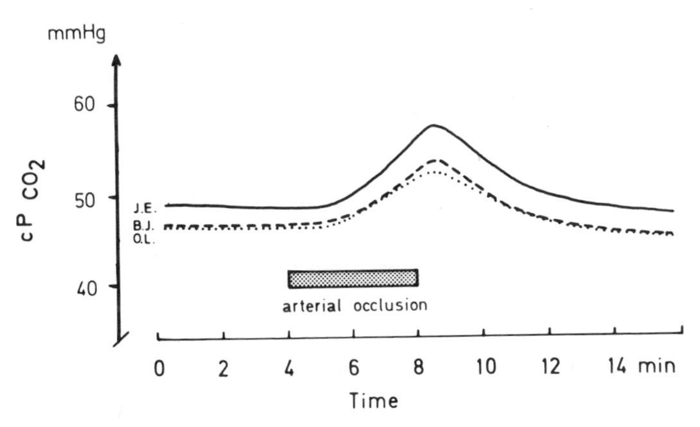


Fig. 4. Three consecutive $cPco_2$ curves from healthy male volunteers showing the increase of $cPco_2$ during stasis.

and later studied by Severinghaus, Stafford and Thunstrom[10]. The latter showed that after stasis there is at first a slow increase in $tcPco_2$ and then a faster one with a rate of 2.2 mmHg/min, corresponding to the lactic acid production in the skin.

As long as oxygen is available in the skin aerobic metabolism is undisturbed, also during stasis, but the $CO_2$ produced cannot escape and a slow increase in $Pco_2$ is seen. When the hypoxia begins and lactic acid is produced, hydrogen ions react with bicarbonate:

$$H^+ + HCO_3^- = H_2O + CO_2 \tag{3}$$

Thus both during aerobic and during anaerobic metabolism $CO_2$ is liberated in the skin and the $Pco_2$ measured on the skin is elevated. For these studies it is best to use an electrode kept at 37° C which eliminates temperature gradients between the blood and the sensor.

Fig. 4 shows the results of three consecutive experiments. The curves are all of the same type, but there are statistical differences between the subjects. From the slope of the curve the anerobic energy production may be calculated and we have expressed this in mmol ATP/l/min. A mean value of $0.33 \pm 10$ was found by taking the average of three measurements on 30 healthy men. This corresponds to about 20 per cent of the aerobic energy production[11].

## Cutaneous $Pco_2$ and metabolic acidosis produced in other organs than the skin

The same chemical reaction as in (3), leading to a $Pco_2$ increase in the skin, occurs when blood with low pH flows into the skin; for instance, when a metabolic acidosis is produced in hypoperfused tissues, or during exercise, or in ketoacidosis.

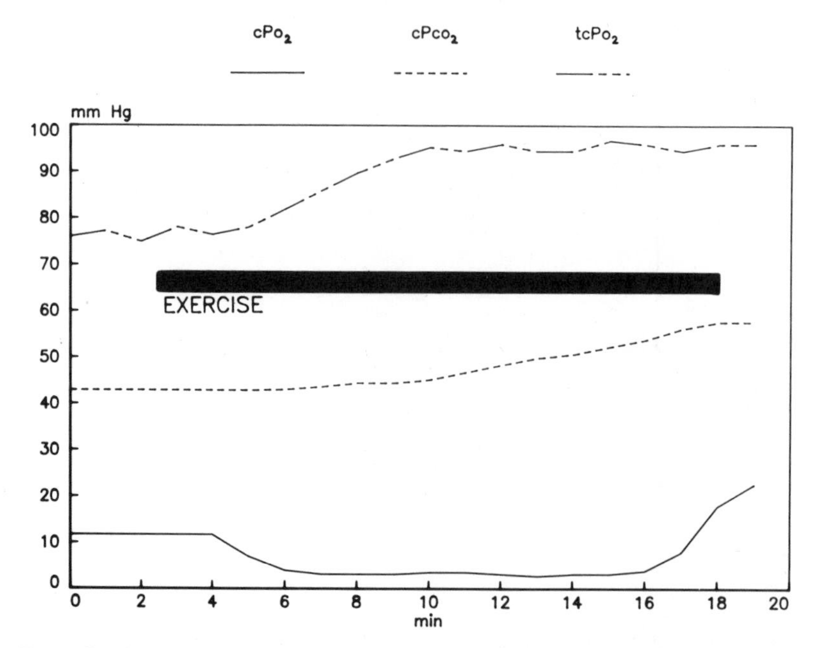


Fig. 5. Changes in transcutaneous $Po_2$, cutaneous $Pco_2$ and $Po_2$ during exercise in a healthy male volunteer.

This is exemplified in Fig 5. As usually observed, $tcPo_o$ increased shortly after the onset of the exercise. Cutaneous $Pco_2$ began to rise at about the same time as $tcPo_2$. In these experiments cutaneous $Po_2$ decreases at the beginning of work, when the blood flow is diverted to the working muscles. This time the subject became warm and $cPo_2$ began to rise only at the end of the experiment. Because of the low flow and stable $cPo_2$, $cPco_2$ was not influenced by changes in skin blood flow.

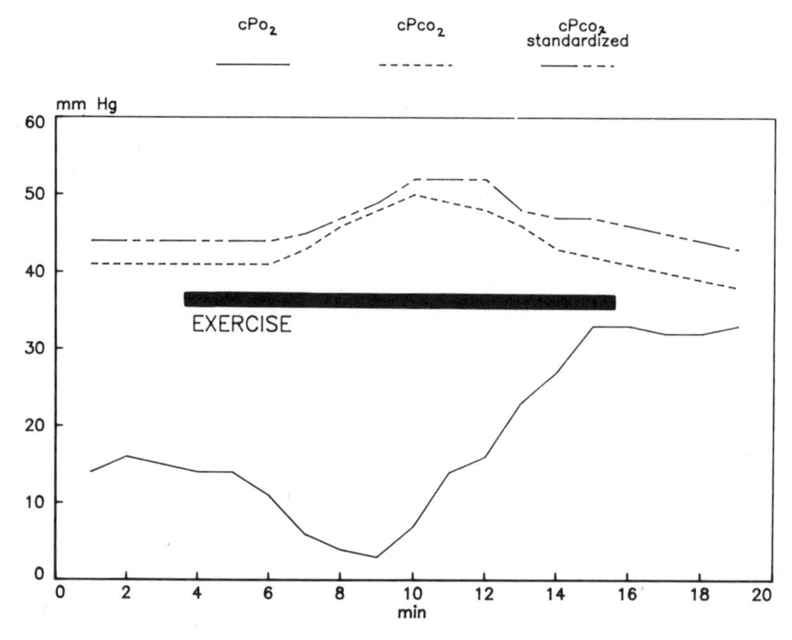


Fig. 6. Standardized and uncorrected cutaneous $Pco_2$ and $Po_2$ during exercise in a healthy male volunteer.

A different example is shown in Fig. 6. Initial skin blood flow was high, and began to fall when the exercise started but rose 5 minutes later to a relatively high value. To evaluate the true $cPco_2$ changes, correction must be made for changes in blood flow.

As always, cutaneous $Pco_2$ began to rise within two min. after onset of work, reached a peak and gradually fell down to a level lower than the inital one. After standardizing $cPco_2$ for a relative blood flow of 1, higher $cPco_2$ values are obtained. A plateau was reached indicating no more changes in the metabolic acidosis of the blood flowing into the skin. The fall in $cPco_2$ 9 min after onset of work indicates that the subject had by that time got his second wind. The uncorrected value would indicate that the subject had already got his second wind 6 min after onset of work.

Except in those who had already got their second wind during work, the $cPco_2$ peak came soon after the end of the work. When there was a rapid increase in $cPco_2$, this correlated significantly with an increase in blood lactate concentration[12]. This is illustrated in Fig.7. After standardization for skin blood flow, there was an increase in $cPco_2$ of 9 mm Hg and lactate concentration reached 11 mmol/l. After the work was over, $cPco_2$ ended with a value 5 mm Hg lower than the initial one - a sign that the metabolic acidosis was disappearing and the bicarbonate concentration in the skin was being restored.

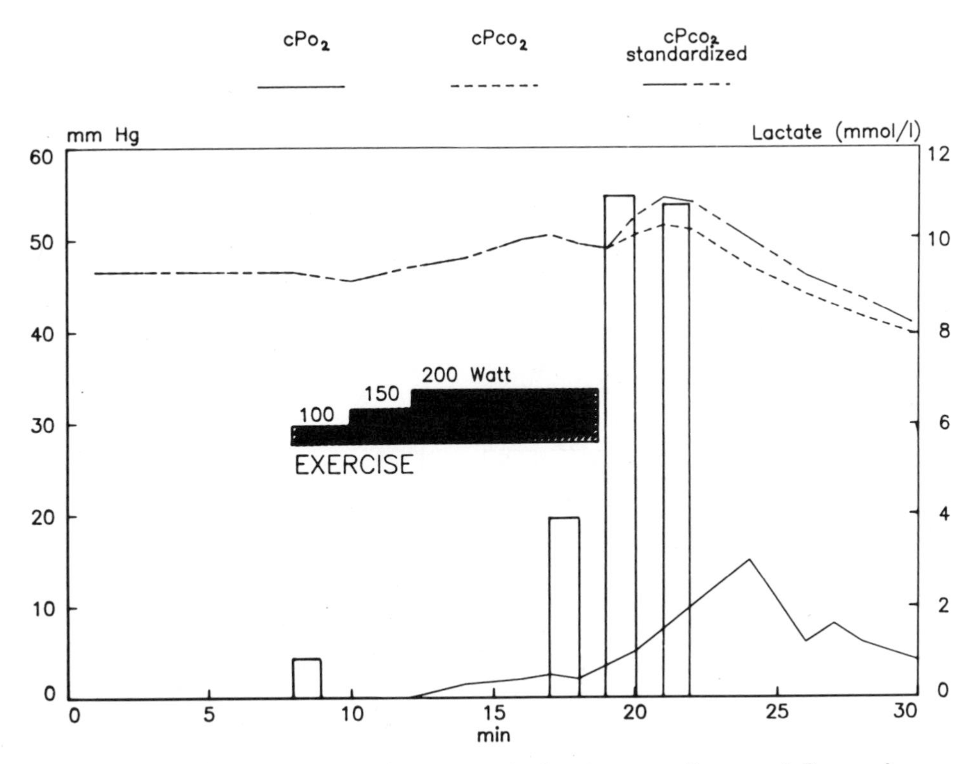


Fig. 7. Standardized and uncorrected cutaneous $Pco_2$ and $Po_2$ and lactate concentration during exercise in a healthy male volunteer.

The details of the buffering action of those tissues, which are not producing the metabolic acidosis, in our case the skin, are complex and basic data such as the $CO_2$ dissociation curve for the whole skin at different levels of base deficit are lacking. However, the underlying principle is simple and similar to that used in the van Slyke technique. Add lactic acid and $CO_2$ is driven out of the bicarbonate. Van Slyke measured the volume or the pressure of the liberated $CO_2$. We monitor $cPco_2$.

A schematic presentation (Fig. 8) shows how normal blood enters the working muscles. The blood leaving the muscles has a base deficit of 10 mmol/l, $Pco_2$ of 60 mm Hg and pH of 7.10. After mixing with blood from other areas and eliminating $CO_2$ in the lungs, arterial blood entering the skin has a base deficit of 2 mmol/l, $Pco_2$ of 40 mm Hg and pH 7.36.

Normally, blood entering the skin absorbs $CO_2$ and (Fig. 9) as a result $Pco_2$ reaches 45 mm Hg and pH 7.35. When instead the blood entering the skin has a base deficit of 2 mmol/l, $Pco_2$ will rise to 50 mm Hg and the resulting pH will be 7.29. The blood leaving the skin will only have a base deficit of 1, i.e. the bicarbonate system of the tissues have contributed to the buffering of the metabolic acidosis.

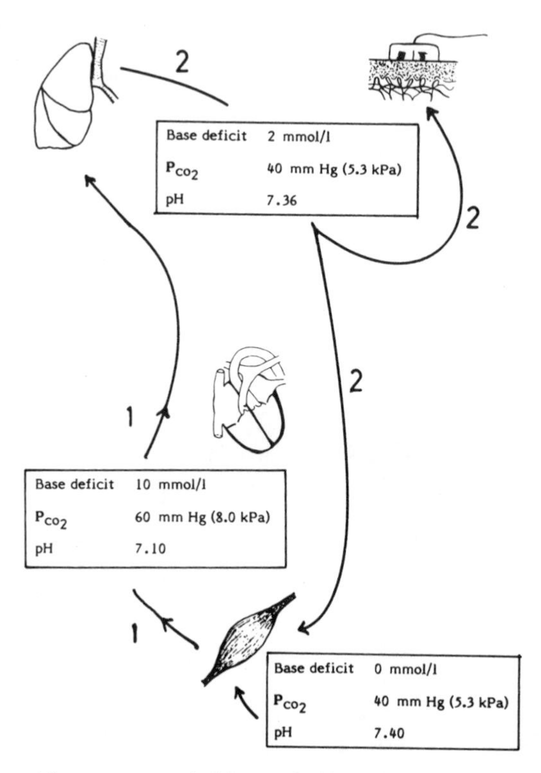


Fig. 8. Schematic representation of the changes in arterial blood after having first perfused a hard working muscle. Read from the bottom right hand corner.

In clinical medicine cutaneous $P_{CO_2}$ monitoring would be indicated any time lactoacidosis is expected to appear. In shock, skin blood flow will be small and no correction for arterialization of the skin capillaries is needed. Whenever treatment becomes effective, cutaneous $P_{CO_2}$ will decrease. We have as yet no systematic studies of our own in this respect but in additon the theoretical aspect there is anecdotal experience from several neonatal intensive care units - although they have usually used $tcP_{CO_2}$. It should be possible to monitor the appearance and disappearance of ketoacidosis in the same way, but fortunately it is so rare today that we have not had the opportunity to test this.

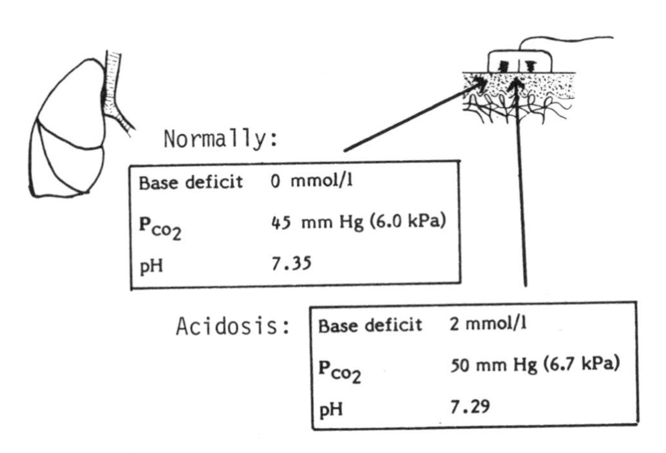


Fig. 9. Schematic representation of changes in arterial blood entering tissues (here the skin). Above,the normal situation, below when the arterial blood has a metabolic acidosis of 2 mmol/l, i.e. the situation in the right upper corner of Fig. 8.

It should be remembered that the $cP_{CO_2}$ indicates changes, not levels, in the metabolic acidosis. As exemplified cutaneous $P_{CO_2}$ is most useful in sport and in occupational medicine. At exhaustion, cutaneous $P_{CO_2}$ increases rapidly. When the second wind comes, it decreases. We have seen how much faster cutaneous $P_{CO_2}$ increases for the same workload when a subject is enclosed in a safety suit as compared to when he is without one.

SUMMARY

The transcutaneous $P_{O_2}$ and $P_{CO_2}$ equipment used at 37$^{\circ}$ gives noninvasive and continuous information about skin blood flow changes and changes in the metabolic acidosis, all of which may be used for experimental, pharmacological or clinical purposes.

Acknowledgement

This study was sponsored by the Joint Reseach Centre of the Commission of the European Communities.

Present address of the author: Ofre Slottsgatan 14 C, S-752 35, Uppsala, Sweden.

REFERENCES

1. A. Huch, R. Huch, K. Meinzer, and D.W. Luebbers, Eine schnelle beheizte Pt-Oberflaechenelektrode zur kontinuierlichen Ueberwachung des $Po_2$ beim Menschen. Med. Tech. Stuttgart, (1972).
2. D. W. Luebbers, Cutaneous and transcutaneous $Po_2$ and $Pco_2$ and their measuring conditions, in: Continuous transcutaneous blood gas monitoring, A. Huch, R. Huch, and J.F. Lucey, eds., A. R. Liss, New York (1979).
3. F. Caligara, G. Rooth, Skin blood flow calculation from transcutaneous gas pressure measurements. This volume.
4. U. Ewald, Evaluation of the transcutaneous oxygen method used at 37°C for measurement of reactive hyperemia in the skin. Clin. Physiol. 4:413 (1984).
5. A. Wennmalm, Cigarette smoking, prostaglandins and reactive hyperemia, Prostaglandins Med. 3:321 (1979).
6. G. Ahlsten, U. Ewald, and T. Tuvemo, Impaired vascular reactivity in newborn infants of smoking mothers, Acta Paediat. Scand. in press.
7. U. Ewald, T. Tuvemo, and G. Rooth, Early reduction of vascular reactivity in diabetic children detected by transcutaneous oxygen electrode, Lancet I:1287 (1981).
8. M. Kobbak, U. Ewald, T. Tuvemo, Vascular reactivity during the first year of diabetes in children, Acta Paed. Scand. Supplement 320:56 (1985).
9. A. Huch, R. Huch, and D.W. Luebbers, Patientueberwachung durch transcutane $Pco_2$-Messung bei gleichzeitiger Kontrolle der relativen lokalen Perfusion, Anaesthesist 22: 379 (1973).
10. J. W. Severinghaus, M. Stafford, and A.M. Thunstrom, Estimation of skin metabolism and blood flow with $tcPo_2$ and $tcPco_2$ electrode by cuff occlusion of the circulation, Acta anaest. scand. Suppl. 68:9 (1978).
11. G. Rooth, U. Ewald, and F. Caligara, Anaerobic skin metabolism in healthy men estimated with transcutaneous $Pco_2$ electrode, Scand. J. Clin. Lab. Invest. 45:393 (1985).
12. U. Ewald, T. Tuvemo, and G. Rooth, Detection of exercise-induced lactic acidosis using transcutaneous carbon dioxide, Crit. Care 13:630 (1985).

# PERFORMANCE OF TRANSCUTANEOUS ELECTRODES

# THE MEASUREMENT OF $TcPO_2$ AND $TcPCO_2$ IN NEWBORN INFANTS AT 44°C, 42°C AND 37°C AFTER INITIAL HEATING TO 44°C

Bent Friis-Hansen, Peter Voldsgaard, Jess Witt, K.G. Pedersen and Peter Steen Frederiksen

Department of Neonatology, University Hospital, Rigshospitalet, Copenhagen, Denmark

## SUMMARY

Prolonged measurement of the trancutaneous $O_2$ and $CO_2$ tension at an electrode temperature of 44 to 45°C often causes a second degree burn of the underlying skin.

To avoid this, we compared the readings at 44°C, 42°C and 37°C, after 2 hours pre-heating of the skin by the electrodes at 44°C. In order to eliminate the electrodes' own temperature coefficients, electrodes with a built-in temperature correction were used. The changes observed therefore represent changes in the $O_2$ and $CO_2$ tension in the skin. The obtained values were compared to repeated arterial samples.

We found that the $TcPO_2$ and $TcPCO_2$ values obtained at 42°C and 37°C were lower than those obtained at 44°C, but when corrected for the in vivo temperature coefficients previously found by us the $TcPO_2$ values at 42 were quite similar to the 44°C values, whereas the 37° values remained lower. $TcPCO_2$ values at 44, 42 and 37° were all similar.

The temperature coefficient of $PO_2$ was calculated to be $0.044 \pm 0.008$ and for $TcPCO_2$ as $0.049 \pm 0.007$.

## INTRODUCTION

It is generally accepted that transcutaneous measurement of the $O_2$ and $CO_2$ tension is best carried out at 44-45°C, but in the newborn infant this often leaves a red spot on the underlying skin, corresponding to a second degree burn. In order to avoid this complication, which is disturbing to the parents, we therefore have tried to measure at lower electrode temperatures, after prolonged pre-heating of the skin. Previous rerults have shown that pre-heating for 2o min. at 44°C did not permit prolonged readings at lower temperature (Voldsgaard et al. 1985). Therefore we have tried now to investigate if pre-heating for 2 hours would make it possible to measure at lower temperatures.

## MATERIAL

12 newborn infants were examined with an average gestational age of 29.9 weeks, and an average birth weight of 1334 g. All were receiving assisted ventilation, and had an umbilical artery catheter inserted.

## METHODS

Each infant was monitored with 3 combi-electrodes (E 527o, Radiometer, Copenhagen, Denmark) (Wimberley et al. 1983), and arterial blood samples were taken at 1, 1 3/4, 2 1/4 and 3 1/2 hour after application. All electrodes were heated at 44$^{o}$C for 2 hours, then one electrode was maintained at 44$^{o}$C, one was lowered to 42$^{o}$C and the third was lowered to 37$^{o}$C. After 2 hours at these temperatures, the electrode temperature was raised again to 44$^{o}$C in both with the lower temperatures. In order to correct the electrodes for internal changes during the temperature changes, an electronic correction of 4.5% was built into the monitor (Wimberley et al. 1985). These temperature changes are illustrated in Fig. 1.

## RESULTS

Fig. 2 shows a typical example. The electrodes showed almost constant values during pre-heating for 2 hours, and the oxygen electrode maintained at 44$^{o}$C, remained constant, whereas a certain decrease was found in the electrodes at 42$^{o}$C, and a marked decrease was observed at 37$^{o}$C. The $TcPCO_2$ values remained almost constant at all temperatures.

If the values observed at 42 and 37$^{o}$C were corrected by a temperature coefficient of o.o44 for $O_2$ and of o.o49 for $CO_2$ (Voldsgaard 1985), the values then obtained are shown in Fig. 3. Here it is seen that the $O_2$ tension at 42$^{o}$C become almost equal to 44$^{o}$C, whereas the 37$^{o}$C value remained low. The $CO_2$ values became even more equal.

Fig. 4 shows simultaneously obtained Tc values from the electrodes of 44, 42 and 37$^{o}$C recorded 2o min. after pre-heating at 44$^{o}$C for 2 hours. Each line represents one patient. When the $TcPO_2$ values are expressed by their ln value, almost all values are remarkably consistent with a constant slope.

When the $TcPCO_2$ values are presented in a similar fashion as ln values, very constant slopes are found (Fig. 5).

All these Tc values were compared to simultaneously obtained arterial samples and expressed as per cent of the true arterial values, as shown in Fig. 6. It is seen that $TcPO_2$ values at 44$^{o}$C (I) increased slightly, after the first 2 hours heating. The corrected $TcPO_2$ of 42$^{o}$C (II) remained identical to the arterial blood tension, whereas the 37$^{o}$C (III) decreased, and cannot be used.

The temperature coefficients calculated from these slopes are presented in Table 1.

## DISCUSSION

The temperature coefficients presented here are quite similar to those described in adults (Wimberley 1985). In blood, however, the temperature coefficient for $O_2$ is somewhat higher in the same range of $O_2$ tension (Siggaard-Andersen et al. 1984). Our results indicate that it is

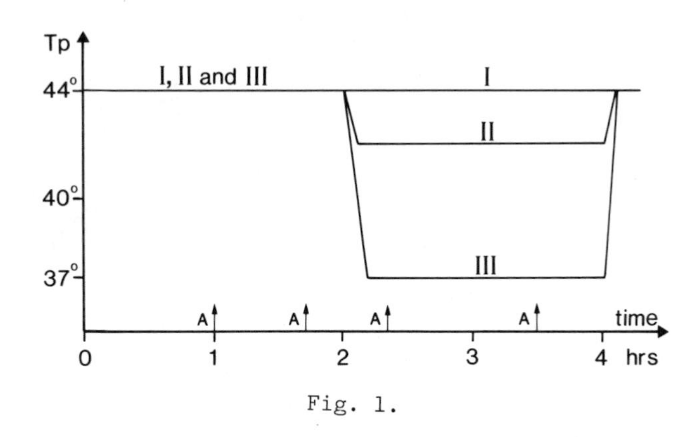
Electrode Tp.
Tp
44°
40°
37°
I, II and III
I
II
III
A
A
A
A
time
0
1
2
3
4 hrs

Fig. 1.

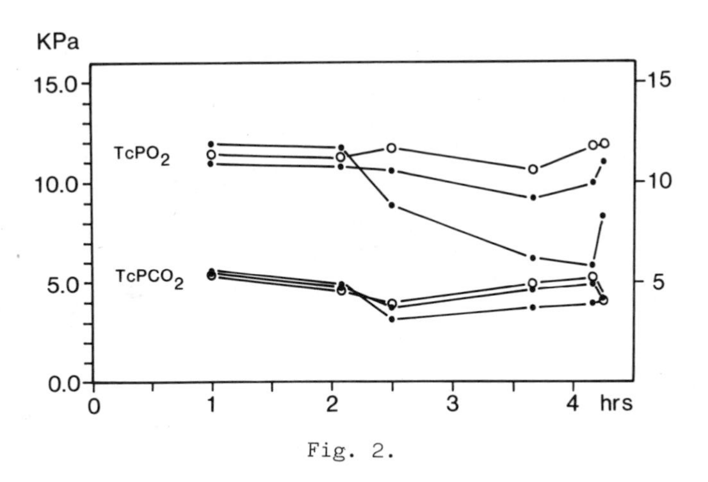
Tc at 44°, 42° and 37.° Uncorrected data
KPa
15.0
10.0
5.0
0.0
TcPO2
TcPCO2
15
10
5
0
1
2
3
4 hrs

Fig. 2.

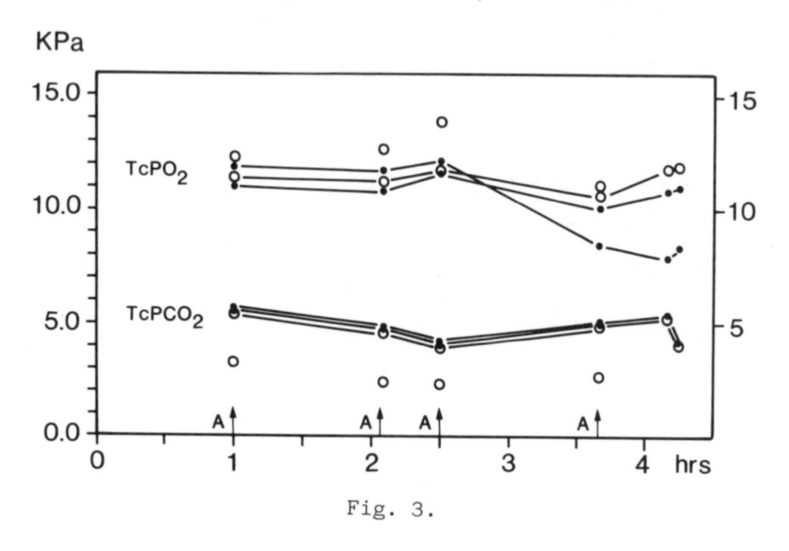
Tc at 44°, 42° and 37°. Tp. corrected
KPa
15.0
10.0
5.0
0.0
TcPO2
TcPCO2
15
10
5
A
0
1
2
3
4
hrs

Fig. 3.

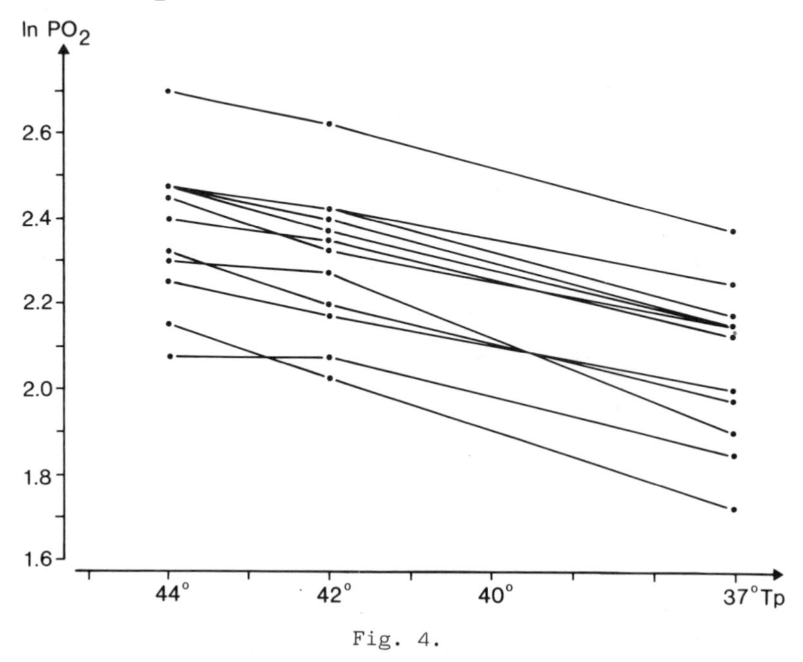
Tc PO2 at 44°, 42° and 37°, 20 min after 2 hrs preheating
ln PO2
2.6
2.4
2.2
2.0
1.8
1.6
44°
42°
40°
37°Tp

Fig. 4.

**Tc $PCO_2$ at 44°, 42° and 37°, 20 min. after 2 hrs preheating**

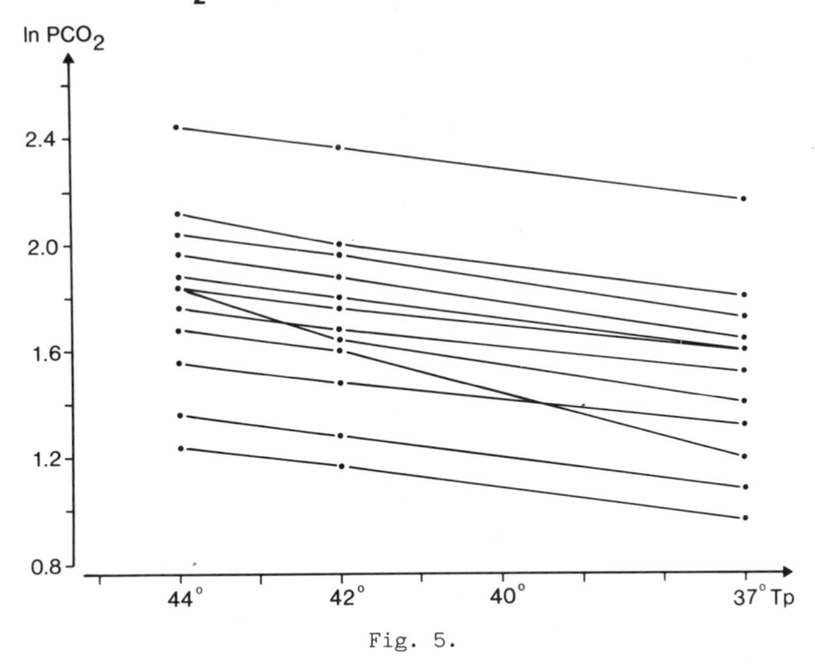


Fig. 5.

**Average Tc $PO_2$ at 44°, 42° and 37°**
**in % of arterial values**

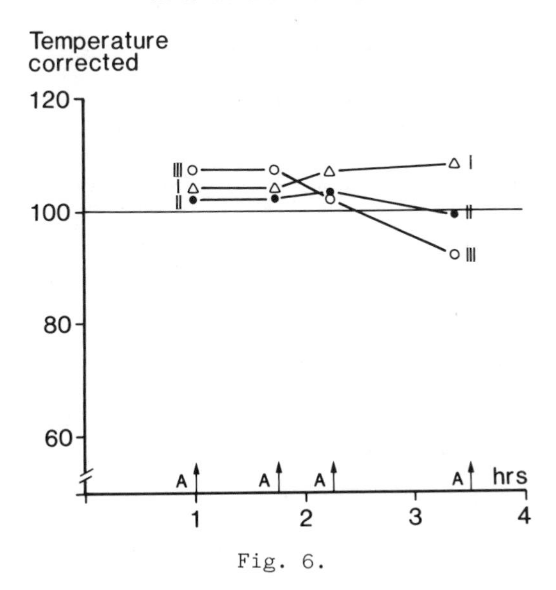


Fig. 6.

possible to monitor transcutaneous $O_2$ and $CO_2$ tensions at $42^oC$ after2 hours pre-heating at $44^oC$. Further investigation is needed to evaluate for how long a time will be possible to get reliable results after 2 hours of pre-heating.

Table 1. Temperature coefficients

| Tp Change ($^oC$) | Pre-heating | dln $PO_2/1^oC$ | dln $PCO_2/1^oC$ |
|---|---|---|---|
| Newborn $44^o - 42^o$ | 2o min. | o.o48 ± o.o19 | o.o48 ± o.o19 |
| Range kPa | | 5.o - 14.1 | 2.9 - 9.8 |
| Newborn $44^o - 37^o$ | 2o min. | o.o53 ± o.oo7 | o.o44 ± o.olo |
| Range | | 6.3 - 12.1 | 2.2 - 7.2 |
| Newborn $44^o - 42^o$ | 2 hrs | o.o41 ± o.oo8 | o.o51 ± o.oo7 |
| Range kPa | | 6.8 - 14.2 | 2.2 - 7.7 |
| Newborn $44^o - 37^o$ | 2 hrs | o.o47 ± o.oo3 | o.o47 ± o.oo4 |
| Range | | 6.8 - 14.2 | 2.2 - 7.7 |

REFERENCES


Siggaard-Andersen, O., Wimberley, P. D., Göthgen, I. and Siggaard-Andersen, M., 1984, A mathematical model of the hemoglobin-oxygen dissociation curve of human blood and of the oxygen partial pressure as a function of temperature, Clin Chem, 3o:1646.

Voldsgaard, P., Witt, J. and Frederiksen, P. S., 1985, Transcutaneous $PO_2$ sampling at $42^oC$ after hyperaeminisatio at $44^oC$. lo. Nord. Congress Of Perinatal Medicine, Turku, Finland.

Wimberley, P. D., Pedersen, K. G., Thode, J., Fogh-Andersen, N., Møller Sørensen, A. and Siggaard-Andersen, O, 1983, Transcutaneous and capillary $PCO_2$ and $PO_2$ measurements in healthy adults, Clin Chem, 29:1471.

Wimberley, P. D., Pedersen, K. G., Olsson, J. and Siggaard-Andersen, O., 1985, Transcutaneous carbon dioxide and oxygen tension measured at different temperatures in healthy adults, Clin Chem, 31:1611.

# DRIFT IN VIVO OF TRANSCUTANEOUS DUAL ELECTRODES

Colm Lanigan, Jose Ponte and John Moxham

Departments of Thoracic Medicine and Anaesthesia
King's College Hospital
London, U.K.

## INTRODUCTION

Transcutaneous oxygen ($tcPO_2$) and carbon dioxide ($tcPCO_2$) monitoring is widely recommended as a non-invasive means of recording trends in arterial blood gases. The stability of $tcPO_2$ and $tcPCO_2$ readings have usually been reported in terms of in vitro gas calibration data with values for drift of 0.1% per hour being commonly quoted (Löfgren, 1978). These results may have little relevance to clinical practice, where local changes in the skin-electrode interface may affect the measured signal to a greater degree (Löfgren, 1978). We therefore undertook a study of the stability of 3 dual electrodes in vivo in normal adults.

## METHODS

We tested three commercially available combined oxygen and carbon dioxide electrodes: the Microgas Combisensor (Kontron Ltd); the E5270 Radiometer electrode (V.A.Howe & Co Ltd); and the Novametrix 850 Commonsensor (Vickers Medical Ltd). Each electrode was remembraned within 5 days of use and had a 2 point dry gas calibration at 45°C before each study according to manufacturers' instructions. Steady state oxygen and carbon dioxide values for a gas calibration standard (5% $CO_2$ + 12% $O_2$ + 83% $N_2$, Corning Medical Gases Ltd) were then recorded both before and after each study: the difference between the two values was defined as the in vitro drift. An increase in the final recorded value was taken as positive drift, and a decrease was taken as negative drift.

Six healthy non-smoking adult volunteers were recruited following informed consent, and were instructed to recline on a couch for 3 hours whilst reading or listening to music during the continuous recordings. The skin on the medial aspect of the subjects' right arms was rubbed until red, then cleaned with alcohol, before fixing the electrodes to the skin with double-sided adhesive rings. All electrodes were operated at 45°C throughout the entire study.

Nasal endtidal oxygen and carbon dioxide ($ETO_2$ & $ETCO_2$) concentrations were measured by a quadripole mass spectrometer (Airspec 2000, drift less than 0.02% per hour), and corrected for barometric and water vapour pressure

Table I. The In Vivo Drift of Dual Electrodes

| Electrode | | Baseline $PO_2$ | Mean differences (+SD) from baseline (torr) 20mins | 40 | 60 | 80 | 100 | 120 | 140 |
|---|---|---|---|---|---|---|---|---|---|
| Kontron | x | 78 | 4.2 | 5.0 | 6.6 | 7.9 | 10.2 | 12.5 | 12.7 |
| | SD | 12.6 | 4.1 | 5.5 | 5.2 | 6.3 | 7.7 | 7.3 | 7.0 |
| Radiometer | x | 79 | 3.2 | 4.8 | 8.4 | 10.4 | 13.4 | 13.6 | 14.3 |
| | SD | 13.0 | 2.2 | 3.1 | 5.8 | 5.8 | 5.6 | 5.5 | 5.6 |
| Novametrix | x | 80 | 4.2 | 5.3 | 6.7 | 10.2 | 13.7 | 16.9 | 16.7 |
| | SD | 14.3 | 4.0 | 4.4 | 6.1 | 8.6 | 12.5 | 11.8 | 12.7 |
| Endtidal | x | 99 | -0.7 | -2.0 | -0.5 | 1.0 | 0.7 | -0.3 | 3.5 |
| | SD | 4.8 | 2.0 | 2.6 | 2.9 | 3.4 | 2.5 | 3.6 | 5.4 |

and the arterio-alveolar gradient for oxygen (taken as 5 torr breathing air). The outputs from the transcutaneous monitors were recorded on a 6 channel chart recorder at a paper speed of 10 mm/min. $TcPO_2$, $tcPCO_2$ and ET values were averaged over 4 minutes after 30 minutes equilibration (= baseline). The four minute averages at seven subsequent 20 minute intervals were then compared with the baseline value (Wilcoxon signed rank test, $P < 0.05$).

## RESULTS

Average $ETPO_2$ and $ETPCO_2$ remained stable throughout the recording period. However all three electrodes recorded a significant positive drift in the $tcPO_2$ signal, with an average rise of 13 to 17 torr during 140 minutes recording ($P<0.05$, see Table 1). No significant drift in the $tcPCO_2$ signal was observed (average difference from baseline less than 3 torr, $P > 0.05$). These changes could not be accounted for by in vitro drift of either the mass spectrometer or the electrodes (see Fig. 1).

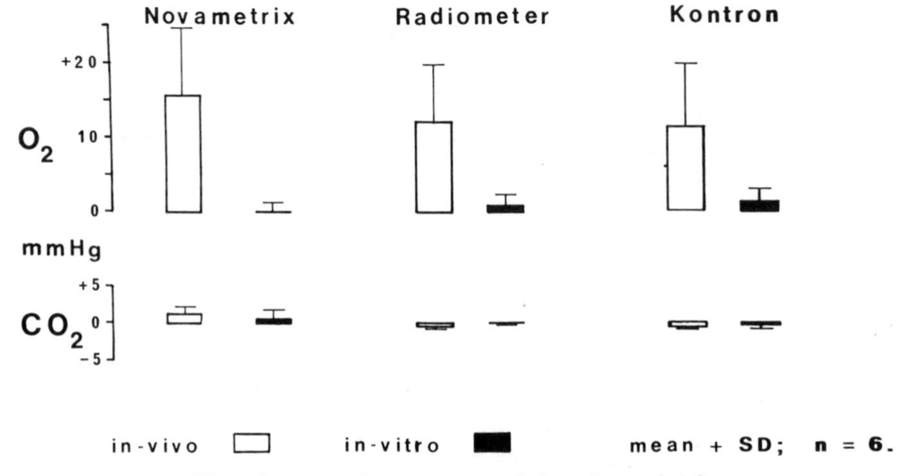


Fig. 1. Total in vivo and in vitro drift.

## DISCUSSION

Our results indicate a positive drift in the $tcPO_2$ signal recorded by transcutaneous dual electrodes in normal adults at rest. Several explanations present themselves:

1) The subjects hyperventilated towards the end of the study. This can be discounted by the stable $ETPO_2$ and $ETPCO_2$ values, the stable $tcPCO_2$ values, and the almost linear rise in $tcPO_2$ values during the study.

2) In vitro drift of the oxygen electrode (Severinghaus, 1981). No significant in vitro drift was observed - steady state in vitro gas calibration values obtained before and after each recording session failed to show any significant average changes (Fig. 1). Although hydroxyl ions produced by a Clark-type polarographic $O_2$ electrode will form $HCO_3^-$ in the presence of $CO_2$ and produce a fall in the recorded $tcPCO_2$ signal, this negative drift was not a problem in this study.

3) Positional changes of the electrodes (Löfgren, 1978). It is well known that changes in the positioning or the pressure applied to a recording electrode can greatly alter the recorded signal: no such changes were detected, and would have been unlikely given that the subjects were at rest, and that both electrodes and cables were securely fixed to the subjects.

4) Increased capillary perfusion (Grønlund, 1985). Skin perfusion is undoubtedly initially increased by local heating of the skin, and is proportional to the electrode operating temperature: a finite maximum to perfusion exists, which one might therefore expect to be better reflected in an exponential, rather than linear rise in the recorded $tcPO_2$ and $tcPCO_2$ values. An analysis of variance indicated that the positive drift in $tcPO_2$ from baseline values can be equally well represented by a linear relationship with time, as by a logarithmic one. Increased capillary perfusion is not the full explanation.

5) Decreased local tissue oxygen consumption (Gøthgen, 1986). The recorded $tcPO_2$ signal reflects the sum of the increased oxygen due to increased capillary perfusion due to local heating, and the increased tissue oxygen consumption caused by increased metabolic demands. Prolonged heating will produce thermal injury and may increase the $tcPO_2$ signal by reducing local tissue oxygen consumption: one might reasonably expect this effect to increase with duration of use, but this was not observed. In addition a small decrease in the $tcPCO_2$ signal should accompany such change, but again this was not observed.

6) Altered skin permeability (Grønlund, 1985, and Gøthgen, 1986). The diffusion coefficient for carbon dioxide is many times that of oxygen, so a change in skin permeability might produce large changes in $tcPO_2$ without noticeable alteration of the $tcPCO_2$ signal. This fact, coupled with the finding of a stable $tcPCO_2$ signal, suggests that the increase in $tcPO_2$ may have been linked to increasing permeability of the skin to oxygen. Inspection of the data reported by Whitehead (1985) indicates that positive drift of the $tcPO_2$ signal, but not the $tcPCO_2$ signal, also occurs when using a single transcutaneous electrode sensor at 44°C in neonates, despite minimal in vitro drift.

We conclude that a large positive drift was observed in the recorded $tcPO_2$ signal in normal adults at rest and that this may be explained in part by increasing skin permeability to oxygen with prolonged heating. The size of this drift makes $tcPO_2$ at 45°C an unreliable trend indicator of arterial oxygen in adults.

SUMMARY

$TcPO_2$ and $tcPCO_2$ monitoring is widely used but the in vitro drift of the new combined sensors is unknown. We tested the in vivo stability of 3 such electrodes in six adults, compared to nasal endtidal values from a mass spectrometer. Each electrode was remembraned within 5 days and had a 2 point dry gas calibration at 45 °C before fixing to the subjects' right arms. $TcPO_2$, $tcPCO_2$ and endtidal values were averaged over four minutes after 30 minutes equilibration, and then at seven subsequent 20 minute intervals. We observed that (1) Endtidal values remained stable (2) $tcPCO_2$ differed from baseline by less than 3 torr (3) $tcPO_2$ rose significantly in all three electrodes by an average of 16 to 21% and, (4) in vivo drift greatly exceeded separately determined in vitro changes. We discuss the possible explanations for the observed results, and conclude that skin permeability changes may play an important role. In the light of the large in vivo $tcPO_2$ drift, transcutaneous dual electrodes are not reliable trend indicators of blood gases in adults.

REFERENCES

Gøthgen, I., 1986, Oxygen tension at the heated skin surface in adults, Acta anaesth.Scan., 30 (80): 1.

Grønlund, J., 1985, Evaluation of factors affecting the relationship between transcutaneous $PO_2$ and probe temperature, J. Appl. Physiol., 59 (4): 1117.

Löfgren, O., 1978, "On transcutaneous $PO_2$ measurements in humans. Some methodological, physiological and clinical studies," Radiometer, Malmö.

Severinghaus, J.W., 1981, A combined transcutaneous $PO_2$-$PCO_2$ electrode with electrochemical $HCO_3^{--}$ stabilization, J. Appl. Physiol., 51 (4): 1027.

Whitehead, M.D., B.V. Lee, T.M. Pagdin, and E.O. Reynolds, Estimation of arterial oxygen and carbon dioxide tensions by a single transcutaneous sensor, Arch. Dis. Child., 1985, 60: 356.

EVALUATION OF SINGLE SENSOR TRANSCUTANEOUS MEASUREMENT

OF $PO_2$ AND $PCO_2$ IN THE NEONATE

J. Messer, A. Livolsi and D. Willard

Service de Néonatologie
Hôpital de Hautepierre
Avanue Molière
67098 Strasbourg Cedex, France

SUMMARY

A single combined transcutaneous sensor for $PO_2$ and $PCO_2$ was evaluated in a neonatal intensive care unit. The values obtained with the combined sensor were compared with the values obtained with two separate electrodes monitoring respectively $PO_2$ and $PCO_2$. Adequate correlations were found. The combined sensor represents an improvement on individual electrodes as it spares available skin surface and needs less handling.

INTRODUCTION

The use of skin surface sensors for the non-invasive measurement of $PO_2$ and $PCO_2$ in neonates is well established. Numerous studies have confirmed their reliability and shown that the results obtained with them correlate well with those obtained by measurements in arterial blood. However, the simultaneous use of two sensors sometimes causes problems, particularly in low birth-weight premature infants. The sensors require two cutaneous sites and these have to be changed every three hours in order to prevent burns. Thus it soon becomes difficult to find an available site ; moreover, the preparation of each sensor takes time. A single, combined sensor, as long as it is as reliable as separate sensors, is bound to be useful. We tested such a sensor (Kontron) and compared it with individual transcutaneous electrodes for $PO_2$ and $PCO_2$. We did not make control measurements of arterial values.

METHODS

Long term recordings were made in 24 children with variable lung disorders : 8 full-term neonates weighing an average of 3240 g, and 16 premature infants (average gestational age 30 weeks, range 28-34 weeks) weighing an average of 1510 g (range 900-2600 g).

Apparatus

A Kontron Microgas 7640 with a combined $PO_2$-$PCO_2$ transcutaneous sensor. Two individual $PO_2$ and $PCO_2$ Kontron sensors with two Kontron 630 monitors. A

Kontron 344 calibrator with a gas 1 calibration gas containing 5 % $CO_2$ and 20.6 % $O_2$ and a gas 2 calibration gas containing 10 % $CO_2$ and 0 % $O_2$.

The study entailed simultaneous recording of $TcPO_2$ and $TcPCO_2$ to compare the values obtained with the single sensor and those obtained with the individual sensors ; the recordings were made on two Kontron 335 recorders at a rate of 0.1 cm per minute.

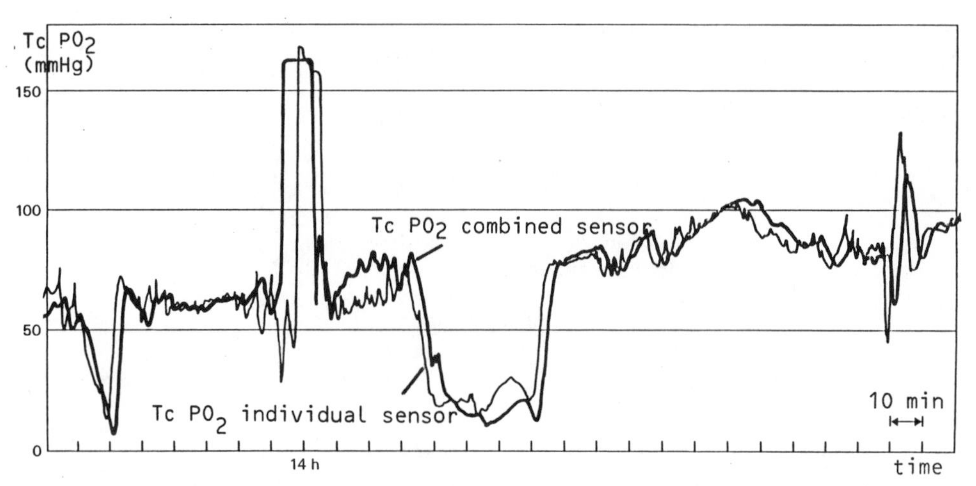


Figure 1 : $TcPO_2$ tracings with combined and individual sensors. Sensor temperature 44 ° C. Distance between pens = 3 mm.

N.B.: During recording, there were 2 hypoxemic episodes in relation to crying. At 14 h, repositioning.

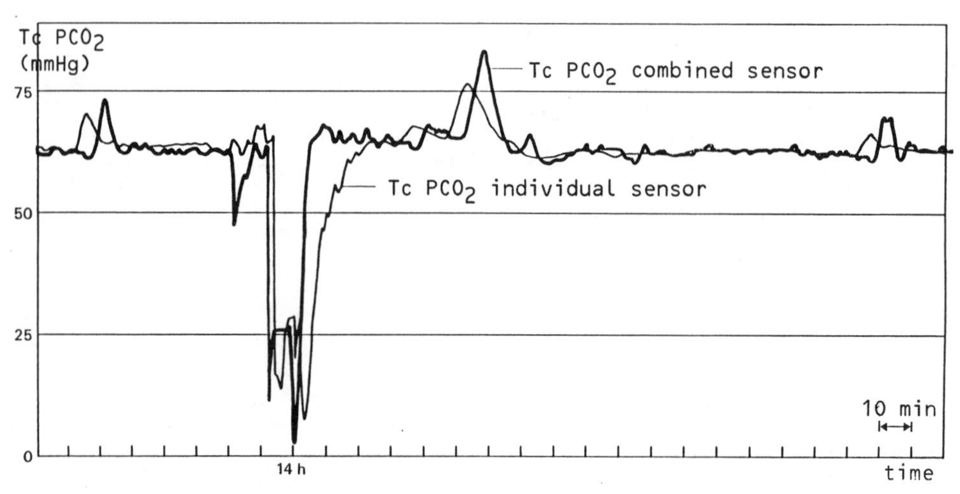


Figure 2 : $TcPCO_2$ tracings with combined and individual sensors. Sensor temperature 44° C.

N.B. : The high $PCO_2$ values are attributable to the child's condition (severe broncho-pulmonary dysplasia).

a) With the combined sensor :

- daily calibrations with gases 1 and 2
- measurement of $PO_2$-$PCO_2$ values in gas 1 every three hours. The values recorded after three hours reveal the drift in the sensor values in relation to the calibration values in gas 1.

The subsequent procedure depended on the instructions for the unit, i. e. either :

- the sensor was reattached (calibration already having been completed)
- or the device required calibration in gas 2 and was repositioned on the child only after the sensor had been calibrated in gases 1 and 2.

The drift was noted down, together with any incident or problem encountered during recording. The sensor temperature was 44° C, and the calibration values were as follows :

- in gas 1 : $PO_2$ : 157 mm Hg ; $PCO_2$ : 28 mm Hg
- in gas 2 : $PO_2$ : 0 mm Hg ; $PCO_2$ : 56 mm Hg, at a barometric pressure of 760 mm Hg.

b) The separate $TcPO_2$ and $TcPCO_2$ sensors were calibrated once daily : the $TcPO_2$ in air, at 157 mm Hg and the $TcPCO_2$ with calibrator, using the same calibration values as for the combined sensor.

c) The three sensors were placed in the same cutaneous zone.

d) The results are based on the period between 08.00 hours on 29.07.85 and 08.00 hours on 03.08.85.

e) Hourly values were obtained for combined sensor $PO_2$ versus individual-sensor $PO_2$ and for combined-sensor $PCO_2$ versus individual-sensor $PCO_2$, i.e. a total of 72 x 2 readings for each pair.

RESULTS

Analysis of the mean values and the standard deviation from the mean revealed a close similarity between the values obtained with the combined sensor and those obtained with the individual sensors.

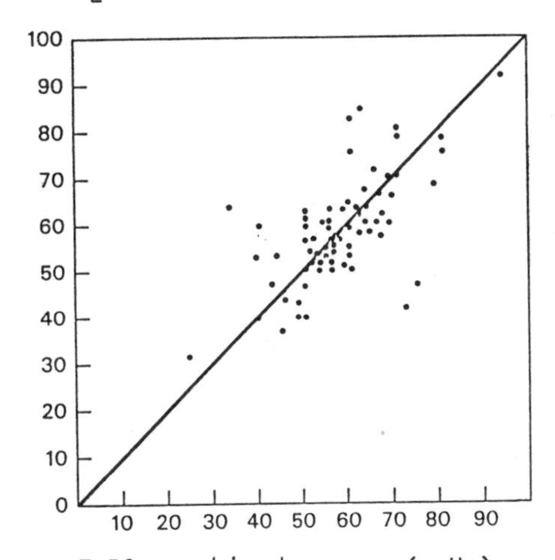


Figure 3 : $TcPO_2$ values recorded with combined and individual sensors, 72 readings

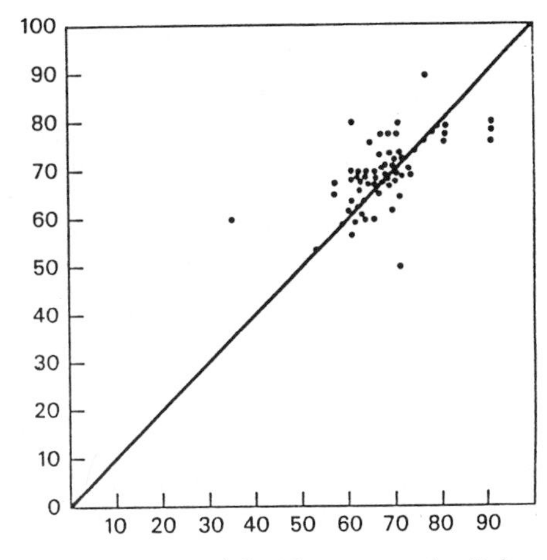


Figure 4 : $TcPCO_2$ values recorded with combined and individual sensors, 72 readings

On the basis of 72 readings, we obtained the following values :

| | Mean value ($\bar{x}$) mmHg | Standard deviation ($O_x$) mmHg |
|---|---|---|
| $PO_2$, individual sensor | 58.1 | 10.9 |
| $PO_2$, combined sensor | 58.1 | 10.8 |
| $PCO_2$, individual sensor | 67.1 | 8.4 |
| $PCO_2$, combined sensor | 69.3 | 7.1 |

Analysis of the drift (D) over three hours revealed that :

- in $PO_2$ monitoring: the mean values for the combined sensor (D = 0.9 mmHg) were more stable than those of the individual sensor (D = 2.1 mmHg) but with a wider scatter for the combined sensor ($O_D$ = 7.5 mmHg) than for the individual sensor ($O_D$ = 1.8 mmHg) ;
- in $PCO_2$ monitoring : readings obtained with the combined sensor showed marked stability : D = 0.5 mmHg, $O_D$ = 0.8 mmHg.

## DISCUSSION

The combined sensor proved to be just as reliable in routine use as the individual sensors. The results are in line with the findings of Whitehead[1] who compared the results obtained with Parker's combined sensor[2] with the arterial values.

### Advantages

The combined sensor spares available skin surface (a significant factor in continuous monitoring of very premature neonates) ; it is easy to use ; preparation and positioning is simpler than with two individual sensors ; the feedback provided by the monitoring of drift in gas 1 improves reliability.

### Disadvantages

The combined sensors so far available must not be exposed to air: they must either be kept in their storage chamber or left in position on the patient. The high degree of reliability presupposes close monitoring of the instrument's functioning. Automatic calibration excludes the option of making manual adjustments, and thus the instrument lacks flexibility. After conducting tests entailing continuous use of the device for twelve hours without calibration, we found wider drift after six hours of use. (The sensor was, of course, repositioned every three hours). In practice, therefore, the following technique should be adopted : calibration in gases 1 and 2 once daily ; verification of drift in gas 1 every six hours ; additional checks on drift whenever the results appear suspect ; systematic checks on the state of membrane and the positioning of the sensor on the child's skin.

## CONCLUSION

We found that the use of a combined sensor offers an improvement on the use of individual sensors. As long as it is used under carefully controlled conditions and by well trained and meticulous personnel, the sensor is of practical value.

## REFERENCES

1. M.D. WHITEHEAD, B.V.W. LEE, T.H. PAGDIN, E.O.R. REYNOLDS, Estimation of oxygen and carbon dioxide tensions by a single transcutaneous sensor. Arch Dis Child 60 : 356 (1985).

2. D. PARKER, D.T. DELPY, E.O.R. REYNOLDS, Single electro-chemical sensor for transcutaneous measurement of $PO_2$ and $PCO_2$, in "Continuous transcutaneous blood gas monitoring", A. Huch, R. Huch, J. Lucey, eds., New-York (1979).

CURRENT CORRECTION FACTORS INADEQUATELY PREDICT THE RELATIONSHIP BETWEEN TRANSCUTANEOUS (tc) AND ARTERIAL PCO2 IN SICK NEONATES

R.J. Martin, A. Beoglos, M.J. Miller, J.M. DiFiore, and W.A. Carlo

Rainbow Babies and Childrens Hospital
Case Western Reserve University, Department of Pediatrics
Cleveland, Ohio

SUMMARY

Despite widespread tcPCO2 monitoring the relationship between tcPCO2 and PaCO2 remains unclear. It has been assumed that after standard temperature correction, a constant metabolic factor can explain the elevation of tcPCO2 over PaCO2. Our data demonstrate a progressive increase in the difference between temperature corrected tcPCO2 and PaCO2 as PaCO2 increases. Thus a constant metabolic factor cannot account for the elevation of temperature corrected tcPCO2 over PaCO2. We speculate that as PaCO2 rises, CO2 production exceeds removal resulting in a progressive gradient between temperature corrected tcPCO2 and PaCO2.

INTRODUCTION

Transcutaneous measurement of carbon dioxide has become a valuable tool in the management of sick neonates. Furthermore, the introduction of combined (O2/CO2) electrodes for transcutaneous monitoring has expanded the role of tcPCO2 monitoring in clinical neonatal care. It has been reported that, after temperature correction, a constant metabolic factor derived from local CO2 production can explain the elevation of tcPCO2 over PaCO2. The purpose of this study was to validate this assumption over a wide range of PaCO2 values in sick infants.

METHODS

The study population comprised 40 normotensive infants with cardiorespiratory disease in whom an umbilical, radial, or tibial arterial line was in place. Mean gestational age was 34±4.3 weeks with a range of 25-42 weeks. Mean birthweight was 2.2±9 kg. with a range of 0.6-4.2 kg. The infants were studied over the first 4±6 days of life with a range of 1-36 days.

Transcutaneous PCO2 was monitored by the Radiometer TCM3 combined transcutaneous electrode. The electrode was freshly membraned prior to calibration. A one-point in vitro calibration was performed with dry gas containing 5.00% CO2 concentration, with the electrode heated to 44$^{\circ}$ C during calibration and throughout each study. The electrode was placed on the infant's abdomen or lower back and 15-30 minutes were required for stabilization of the electrode on each infant. After stabilization (≤2mmHg fluctuations in tcPCO2) was achieved an arterial sample (0.3 mL) was obtained. Arterial blood gases were measured immediately (<5min.) on a Radiometer ABL-2 analyzer.

Uncorrected tcPCO2 was correlated with the corresponding arterial PCO2. In addition, uncorrected tcPCO2 was temperature corrected and the difference between temperature corrected tcPCO2 and PaCO2 was compared to the corresponding PaCO2.

## RESULTS

An excellent correlation existed between the 40 paired arterial and uncorrected tcPCO2 points, r=.95. (Fig.1) However, as temperature corrected tcPCO2 increased, the difference between temperature corrected tcPCO2 and PaCO2 showed a progressive increase. (Fig.2)

## DISCUSSION

These data demonstrate that a constant metabolic factor cannot explain the elevation of temperature corrected tcPCO2 over PaCO2. These findings are in contrast to the data of Hansen and Tooley from 17 sick infants, who reported a relationship between tcPCO2 and PaCO2 that could be accounted for by temperature correction and a constant metabolic factor of 2-3mmHg. Although this disparity of results is not clearly understood, we speculate that at high PCO2 levels, tissue CO2 production exceeds removal, resulting in a progressive increase in the difference between tcPCO2 and PaCO2. Our clarification of the relationship between tcPCO2 and PaCO2 should allow a better understanding and interpretation of tcPCO2 measurements in sick infants.

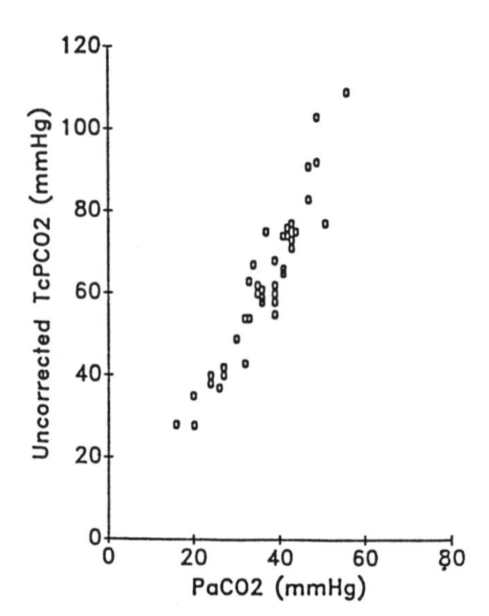


Fig.1 Relation between uncorrected transcutaneous PCO2 and arterial PCO2, r=.95.

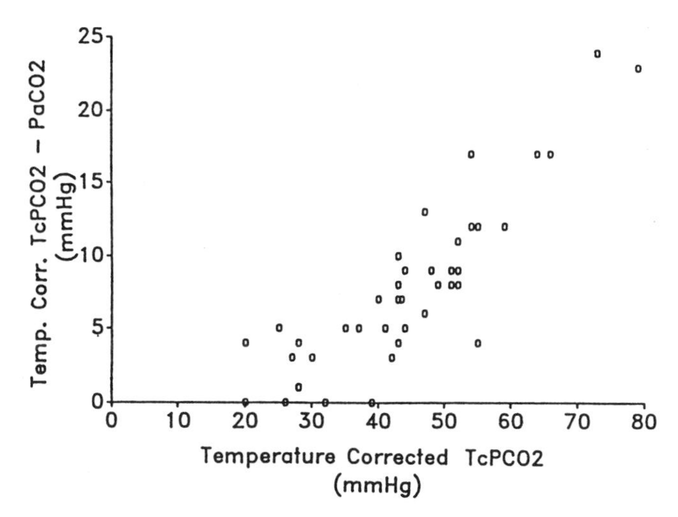


Fig.2 Comparison of the difference between temperature corrected transcutaneous PCO2 and arterial PCO2 versus temperature corrected transcutaneous PCO2.

ACKNOWLEDGEMENTS

This research was supported by NIH (HL25830) and Radiometer Corp.

REFERENCES

1. O. Siggard-Andersen, "The Acid-Base Status of Blood," Munkgaard, Copenhagen(1976).
2. J. Severinghaus, M. Stafford, and A.F. Bradley, TcPCO2 electrode design, calibration and temperature gradient problems, Acta Anesthesiol Scand (Suppl) 68:118-122 (1978).
3. T.N. Hansen and W.H. Tooley, Skin surface carbon dioxide tension in sick infants, Pediatrics 64:942-945 (1979).

# COMPUTING THE OXYGEN STATUS OF THE BLOOD FROM HEATED SKIN $pO_2$

Ivar H. Gøthgen and Erik Jacobsen

Department of Anesthesia
Gentofte Hospital and Rigshospitalet
University of Copenhagen
Denmark

## INTRODUCTION

The non-invasive measurements of $pO_2$ (transcutaneous oxygen) has been shown to be different from arterial $pO_2$ due to dependence upon various factors in the skin.[1] Recently a mathematical model[2] has been confirmed in adult ICU patients.[3] By means of this model, describing the relation between arterial- and skin $pO_2$, other oxygen parameters of the blood can be calculated from skin $pO_2$ ($pO_2(S)$) measurements and thereby the oxygen status of the blood may be evaluated by three parameters, i.e. the partial pressure of oxygen ($pO_2$), saturation of hemoglobin with oxygen ($sO_2$) and the concentration of total oxygen ($ctO_2$) all obtained non-invasively.

The clinical controversy concerning pulse-oximetry versus transcutaneous $pO_2$[4] may be elucidated by computing oxygen saturation from skin $pO_2$. In the present study calculated oxygen parameters are compared to measured values of the same parameters.

## METHODS

In 16 ICU patients (25-81 years) $pO_2(S)$ was measured with an E5240 skin electrode heated to 43°C and a TCM-1 monitor (Radiometer A/S, Copenhagen). Arterial $pO_2$ ($pO_2(aB)$) was measured in ABL-3 blood gas instrument (Radiometer A/S, Copenhagen). Oxygen saturation ($sO_2(aB)$) and hemoglobin concentration were measured in OSM-2 hem-oxymeter (Radiometer A/S, Copenhagen). Calculations of oxygen parameters were done by the previous described technique[3] assuming electrode heating efficiancy = 0.7, arterio-capillary oxygen difference = 0.17 $mmol \cdot l^{-1}$ and capillary-skin $pO_2$ difference = 5 kPa as determinants in the $pO_2(S)$ algorithm.

PARTIAL PRESSURE OF OXYGEN

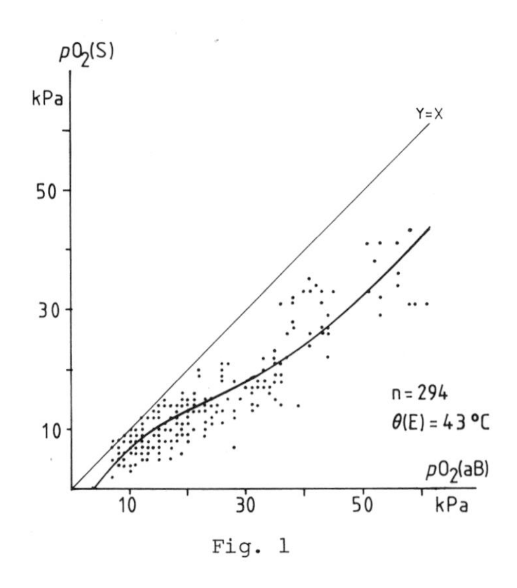


Fig. 1

RESULTS

A total of 294 pairs of data were obtained. Figure 1 shows the plot of skin $pO_2$ versus arterial $pO_2$. The sigmoid curve is the mathematical model with the above given values for the skin factors. Linear regression analysis shows Y = 0.72·X - 0.98 and r = 0.98. In figure 2 measured and calculated oxygen saturation are compared, and the linear regression is Y = 0.96·X + 0.03 with r = 0.70. The mean difference was 0.005 (0.5%), s.d. 0.02. The oxygen content is the sum of hemoglobin bound - and free dissolved oxygen,

CONCENTRATION OF TOTAL OXYGEN

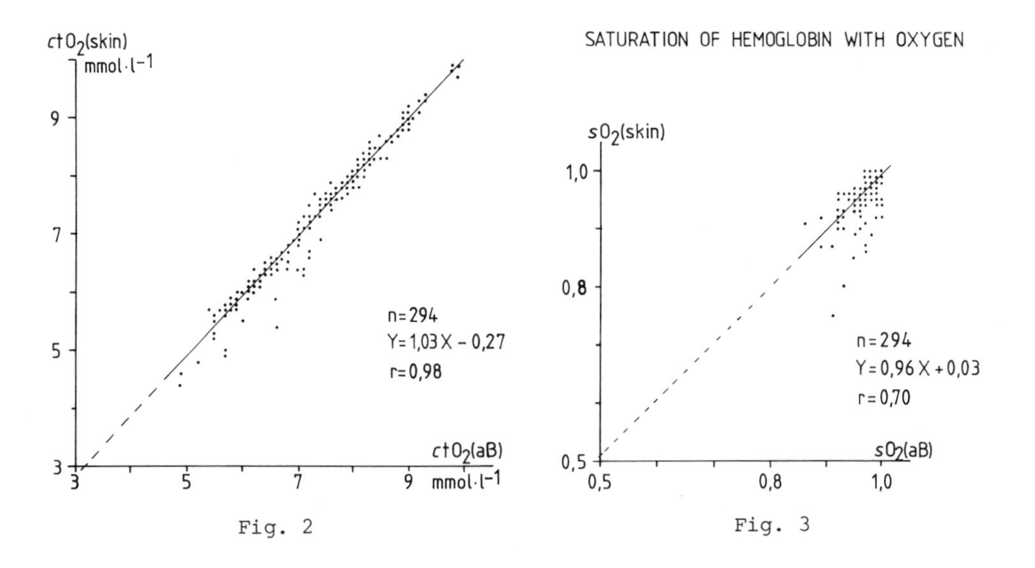


Fig. 2

Fig. 3

i.e. the concentration of total oxygen ($ctO_2$). Comparison of calculated and measured $ctO_2$ is shown in figure 3. The linear regression is $Y = 1.03 X - 0.27$ with $r = 0.98$, and the mean difference was 0.03 $mmol \cdot l^{-1}$.

DISCUSSION

By application of the mathematical model for the relation between skin $pO_2$ and arterial $pO_2$, the oxygen status of arterial blood can be evaluated non-invasively. The $pO_2$ plot (fig. 1) shows a large scattering and non-identity between the two oxygen tensions. In the oxygen-saturation plot and especially in the oxygen-content plot (fig. 2 and 3) the scattering is reduced and almost identity can be demonstrated. Thus the present study demonstrates that the information given by skin $pO_2$ measurement can be increased substantially by calculating other oxygen parameters.

The clinical controversy between skin $pO_2$ and pulse oximetry may also be elucidated by this study. The pulse oximeter discriminates arterial from capillary and venous blood hemoglobin by measuring light absorbance changes coherent with pulsation. The pulse-added light absorbance signal is empirically correlated with arterial hemoglobin saturation, and this empirical relationship is programmed into the oximeter.[4,5] No further calibration is necessary and the calibration cannot be checked during clinical measurements. The skin $pO_2$ technique is based upon electrochemical measurement and needs calibration with gas mixtures before clinical measurements, but the measured $pO_2$ is the real $pO_2$ on the skin surface without any empirical correction. The $pO_2(S)$ algorithm is thus the equivalent of the empiric computer program for pulse-oximetry and therefore direct comparison of skin $pO_2$ and pulse-oximetry are possible. Table 1 shows such a comparison and indicates that the precision and accuracy in esti-

Table 1. Comparison of pulse-oximetry and skin $pO_2$ by linear regression of arterial oxygen saturation (X-axis) versus saturation measured by pulse-oximetry or computed from skin $pO_2$ (Y-axis)

| Type of measurements | n | slope | intercept | r |
|---|---|---|---|---|
| $sO_2$ (BIOX III) Anesthesiology 1985: 63: A 175 | 326 | 0.93 | 5.2 | 0.57 |
| $sO_2$ (NELCOR) J Clin Monit 1985: 1: 156-160 | 277 | 0.94 | 5.1 | 0.98 |
| $pO_2$ (TCM-1) Present study | 294 | 0.96 | 3.0 | 0.70 |

mating arterial oxygen saturation are equal for the two non-invasive methods of oxygen monitoring.

The disadvantages of skin $pO_2$ measurements are the calibration, the warm-up time and the risk of skin burns. The disadvantages of pulse-oximetry are the empiric calibration and no trend of $pO_2$ during normo- and hyperoxi.[4] If the two methods are combined, the disadvantages of the individual methods may be minimized, e.g. low skin $pO_2$ and high $sO_2$ may indicate low skin blood flow and/or low concentration of oxyhemoglobin, high $pO_2$ and high $sO_2$ indicate significant hyperoxi and high skin $pO_2$ and low $sO_2$ indicate sensor error etc.

CONCLUSION

The high correlation and almost identity in oxygen saturation and oxygen content indicate that the value of skin $pO_2$ can be increased substantially by computing other oxygen parameters. The $sO_2$ correlation is comparable in accuracy with results obtained by pulse-oximetry, and in this way the two methods of non-invasive oxygen monitoring may be compared. If the two methods are combined, ideally in one sensor, measured and computed $sO_2$ can be used to predict arterial $pO_2$ more precisely and the drawbacks of the individual methods may be minimized.

References


1. R. Huch, A. Huch, and D. W. Lübbers, "Transcutaneous $pO_2$", Georg Thime Verlag, Stuttgart, New York (1981).
2. J. W. Severinghaus, M. Stafford, and A. M. Thunstrom, Estimation of skin metabolism and blood flow with $tcpO_2$- and $tcpCO_2$ electrodes by cuff occlusion of the circulation, Acta Anaesthesiol Scand, Suppl 68:9 (1978).
3. I. H. Gøthgen, Oxygen tension on the heated skin surface in adults, Acta Anaesthesiol Scand, Suppl 83:1 (1986).
4. Clinical Controversy: New/Barker and Tremper, Pulse oximetry versus measurement of transcutaneous oxygen, J Clin Monit, 1:126 (1985).
5. N. Mackenzie, Comparison of a pulse oximeter with an ear oximeter and an in-vitro oximeter, J Clin Monit, 1:156 (1985).

# APPLICATION IN ADULTS

TRANSCUTANEOUS MONITORING OF $PO_2$ AND $PCO_2$ DURING RUNNING -

A NONINVASIVE DETERMINATION OF GAS TRANSPORT

Jürgen M. Steinacker and Kai Röcker

Department of Applied Physiology
University of Ulm
D-7900 Ulm, Fed. Rep. of Germany

SUMMARY

Transcutaneous $pO_2$ and $pCO_2$ ($tcpO_2$ and $tcpCO_2$) were measured during running with stepwise increased velocities and with constant speed, under both aerobic and anaerobic conditions, for the determination of blood gas transport during exercise.

Arterial and transcutaneous blood gas values correlated significantly ($pO_2$ $r=0.87$, $p<0.001$, $pCO_2$ $r=0.91$, $p<0.001$ respectively). Transcutaneous $pCO_2$ is a noninvasive method of monitoring arterial $pCO_2$ and lactate formation during exercise. When athletes run, arterial $pO_2$ falls to a specific limit depending on the intensity of work. This seems to be characteristic for maximum oxygen transport capacity. The aerobic endurance measured by the aerobic-anaerobic threshold may be dependent on the possibility of sustaining low arterial $pO_2$ during high working levels at high oxygen consumption.

INTRODUCTION

Arterial $pO_2$ and $pCO_2$ during exercise can be monitored continuously by means of transcutaneous measurements[3,4]. We have shown that the changes of arterial $pO_2$ and $pCO_2$ during exercise are related to the intensity of work and to the physical fitness of the subject tested[3,4]. A fall in $p_aO_2$ in trained subjects is related to an increase in $\dot{V}O_2$. During incremental exercise, the increase in blood lactate concentration above the aerobic-anaerobic threshold (AAS) is explained by the assumption that anaerobic ways of cellular metabolism are used above the AAS because the aerobic mitochondrial metabolism reach their maximum capacity at the AAS. It has been shown that lactate increase is influenced by $pO_2$-delivery, exspecially in the case of hypoxia[1,2].

In this study, transcutaneous measurements of $pO_2$ and $pCO_2$ are examined during running in well-trained athletes. It would be of interest to determine whether metabolism during strenuous exercise is influenced by gas transport and tissue hypoxia.

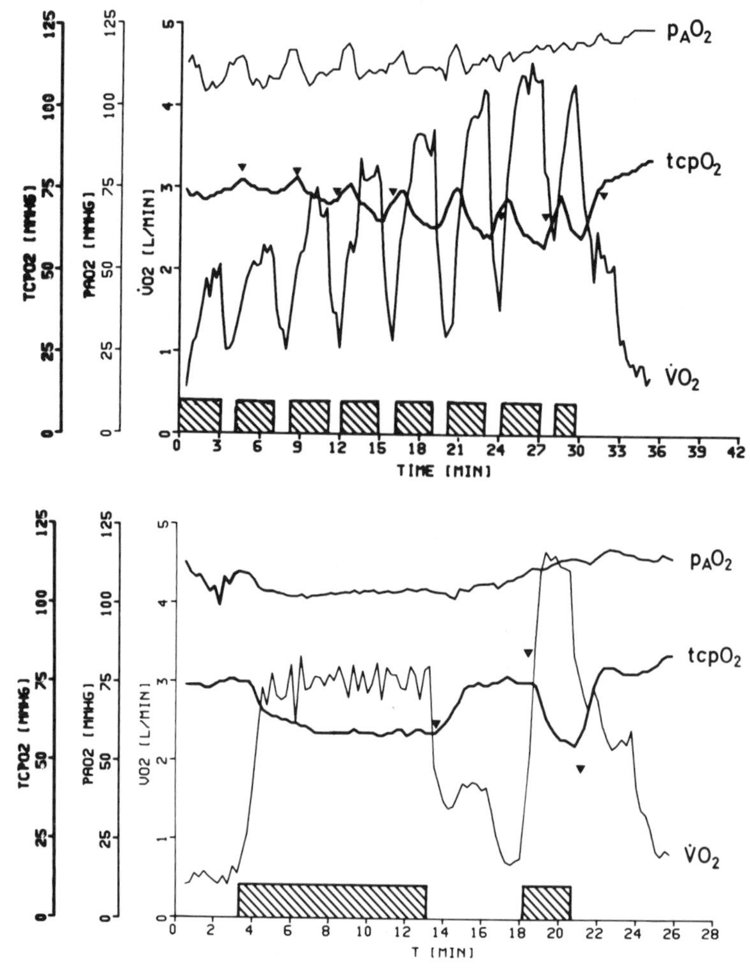


Fig.1: $tcpO_2$, $p_aO_2$ (▼), $p_AO_2$ and $\dot{V}O_2$ during the MST (upper panel) and AT and ANT (lower panel) on the treadmill of a well-trained subject. The bars indicate the duration of running. For explanation, see text.

METHODS

Transcutaneous $pO_2$ and $tcpCO_2$ were measured during running on a treadmill starting with a multi-stage-test (MST) with stepwise increased velocities beginning with 8 km/h and an increase of 2 km/h every 3.5 min. Two hours after the MST, an aerobic test (AT) was performed at a velocity near the aerobic-anaerobic threshold (AAS) for 10 min followed by an anaerobic test (ANT) at maximum speed for 1-2 min. Eleven well-trained middle-distance runners were studied. Their mean maximum velocity at the MST was 21.6 km/h (s=±1.62), their mean AAS was at 18.8 km/h (s=±0.81), which indicates a high aerobic endurance. $\dot{V}_E$, $\dot{V}O_2$ and the expiratory gas pressures were measured with an open spirometric system. Endexspiratory alveolar $pO_2$ ($p_AO_2$) was monitored by a fast polarographic electrode. Blood gases, lactate and hemoglobin concentrations were measured in capillary blood samples taken between intervals of MST and before and after AT and ANT. The electrodes were fixed on the upper back of the athletes. The electrode temperature for $tcpO_2$ was 45°C and for $tcpCO_2$ 44°C.

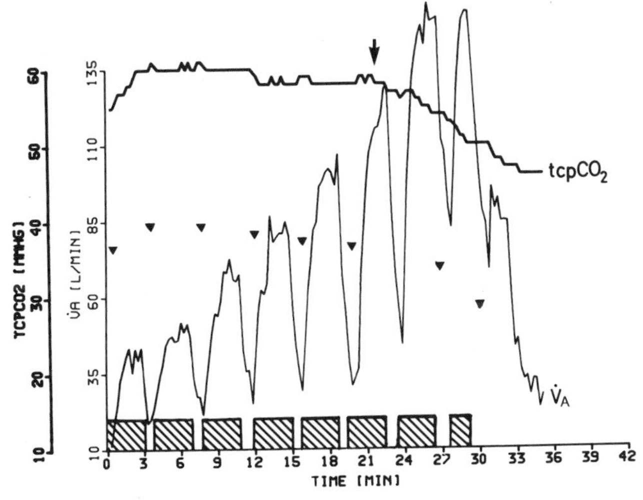


Fig.2: $tcpCO_2$, $p_aCO_2$ (▼) and $\dot{V}_A$ during the MST on the treadmill (↓ =AAS).

The change of $tcpO_2$ and $tcpCO_2$ during exercise from the values at rest was calculated as $\Delta pO_2$ and $\Delta pCO_2$.

## RESULTS AND DISCUSSION

### Methodological Aspects

The results found in the different tests are illustrated by typical records of a well-trained subject in Fig.1 and 2. Transcutaneous and capillary values of $pO_2$ from all subjects were found to correlate highly significantly ($r=0.87$, $p<0.001$, for $pCO_2$ $r=0.91$, $p<0.001$). We have previously shown that $tcpO_2$ during exercise reflects arterial $pO_2$[4]. Therefore, reference in the discussion will be arterial $pO_2$ although $pO_2$ was measured transcutaneously.

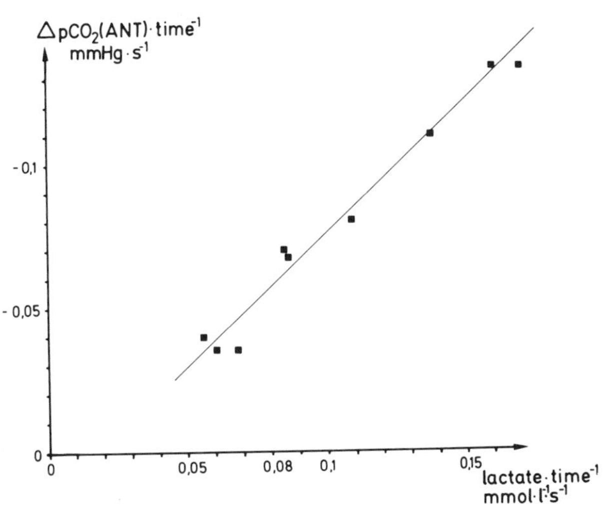


Fig.3: $\Delta pCO_2$ per time versus lactate increase in the anaerobic test (ANT) (regr.line $y=0.96x-0.02$, $r=-0.97$, $p<0.001$, $n=10$)

Transcutaneous monitoring is superior to invasive methods of monitoring arterial blood gases in physiological exercise studies because the error may be calculated. This is more uncertain for blood samples, which have to be handled with care and are difficult to obtain in exercising men. It seems evident that the discontinuous methods of measurement, involving arterial blood sampling und subsequent gas analysis, are not sufficient for exact determination of rapid changes of blood gases. Transcutaneous measurements allow the use of on-line computer monitoring.

No correction was made for the anaerobic change of $p_aCO_2$ by the heating temperature. Therefore, $tcpCO_2$ overestimates $p_aCO_2$ by a constant factor. A significant correlation between $\Delta pCO_{2max}$ and the maximum lactate concentrations $La_{max}$ was calculated for the MST ($r=-0.89$, $p<0.001$) and for the ANT ($r=-0.73$, $p<0.01$). Calculating the change per time (Fig.3), a highly significant correlation was found between $\Delta pCO_{2max}$ and $La_{max}$ ($r=-0.97$, $p<0.001$) at the ANT. Thus, $tcpCO_2$ seems to be a good indicator of lactate, not only for $p_aCO_2$. In further investigations it may be possible to monitor lactate non-invasively by $tcpCO_2$.

## Oxygen Tension During Running

In the MST, $tcpO_2$ decreased at every stage of running and increased during the breaks between the stages. A reciprocal time course was seen in $p_AO_2$, the $AaDO_2$ was greater with higher $\dot{V}O_2$. The $AaDO_2$ at the AT and at the ANT corresponds to the values at the MST (Fig.1), which means that the $AaDO_2$ is related to the intensity of work.

Alveolar ventilation is not the limiting factor for $O_2$-exchange. This can be concluded from the increase of $p_AO_2$ during incremental exercise. The influence of venous shunts and pulmonary circulation in the lung is diminished at high work load[5]. The increase in $AaDO_2$ is probably due to diffusion limitation of $O_2$ and can be explained by a reduced contact time for gas exchange in pulmonary capillaries at high cardiac output[3,4,5]. In AT and ANT, $pO_2$ adjusted at a specific level (Fig.1). The maximum changes in $pO_2$ during MST and ANT did not correlate significantly ($r=0.42$). Better results were found for $\Delta pO_2$ at the AAS (MST) and at AT ($r=0.58$, $p<0,05$), where a steady state was attained.

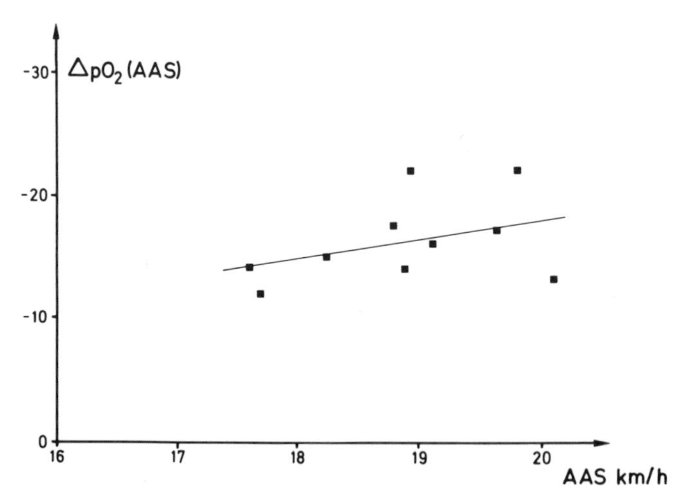


Fig.4: Running velocity at the aerobic-anaerobic threshold (AAS) and $\Delta pO_2$ at the AAS (regr.line $y=-1.6x+14.5$, $r=-0.38$, $n=10$)

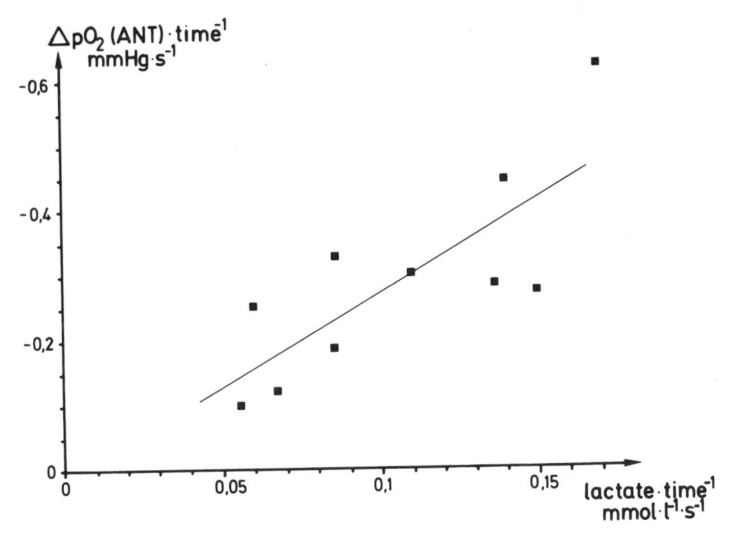


Fig.5: $\Delta pO_2$ per time versus lactate increase in the ANT (regr.line $y=3.01x-0.04$, $r=-0.79$, $p<0.01$, $n=10$).

This dependency of $\Delta pO_2$ on performance was shown in our previous studies[3,4]. The decrease in $pO_2$ also seems to be related to the aerobic capacity of the athlete. The running velocity at the AAS, which is specific for the aerobic endurance, tends to correlate to $\Delta pO_2$ at the AAS in this homogeneous group of well-trained athletes (Fig.4), but this correlation is not significant.

At maximum work in the ANT it was found that $\Delta pO_{2max}$ had a significant correlation to maximum lactate concentrations calculating the change per time ($r=-0.79$, $p<0.01$). The duration of ANT and the increase in lactic acid correspond to the anaerobic capacities. The decrease in $pO_2$ correlated linearly to the increase in lactate concentration and therefore also to anaerobic metabolism (Fig.5).

These results support a hypothesis that the AAS is related not only to the power of the mitochondrial substrate turnover but also to the capacity of oxygen transport system[4]. The AAS may depend on the possibility of sustaining low arterial $pO_2$ during hard work at high oxygen-turnover. Athletes, trained for endurance and therefore for increased AAS, are able to maintain these turnover-rates by special adaptions: Increased cardiac output[4,5,6], increased hemoglobin[1,5] and myoglobin concentrations[1], increased capillarisation and decreased vascular resistance in trained muscles and enlarged mitochondrial volume[1].

Earlier results indicate that $pO_2$ is dependent on hemoglobin concentrations and that $\Delta pO_2$ is correlated with the decrease in venous oxygen content and the increase in heart rate and cardiac output during exercise[3,4]. The decrease in $p_aO_2$ at high $\dot{V}O_2$ may effect adaptive mechanisms in well-trained athletes and may result in an improved tolerance to cellular hypoxia[3,4]. Lactate concentrations during exercise are significantly higher during hypoxia compared with hyperoxia[2]. These differences in lactate accumulation might be explained by different tissue $pO_2$. The findings in our study might have the same explanation. As there was a mean decrease of 21.5 mmHg (±8.33) of $pO_2$ during anerobic exercise in the ANT, the reduced arterial $p_aO_2$ level could be a contributing factor to lactate formation in these experiments. The influence of $p_aO_2$ on lactate formation during work needs further investigations.

Carbon Dioxide Tension and the Regulation of Breathing During Running

During moderate intensity exercise, at any level of $p_aCO_2$, $\dot{V}_A$ changes as a linear function of $\dot{V}CO_2$ in response to steady state increases in work rate[6]. The time constants of $\dot{V}CO_2$ and $\dot{V}O_2$ responses to steady state increases in work are different. The slower kinetics of $\dot{V}CO_2$ reflects the considerable capacity for tissue $CO_2$ storage. As a consequence, at the beginning of moderate exercise, a transient fall in $p_aO_2$ is evident, the slightly slower time course of $\dot{V}_E$ elicits a transient rise in $p_aCO_2$ (Fig.1, MST, Fig.2)[6].

With increasing steady state work intensity, lactic acid production is elevated. The hydrogen ions of dissociated lactic acid are buffered by bicarbonate both in the muscle and in the blood. The concentration of bicarbonate in the blood decreases, $CO_2$ is liberated and exhaled by the lungs. The change in lactate and pH is inversely related to the change in $pCO_2$. At the AAS, $pCO_2$ decreases, $\dot{V}CO_2$ and $\dot{V}_A$ increase exponentially[1,2,6]. The aerobic-anaerobic threshold can be determined non-invasively from the point of beginning $tcpCO_2$ decrease (Fig.2). It is interesting that in another study $tcpCO_2$ measured at 37°C increased during lactate accumulation[7]. This might be explained by the fact that at 37°C, $tcpCO_2$ is more dependent on local metabolism than on arterial $pCO_2$.

The exponential increase of $\dot{V}_A$ at the AAS cannot be fully explained by compensatory hyperpnea for acute metabolic acidosis due to changes in lactate, pH and $\dot{V}CO_2$[1]. It was found that the ventilatory response to exercise is modified by hypoxia or hyperoxia[6]. It should be discussed whether parts of ventilatory response to exercise are due to tissue hypoxia.

SPECIAL ABBREVIATIONS

| | | | | | |
|---|---|---|---|---|---|
| AAS | = | aerobic-anaerobic threshold | AT | = | aerobic test |
| MST | = | multi stage test | ANT | = | anaerobic test |

ACKNOWLEDGEMENT

This study was supported by the Deutsche Forschungsgemeinschaft. We thank Dr.P.Meier for her help in preparation of this manuscript.

REFERENCES

1.) L.B. Gladden, Current "anaerobic threshold" controversies, Physiologist 27:312 (1984)
2.) M.C. Hogan, R.H. Cox, H.G. Welch, Lactate accumulation during incremental exercise with varied inspired oxygen fractions, J Appl. Physiol. 55:1134 (1983)
3.) J.M. Steinacker, R.E. Wodick, Transcutaneous $pO_2$ during exercise, Adv. Exp. Med. Biol. 169:763 (1984)
4.) J.M. Steinacker, R.E. Wodick, Möglichkeiten und Grenzen der transcutanen Bestimmung des arteriellen $pO_2$ und $pCO_2$ bei der Ergospirometrie. Med. Welt 37:193 (1986)
5.) G. Thews, Theoretical Analysis of the pulmonary gas exchange at rest and during exercise, Int. J. Sports Med. 5:113 (1984)
6.) B.J. Whipp, S.A. Ward, Ventilatory control dynamics during muscular exercise in man, Int. J. Sports Med. 1:146 (1980)
7.) U. Ewald, T. Tuvemo, G. Rooth, Detection of exercise-induced lactic acidosis using transcutaneous carbon dioxide, Crit. Care Med. 13:630 (1985)

TRANSCUTANEOUS OXYGEN TENSION DURING EXERCISE IN PATIENTS WITH PULMONARY EMPHYSEMA

D.C.S. Hutchison, B.J. Gray, J.M. Callaghan and R.W. Heaton

Department of Thoracic Medicine, King's College School of Medicine, London SE5

SUMMARY

Transcutaneous $PO_2$ ($tcPO_2$) and arterial $PO_2$ ($PaO_2$) were compared during exercise in six patients with pulmonary emphysema. For calibration purposes, the $tcPO_2$ electrode was first attached to the skin and after stabilisation its reading was adjusted to correspond to the $PO_2$ of an initial arterial blood sample. It was shown that $tcPO_2$ measurement could follow accurately the rapid changes in $PaO_2$ occurring during exercise. Sixty-eight paired comparisons of $PaO_2$ and $tcPO_2$ were available and the regression equation was given by: $tcPO_2$ (mmHg) = 0.98 $PaO_2$ + 0.7 (correlation coefficient, 0.985; 95% confidence limits, 5.7 mmHg).

INTRODUCTION

The changes in arterial $PO_2$ ($PaO_2$) which take place during exercise in patients with pulmonary disease are of physiological and clinical interest. $PaO_2$ has been shown (1) to fall during exercise in emphysematous patients; repeated sampling of arterial blood, however, requires insertion of an arterial cannula, an invasive method with possible complications.

Transcutaneous oxygen electrodes are commonly calibrated by a two point procedure before application of the electrode to the skin. The relationship between $PaO_2$ and transcutaneous $PO_2$ ($tcPO_2$) obtained by this method was found to be unsatisfactory (2), presumably due to unpredictable variations in skin metabolism or blood flow. Since we were primarily interested in the function of the lungs we aimed to carry out a direct comparison of $tcPO_2$ and $PaO_2$ during exercise in patients with pulmonary emphysema using a preliminary estimation of $PaO_2$ as the upper calibration point.

PATIENTS AND METHODS

$tcPO_2$ and $PaO_2$ were compared in six male patients (mean age 58.5 years: range 52-67) with chronic airflow obstruction and radiological evidence of pulmonary emphysema. Their mean $FEV_1$ (as % of predicted)

was 35% (SD 10) and mean CO transfer factor (as % of predicted) 41% (SD 12). The patients gave informed consent and the project was approved by the King's College Hospital ethical committee.

Transcutaneous oxygen tension ($tcPO_2$) was measured by means of a Radiometer E5240 electrode heated to 45°C and applied to the skin surface over the biceps muscle. The electrode was used in conjunction with a TCM1 monitor and its output could be read digitally on the monitor or displayed on a chart recorder. An in-dwelling arterial cannula was inserted into the radial or brachial artery under local anaesthesia. When the $tcPO_2$ reading had stabilised (not more than 5 mmHg change over 5 min) an arterial blood sample was obtained and the $PaO_2$ measured with a Radiometer ABL2 analyser. The $tcPO_2$ reading was then adjusted to correspond to the $PaO_2$, using the monitor gain control.

The patients then undertook a progressive exercise test on an electrically braked cycle ergometer. The $tcPO_2$ was monitored for at least 1 min while the patient was seated at rest on the ergometer until a stable reading was obtained and was monitored throughout the exercise study and for 30 min after the end of exercise. Arterial blood samples were taken as often as possible during this period and $tcPO_2$ was read from the digital meter at the mid-point of arterial sampling. During the exercise period, the load was increased by 10 watts every min until the maximum load was achieved. The electrocardiogram was displayed throughout the procedure but no dysrhythmic or ischaemic episodes were observed.

## RESULTS

In all six patients a similar biphasic pattern in $tcPO_2$ was observed (Fig. 1). After the start of exercise there was a progressive fall in $tcPO_2$; immediately after the end of exercise there was a rapid rise in $tcPO_2$ which greatly exceeded the resting values. The $PaO_2$ values corresponded closely to the $tcPO_2$ values at each time point.

Sixty-eight paired comparisons of $PaO_2$ and $tcPO_2$ were available and the regression equation was given by: $tcPO_2$ (mmHg)=0.98 $PaO_2$ + 0.7. The correlation coefficient was 0.985 and the 95% confidence limits 5.7 mmHg.

## DISCUSSION

In this study we have shown that $tcPO_2$ measurement can accurately follow the rapid changes in $PaO_2$ which take place during and after exercise in pulmonary emphysema. The regression line relating $PaO_2$ and $tcPO_2$ was not significantly different from the line of identity and the 95% confidence limits (5.7 mmHg) were satisfactory. One might have expected that changes in skin blood flow during exercise would influence the relationship but this did not prove to be the case. Extrapolation of these results to patients with other disorders, however, would be unwise.

These observations confirm the validity of our previous findings (3) in 23 emphysematous patients; in that study it appeared that the magnitude of the fall in $tcPO_2$ during exercise was related to the severity of lung function impairment. In emphysema, the destruction of the lung parenchyma is likely to be associated with a reduction in the pulmonary capillary volume which would result in a reduction in the transit time of blood through the capillary. The oxygen tension of the pulmonary capillary blood may, therefore, not have time to reach

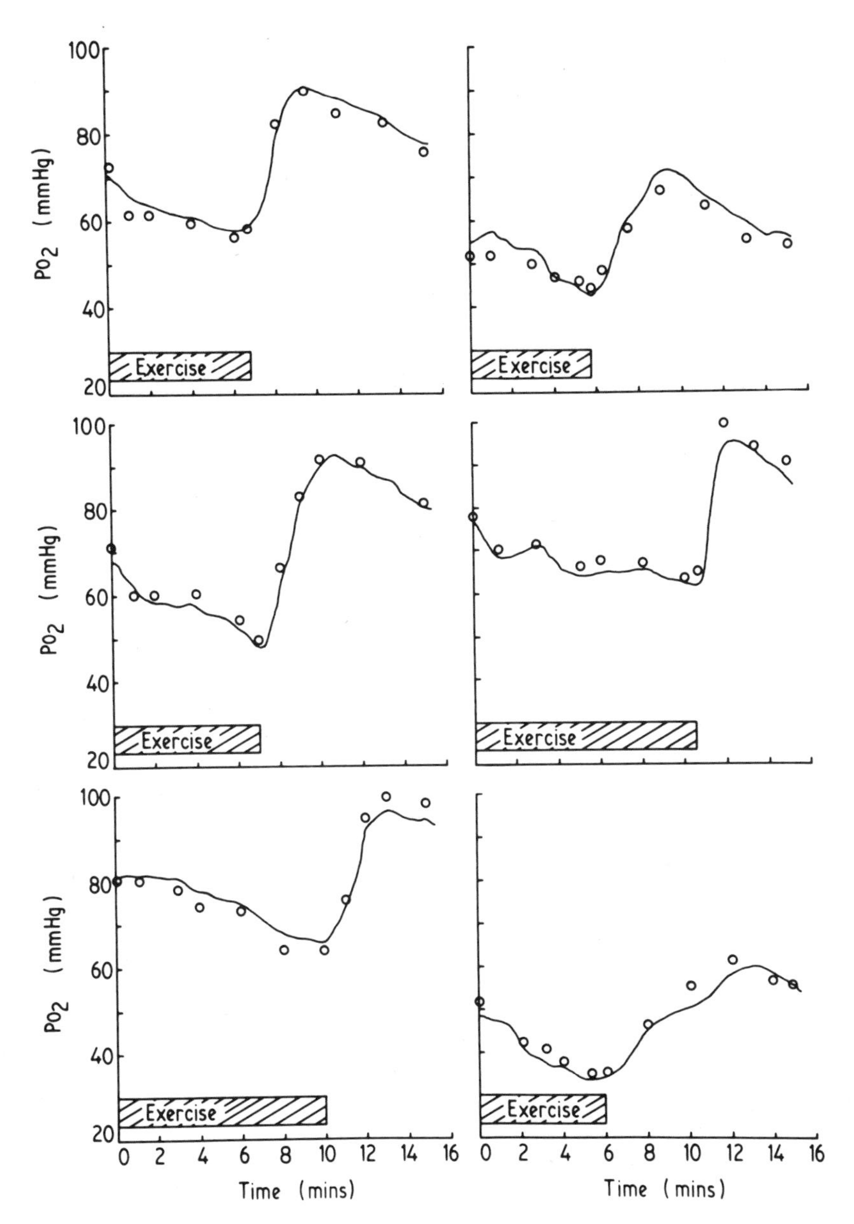


Fig 1: Simultaneous measurements of $tcPO_2$ and $PaO_2$ in six emphysematous patients during progressive exercise tests. Continuous line: $tcPO_2$. Circles: $PaO_2$. Reproduced by permission of the Editor of 'Breath'.

equilibrium with alveolar gas, an effect which would be enhanced during exercise.

At the end of the exercise period, $tcPO_2$ underwent an immediate and rapid increase in all six patients; this is likely to be due to the sudden reduction in oxygen consumption leading to a sharp rise in central venous $PO_2$, while ventilation and cardiac output remain considerably above normal.

We conclude that measurement of $tcPO_2$ can provide a valid estimate of $PaO_2$ during exercise in emphysematous patients. The $tcPO_2$ electrode could be calibrated using a single arterial sample which under local anaesthesia is normally a safe and painfree procedure.

REFERENCES

1. Jones N L. Pulmonary gas exchange during exercise in patients with chronic airway obstruction. Clin Sci 1966; 31: 39-50.
2. Hutchison D C S, Rocca G, Honeybourne D. Estimation of arterial oxygen tension in adult subjects using a transcutaneous electrode. Thorax 1981; 36: 473-77.
3. Hughes J A, Gray B J, Hutchison D C S. Changes in transcutaneous oxygen tension during exercise in pulmonary emphysema. Thorax 1984; 39: 424-431.

THE EFFECT OF INHALED BRONCHOCONSTRICTORS ON TRANSCUTANEOUS GAS TENSIONS IN NORMAL ADULT SUBJECTS

Barry Gray and Neil Barnes

Department of Thoracic Medicine
King's College School of Medicine & Dentistry
London, U.K.

SUMMARY

The administration of histamine and leukotriene $D_4$ ($LTD_4$) by nebulised aerosol in logarithmically increasing doses to normal subjects resulted in significant bronchoconstriction. Transcutaneous oxygen tension ($tcPO_2$) was monitored during and after the bronchial challenge tests. Following histamine challenge there was significant hypoxaemia in all subjects (mean fall in $tcPO_2$, 20mmHg). However, following $LTD_4$ administration, there was a small and insignificant fall in $tcPO_2$. Transcutaneous carbon dioxide tension ($tcPCO_2$) was also monitored throughout bronchial challenge, but showed no significant change. We suggest that the hypoxaemia following histamine challenge was due to increased ventilation/perfusion (V/Q) mismatching in the lung induced by histamine deposition.

INTRODUCTION

Bronchial challenge testing by the administration of nebulised bronchoconstrictor substances has typically measured responses in terms of changes in airway calibre and lung volumes. Since gas exchange is the prime function of the human lung it is remarkable that so little work has investigated the changes in arterial blood gases (ABGs)) during and after bronchial challenge, especially as changes in ABGs are important considerations in the investigation of asthma. We have investigated the effects of administering by nebuliser, solutions of the bronchoconstrictors histamine acid phosphate and leukotriene $D_4$ ($LTD_4$) on the blood gases of five normal adult subjects.

METHODS

Five normal subjects (four male, one female), age range 24-36, were studied. All were non-atopic non-smokers with normal lung function. There were two study days for each subject separated by at least one week. On one day histamine acid phosphate was given in buffered normal saline in increasing concentrations (1/2 log 10 increases in dose) at 10 minute intervals up to a maximum of $10^{-1}$ moles per litre. On the other day, $LTD_4$ was administered in increasing concentrations (1/2 log 10 increases in dose) at 10 minute intervals up to a maximum dose of 2 x $10^{-4}$ moles/l. On each day, baseline measurements of airway calibre

($FEV_1$ and flow at thirty percent of vital capacity, $Vmax_{30}$) were performed. Blood gases were monitored during the study by a Radiometer E5243/0 $tcPO_2$ electrode and a Radiometer E5230 $tcPCO_2$ electrode attached to the skin over the upper arm at an electrode temperature of $45^oC$. Calibration of the $tcPO_2$ electrode was by an electrical zero point, and an air calibration. The $tcPCO_2$ electrode was calibrated in 5% and 10% carbon dioxide gas solution using a Radiometer A7405TC calibration unit. After application, the electrodes were allowed to reach a stable output in situ (time taken usually about 20 minutes) and then re-calibrated by an in vivo method[1]. The reading of each electrode was calibrated to correspond to the $PO_2$ and $PCO_2$ of an end tidal gas sample measured on a mass spectrometer. Re-calibration of $tcPO_2$ and $tcPCO_2$ readings was achieved by altering the gain control on the respective monitors until $tcPO_2$ was equal to end-tidal $PO_2$ (with an allowance of 3 to 5 mmHg deducted for alveolar-arterial difference) and $tcPCO_2$ equalled end-tidal $PCO_2$.

Following these baseline measurements and calibrations, a saline aerosol generated by a Wright nebuliser was administered as a control and the measurement of $FEV_1$ and $Vmax_{30}$ repeated. At 10 minute intervals logarithmically increasing doses of nebulised histamine or $LTD_4$ were administered to the maximum tolerated dose, and the measurements of airway calibre repeated. The $tcPO_2$ and $tcPCO_2$ were monitored throughout each study, after in vivo calibration.

RESULTS

Baseline $tcPO_2$ $\pm$ SEM was 97 $\pm$ 2.5 mmHg before histamine challenge and 100 $\pm$ 3 mmHg before $LTD_4$ challenge. The respective mean baseline $tcPCO_2$ measurements were 39.2 mmHg before histamine and 40.1 mmHg before $LTD_4$. The baseline $Vmax_{30}$ in each individual subject was well matched on the two study days with less than 10% variation between days. The inhalation of both histamine and $LTD_4$ resulted in significant dose-dependent falls in $Vmax_{30}$. The maximum concentration of histamine inhaled by all subjects ($3.2 \times 10^{-2}$ moles/l) resulted in a mean fall in $Vmax_{30}$ to 75 $\pm$ 7.4% of baseline and a significant fall in mean $tcPO_2$ to 77 $\pm$ 7.7 mmHg. For $LTD_4$ the pattern of response was quite different. At the maximum concentration of $LTD_4$ inhaled by all subjects ($6.4 \times 10^{-5}$ moles/l) mean $Vmax_{30}$ fell to 63 $\pm$ 7% of baseline but the mean fall in $tcPO_2$ was small and not statistically significant (to 93.1 $\pm$ 2 mmHg). Thus in the case of $LTD_4$, there was a small and insignificant fall in $tcPO_2$ despite a greater fall in $Vmax_{30}$. The $tcPCO_2$ did not vary significantly from baseline values throughout either study.

SIGNIFICANCE

Transcutaneous gas measurements was the method selected to monitor ABGs because it has notable advantages over other possible methods in this type of study. Direct arterial sampling would have required repeated stabs or arterial cannulation. These methods give discontinuous data which may miss important changes in ABGs, and were also felt to be unacceptably invasive and not without risk.

Ear or finger oximetry provide a continuous measurement of oxygenation but give no information on $PaCO_2$. In addition, resting oxygen saturation in normal adults is above 97% and because of the sigmoid nature of the oxy-haemoglobin dissociation curve, the use of oxygen saturation to derive $PaO_2$ lacks sensitivity on the upper, flatter part of the curve.

End-tidal gas analysis, using a mass spectrometer, can give a continuous and accurate estimation of $PaO_2$ and $PaCO_2$ in the normal adult, but becomes increasingly inaccurate in the presence of airflow obstruction, with changes in the distribution of inspired gas[2], and these changes would occur during airflow obstruction induced by the administration of bronchoconstrictors.

We have used an in vivo calibration technique on skin electrodes in this study. In vivo calibration of $tcPO_2$ in adults improves the accuracy of estimation of $PaO_2$ from $tcPO_2$[1]. There is no evidence to show that in vivo calibration of $tcPCO_2$ electrodes increases the accuracy of estimation of $PaCO_2$ but there is no theoretical reason to suppose that it does not. Skin blood flow is a major determinant of $tcPO_2$ and $tcPCO_2$. However, at electrode temperatures of $45^{o}C$ used in this study, in normal adults with stable circulations, there is likely to be maximal dilatation of vessels below the electrode with stable skin blood flows. It is also extremely unlikely that the very small doses of histamine and $LTD_4$ administered by inhalation would have any effect on skin blood flow. Significant hypoxia occurred in all subjects following the inhalation of the highest dose of histamine (mean fall in $tcPO_2$ 20 mmHg) but following the inhalation of $LTD_4$ there was a small and insignificant fall in $tcPO_2$ (mean fall 7 mmHg). It is unlikely that this differential effect on $tcPO_2$ was due to changes in ventilation ($tcPCO_2$ was unchanged) or changes in skin blood flow. The most likely explanation is that histamine and $LTD_4$ deposited in the airways are having different effects on local V/Q matching in the lung. Both histamine and $LTD_4$ are potent bronchoconstrictors. $LTD_4$ is a pulmonary vasoconstrictor, and has the effect of preserving V/Q matching in areas where it is deposited. Histamine has variable effects on the pulmonary vasculature but it is possible that vasodilator effects are mediated via the $H_2$ receptor[3]. It is therefore possible that deposition of histamine in the airways and alveoli results in local bronchoconstriction coupled with vasodilation leading to increased local V/Q mismatching with a low V/Q ratio. This explanation would fit with the observed hypoxaemia following histamine challenge seen in this study.

REFERENCES


1. BJ Gray, RW Heaton, AF Henderson and DCS Hutchison. Single point in vivo calibration improves the accuracy of transcutaneous oxygen tension estimations in adult patients. Thorax. 39: 712 (1983).
2. BL Hoffbrand. The expiratory capnogram: a measurement of ventilation-perfusion inequalities. Thorax. 21: 518 (1966).
3. N. Chand. Distribution and classification of airway histamine receptors: the physiological significance of $H_2$ receptors. Adv Pharmacol Chemotherap. 17: 103. (1980).

# IN VIVO CALIBRATION OF A TRANSCUTANEOUS OXYGEN ELECTRODE IN ADULT PATIENTS

B.J. Gray, R.W. Heaton, A. Henderson and D.C.S. Hutchison

Department of Thoracic Medicine
King's College School of Medicine and Dentistry
London, U.K.

## SUMMARY

Transcutaneous oxygen tension ($tcPO_2$) has been compared with arterial oxygen tension ($PaO_2$ in 14 haemodynamically stable patients in an intensive care unit. Two calibration methods have been compared: (1) "In vitro" calibration, a two point calibration procedure carried out before attachment to the skin. (2) "In vivo" calibration, calibration using a single arterial sample, to recalibrate the upper point after attachment of the electrode to the skin and stabilisation of the electrical output. After "in vitro" calibration the regression equation was given by $tcPO_2$ (mmHg) = 0.58 $PaO_2$ + 13.4 (95% confidence limits $\pm$ 19.6). After "in vivo" calibration, the regression equation for 55 comparisons over the range 50 to 120mmHg was given by: $tcPO_2$ (mmHg) = 0.98 $PaO_2$ + 1.6 (95% confidence limits $\pm$ 6.6). The "in vivo" calibration method therefore allows a close estimate of $PaO_2$ to be made from $tcPO_2$ values in adult patients providing strict operating criteria observed.

## INTRODUCTION

The method of calibration of transcutaneous oxygen electrodes is clearly of considerable importance if accurate results are to be obtained. As commonly practised a two point "in vitro" calibration is carried out before application of the electrode to the skin, employing for the low point electrical zero, or an oxygen free mixture, and for the upper point calibration, atmospheric air. This method has been widely used in neonates. However, in adults the accuracy with which $tcPO_2$ can be used to predict $PaO_2$ has been extremely variable and not of a high order, perhaps due to variations in skin metabolism or in the state of the local capillary circulation.

We have proposed that an in vivo calibration of the upper calibration point would be likely to improve the accuracy of the $tcPO_2$ in adults, and the purpose of this study was to test that proposition by ascertaining to what extent increased accuracy can be achieved.

METHODS

## Patients

Fourteen patients in the intensive care unit at King's College Hospital were studied. No patient was hypothermic and all had stable circulatory indices with no evidence of central or peripheral vascular insufficiency. All patients were receiving supplemental oxygen, five by mechanical ventilation, three by spontaneous ventilation via a T-piece and endotracheal tube and six by spontaneous ventilation via a face mask. Every patient had an indwelling arterial line in place as part of their management.

## Electrode

A Radiometer E5243/0 transcutaneous oxygen electrode with a 25 micron platinum microcathode was used in this study. The temperature was set at 45°C and positioned on the skin over the biceps away from major vessels. A Radiometer TCM1 monitor was used in conjunction with the electrode and its output displayed digitally and on a chart recorder.

## In vitro calibration

A two point calibration was carried out at an electrode temperature of 45°C before each study. Electrical zero was used for the lower point and atmospheric air used for the upper point.

## In vivo calibration

When the $tcPO_2$ had stabilised (usually after 20 to 30 minutes), an arterial blood sample was taken and the $PaO_2$ measured with a Radiometer ABL2 blood gas analyser. The $tcPO_2$ reading was then adjusted to correspond to the $PaO_2$ using the monitor gain control. Thus the upper point was re-calibrated to correspond to $PaO_2$.

## Comparison

After in vivo calibration the $FiO_2$ was increased or decreased a number of times in each patient. The resulting changes in $tcPO_2$ were recorded until a new stable plateau was reached. An arterial blood sample was then taken and analysed and the simultaneous $tcPO_2$ and $PaO_2$ values noted. It was thus possible to compare $tcPO_2$ and $PaO_2$ over a wide range (at least five points) in each subject.

## Drift

To calculate drift the "in vitro" calibration was repeated at the end of each study when the oxygen electrode was carefully detached from the skin. When corrected for any alteration in the gain control involved in the re-calibration of the upper point, drift was never greater than 5% of the original upper point value. Zero drift was never greater than $\pm$ 2mmHg.

RESULTS

## In vitro calibration

The $tcPO_2$ was compared with $PaO_2$ in the 14 patients following in vitro calibration of the electrode, when the electrode output had reached a steady state. The fourteen data points gave the following regression equation: $tcPO_2$ (mmHg) = 0.58 $PaO_2$ + 13.4. 95% confidence limits = $\pm$ 19.6.

In vivo calibration

Following in vivo calibration a total of 77 comparisons were obtained in the 24 patients at various levels of $FiO_2$. Overall these 77 data points gave the following regression equation in the range 50 to 294 mmHg: $tcPO_2$ (mmHg) = 0.74 $PaO_2$ + 21.8. 95% confidence limits= $\pm$ 13. It was obvious that the greatest inaccuracy occurred at the higher values of $PaO_2$, i.e. outside the calibration range. If the 55 comparisons in the range 50 to 120 mmHg are considered (each patient contributing at least 3 data points) the equation becomes: $tcPO_2$ = 0.98 $PaO_2$ + 1.6. 95% confidence limits = $\pm$ 6.6.

SIGNIFICANCE

The results of this study demonstrate that in adult patients, $PaO_2$ can be estimated from $tcPO_2$ with much greater accuracy when "in vivo" calibration using a single arterial sample is employed rather than the more commonly used "in vitro" method. The regression equation obtained by the "in vivo" method within the range 50-120 mmHg was close to the line of identity with acceptable 95% confidence limits ($\pm$ 6.6 mmHg) for clinical purposes.

In contrast, the results of the "in vitro" procedure on the same patients are similar to results already reported[1], i.e. the slope of the regression equation considerably less than unity, a large positive intercept on the $tcPO_2$ axis and unacceptable 95% confidence limits for clinical purposes.

In the intensive care unit the "in vivo" calibration technique offers no further problems if an arterial cannula is in situ. This is seldom the case in ambulant patients or normal subjects undergoing physiological studies and it can be argued that arterial sampling to obtain "in vivo" calibration is ethically unjustified. However, in our hands and those of several other investigators, a single arterial puncture under local anaesthesia is painless, and amply justified by the increased accuracy which it affords.

REFERENCES

1. DCS Hutchison, G Rocca and D Honeybourne.
Estimation of oxygen tension in adult subjects using a transcutaneous electrode. Thorax. 36: 473. (1981)

# INFLAMMATION AND TRANSCUTANEOUS MEASUREMENT OF OXYGEN PRESSURE IN DERMATOLOGY

Anneliese Ott

Hautklinik und Poliklinik des Klinikum Charlottenburg
der freien Universität Berlin
Direktor: Prof. Dr. med. G. Stüttgen

## SUMMARY

In this study the influence of local inflammation on $tcPO_2$ values at different electrode temperatures has been investigated. The measurements were performed on UV-erythema, contact dermatitis and other inflammatory reactions of the skin. Different types of inflammation and different degrees of damage can lead to very different reactions in the microcirculation and this will have an effect on the $tcPO_2$ levels.

## INTRODUCTION

In the last few years the transcutaneous $PO_2$ technique which was introduced by Huch et al (1,2,3) for estimating arterial $PO_2$, has been widely used at different electrode temperatures as a noninvasive tool in clinical and experimental angiology. For the interpretation of the results of the $tcPO_2$ measurement it is important to take into account that $tcPO_2$ values are dependent on several local factors. Acute, subacute or chronic dermatitis can substantially influence $tcPO_2$ values which we use for the diagnosis and therapy control of a basic disease. Due to the complexity of this subject this paper will concentrate on two types of inflammation a physical form, the UV-Erythema, and an immunological form, the allergic contact dermatitis (type IV reaction according to Coombs and Gell). Some examples of other forms of local inflammation will also be given.

## METHODS

In 16 healthy volunteers an area of 2cm x 3cm was irradiated with the Ultravitalux lamp (Osram) for a period of 4 minutes at a distance of 30 cms. After 24 hrs $tcPO_2$ was simultaneously registered on the exposed area and on a comparable unexposed area at the electrode temperatures 37°C and 43°C. These measurements were repeated under the same conditions during an $O_2$ respiration of 4l $O_2$/min lasting 3 mins using a nasal tube.

Under similar conditions 8 patients with proven contact sensitisation to different contact allergens were examined on an epicutaneous test area. 72 hrs after the contact allergen had been applied, $tcPO_2$ was monitored on the test area and an untreated area. For $tcPO_2$ measurement we used the TCM 2 monitor in conjunction with the two-channel recorder both manufactured by Radiometer Copenhagen.

## RESULTS

The results are summarized in Table 1.

Table 1

| | | initial $tcPO_2$ (mm Hg) | | $tcPO_2$ (mm Hg) during $O_2$ respiration | |
|---|---|---|---|---|---|
| | | mean | ± sd | mean | ± sd |
| Normal skin | at 37°C: | 4 | ± 4 | 4 | ± 4 |
| UV exposed skin | " : | 32 | ± 12 | 62 | ± 28 |
| Contact dermatitis | " : | 2.4 | ± 3.2 | 11.6 | ± 17.3 |
| Normal skin | at 43°C: | 68 | ± 8 | 121 | ± 17 |
| UV exposed skin | " : | 60 | ± 13 | 119 | ± 22 |
| Contact dermatitis | " : | 14 | ± 16 | 50 | ± 35 |

It will be seen that:

I. On normal skin at an electrode temperature of 37°C $tcPO_2$ values did not rise during $O_2$ respiration.

II. At 37°C the skin exposed to UV-rays showed a high initial value, which was doubled during $O_2$ respiration.

III. On normal skin at an electrode temperature of 43°C there was a significant rise during $O_2$-respiration.

IV. At an electrode temperature of 43°C there was no significant difference between the UV-exposed and unexposed area whether with or without $O_2$-respiration.

V. On the positive test areas there was no significant difference to the normal skin for $tcPO_2$ at 37°C in contrast to the skin exposed to UV-rays (compare II). $O_2$ respiration caused a moderate rise in only a few cases.

VI. At the temperature of 43°C the $tcPO_2$ average values for contact dermatitis both with and without $O_2$ respiration were considerably lower than the comparable values for normal skin as well as those reached for UV-Erythema.

To what degree various local factors can influence $tcPO_2$ values will be illustrated by some individual cases:

Case 1: ♀, 67 years old housewife, tentative diagnosis: Photodermatosis.

$tcPO_2$ at 37°C: 2mmHg, after $O_2$ respiration: 24mmHg.
43°C: 10mmHg, " " : 26mmHg.

$TcPO_2$ values indicated a case of contact dermatitis. This was later confirmed by a patch-test. The anamnesis showed that she had used a pyrethrum spray. Pyrethrum which is derived from the Chrysanthemum species acts by paralysing insects with low toxicity to mammalians but it is a well-known allergen. Its allergens are the sesquiterpene lactone Pyrethrosin and the insecticidal Pyrethrin II (4).

Cases 2 and 3: ♀, 83 and 65 years old respectively, diagnosis: allergic vasculitis. The electrodes were attached to the reddened area surrounding the necroses.

| | affected skin | | non-affected skin | |
|---|---|---|---|---|
| | case 2 | case 3 | case 2 | case 3 |
| $tcPO_2$ at 37°C : | 1 mmHg | 1 mmHg | 4 mmHg | 1 mmHg |
| " " 37°C+$O_2$ respiration: | 1 " | 1 " | 4 " | 1 " |
| " " 43°C : | 1 " | 1 " | 40 " | 24 " |
| " " 43°C+$O_2$ respiration: | 1 " | 1 " | 64 " | 44 " |

The microcirculation of the affected skin had obviously almost completely broken down.

Case 4: ♂, 22 years old, diagnosis: Phyto-phototoxic dermatitis. He had been pruning a giant heracleum shrub 5 days previously.

| | normal skin | 5 day old reddened contact area | test area UVA exposed |
|---|---|---|---|
| $tcPO_2$ at 37°C : | 6 mmHg | 6 mmHg | 50 mmHg |
| " " 37°C +$O_2$ resp. : | 10 " | 36 " | 91 " |
| " " 43°C : | 60 " | 8 " | 70 " |
| " " 43°C +$O_2$ resp. : | 112 " | 40 " | 140 " |

Because of the warm summer weather he had been only lightly dressed which meant that his skin came into extensive contact with the juice of the huge bush. His skin showed extensive reddened areas and bullous reactions. Heracleum is a plant containing psoralens. The presence of psoralens makes the skin susceptible to the longer wavelengths (320 - 360 nm). The severity of the reaction depends upon the concentration of the psoralens in the skin, its potency and the dose of irradiation. Thus the exposed patch-test (exposed to 10J UVA) shows a sunburn-type reaction. The 5 days old reaction to an unknown amount of exposure to the sun's rays and contact with an unknown quantity of the plant juice was obviously severely damaging the microcirculation in a way which does not conform to any of the previous reaction patterns.

## DISCUSSION

Macroscopically, a reddening and swelling was present both in the case of UV-erythema and of contact dermatitis. The epicutaneous test areas also showed a papulous pattern. Bullous reactions were excluded in both cases.

As a result of UV-radiation keratinocytes and Langerhans cells are primarily injured, releasing inflammatory mediators which penetrate the vessel-carrying layers and induce vasodilation. A biopsy of a first degree sunburn shows extensive vasodilation and only a sparse perivascular infiltrate.

It is of practical interest that the increased $tcPO_2$ values after UV exposure were detectable even without visible erythema. The direct release of inflammatory mediators induced by UV radiation is in contrast to the indirect one in contact dermatitis. Specifically sensitized T-lymphocytes penetrate the epidermis releasing various lymphokines which damage Langerhans cells and this probably results in their discharging inflammation producing mediators. The compound of the various mediators greatly determines whether vasodilation or increased permeability dominates. Histologically, after 72 hrs one finds, apart from the vasodilation a pronounced perivascular infiltate in the upper dermis, edema of the papillary dermis, exocytosis, inter- and intracellular edema of the epidermis. The perivascular infiltrate, the edema of the papillary dermis as well as the spongiosis influence the microcirculation by compression and prevent $O_2$ diffusion. Thus the $tcPO_2$ technique in these cases is a non-invasive test to indicate the type of microcirculatory disturbance present in different types of dermatitis.

REFERENCES


1. A. Huch, R. Huch, D. W. Lübbers, Quantitative polarographische Sauerstoffdruckmessung auf der Kopfhaut des Neugeborenen, Arch Gynäkol 207: pp 443-451 (1969).

2. R. Huch, A. Huch, D. W. Lübbers, Transcutaneous $PO_2$, Georg Thieme Verlag, Stuttgart New York (1981).

3. D. W. Lübbers, Cutaneous and transcutaneous $PO_2$ and $PCO_2$ and their measuring conditions, Birth defects: Original article series, Vol IV, 4: pp 13-31 (1979).

4. E. Cronin, Contact Dermatitis, Churchill Livingstone, New York (1980).

# DIAGNOSTIC ASSESSMENT OF DIABETIC MICROANGIOPATHY BY $TCPO_2$ STIMULATION TESTS

Norbert Weindorf, Ulrich Schultz-Ehrenburg, and Peter Altmeyer

Department of Dermatology, St. Josef-Hospital, Ruhr-University Bochum
Gudrunstr. 56, D-4630 Bochum

## INTRODUCTION

The diagnosis of diabetic microangiopathy is usually secured by an ophthalmological examination of the eye fundus. But the vessels which are viewed by funduscopy are small arteries and no terminal vessels. Thus even funduscopy is only an indirect method for diagnosing microangiopathy which beyond that gives qualitative information only. For the first time, Ewald et al. (1981) showed that the $tcPO_2$ method can successfully be used for the assessment of microcirculation in diabetic children. But the vascular changes of diabetic adults are much more difficult to evaluate by this technique because of two reasons, i. e., thickening of the horny layer and overlapping of microvascular and macrovascular disturbances.
In the present work, special stimulation tests have been used in order to improve the applicability of the $tcPO_2$ technique. The investigations aimed for developing a method of examination which is suitable for a quantitative assessment of microangiopathy in adult diabetic patients.

## MATERIAL AND METHODS

The study comprises 21 diabetic patients, twelve of type 1 and nine of type 2 who needed insulin therapy. The duration of disease was 0.5 - 50 years, the mean age 58.1 years (range 28 - 78 years). The control group consisted of 15 healthy volunteers with a mean age of 48.3 years (range 26 - 76 years). We used the apparatus of Radiometer, Copenhagen, Danmark, with the Clark electrode E 5242. The electrode was heated to 37 °C and placed on the volar aspect of the forearm after 10 times of stripping the horny layer. All subjects were studied in supine position at a room temperature of 24 - 28 °C. The following stimulation tests were applied:

- resting value (RV)
- postocclusive value (POV) after 4 minutes of ischemia using a sphygmomanometer cuff
- rubefacient-induced value (RIV) after local application around the electrode of a commercially available ointment (Finalgon[R], Thomae company, F.R.G.) containing nicotinic acid butyl ester and nonylic acid vanillylamide.
- 45 °C heating value (HV) after heating the electrode from 37 °C to

45 °C. This measurement was performed on the contralateral forearm because of the long lasting hyperemia after application of the rubefacient.

## RESULTS

The results of all $tcPO_2$ examinations are shown in Table 1 and Figure 1. Diabetic patients showed elevated resting values (1.58 ± 1.05 kPa), diminished rubefacient-induced values (4.06 ± 1.75 kPa) and diminished 45 °C heating values (11.98 ± 2.15 kPa), whereas the postocclusive $PO_2$ values were nearly identical with those of the control group. Even more striking were the changes concerning the $PO_2$ increase, i. e. the difference between the stimulated and resting values, which was markedly reduced in the rubefacient test as well as in the heating test (Fig. 2).

Table 1. Results of $PO_2$-measurements ($\bar{x}$ ± SD in kPa)

| Stimulation test | Diabetics (n = 21) | Controls (n = 15) |
|---|---|---|
| RV | 1.58 ± 1.05 | 0.67 ± 0.54 |
| POV | 2.74 ± 0.74 | 2.72 ± 1.14 |
| RIV | 4.06 ± 1.75 | 5.80 ± 1.20 |
| HV | 11.98 ± 2.15 | 13.50 ± 1.86 |
| Δ POV-RV | 1.16 ± 0.31 | 2.04 ± 0.60 |
| Δ RIV-RV | 2.48 ± 1.89 | 5.12 ± 1.30 |
| Δ HV-RV | 10.79 ± 2.39 | 13.60 ± 2.13 |

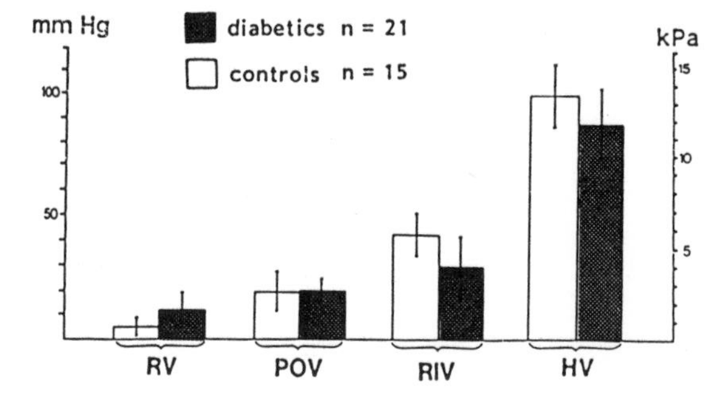


Fig. 1. Resting and stimulated $PO_2$ values in diabetics and healthy controls
RV = resting value
POV = postocclusive value
RIV = rubefacient-induced value
HV = heating value

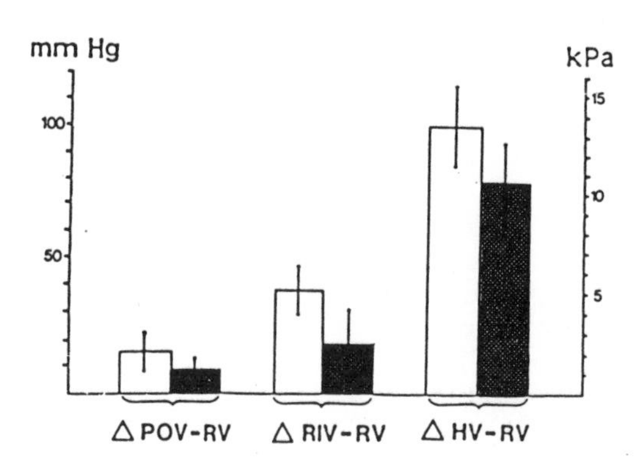


Fig. 2. $PO_2$ increase after stimulation tests in diabetics and healthy controls. Abbreviations see Fig. 1

DISCUSSION

Among the various function tests already the resting $PO_2$ measurement gave different results in the diabetic group and in the control group. But the resting value alone is a rather unstable parameter depending on many variables like kind of stripping of the horny layer, room temperature, motion of the patient etc. The differences between the two groups stood out more clearly when the values of the stimulation tests - especially the rubefacient-induced value and the 45 °C heating value which proved to be the most reliable parameters - were compared. The $PO_2$ increase after application of the rubefacient and after heating the electrode to 45 °C was significantly diminished in the diabetic group ($p \leq 0.001$). Nevertheless, the absolute $PO_2$ values should be preferred, because of the uncertain behaviour of the resting $PO_2$ as mentioned above.

The various stimulation tests represent different vascular functions. First the resting $PO_2$ depends on the vasoconstrictor tonus of the autonomous nervous system. The high resting $PO_2$ value, therefore, indicates that an autoregulatory sympathicolysis has taken place to a certain extent which has also been reported from other studies. Alexander et al. (1968) using venous occlusion plethysmography found an elevated resting flow in the diabetic group in comparison to that of the control group. But in the $tcPO_2$ technique, in particular the resting value is susceptible to trouble.

The postocclusive $PO_2$ value first of all reflects the macrovascular function, but the transcutaneously measured value can be influenced by microcirculation, too. In our observations, the macrovascular function is predominating in the postocclusive outcome. Our findings are in contrast to the results of Ewald et al. who found a diminished postocclusive $PO_2$ increase in diabetic children ($1.7 \pm 1.1$ kPa in the diabetics and $3.7 \pm 0.7$ kPa in the controls). There might be following explanations for this discrepancy: the group of patients is different concerning the age (28-78 years in our group -nd 9 - 18 years in Ewald's group), and also the type of diabetes (in our study type 1 and 2, in Ewald's study type 1 only). There are also slightly different examination conditions of room temperature and of the stripping of the horny layer.
Railton et al. (1983) found less diminished values for the postocclusive $PO_2$ increase in diabetic patients than those reported by Ewald et al. The mean postischemic change in diabetics was $2.0 \pm 1.0$ kPa and $2.1 \pm 1.0$ kPa respectively after a second occlusion, in the control group the difference was $2.9 \pm 1.2$ and $2.2 \pm 1.0$ kPa respectively. The mean age of their patients was 28 years (range 18 - 39 years). These findings speak in favour of the assumption that there are different reaction patterns in children and adults. According to our experience the postocclusive value ist not sufficient for diagnosing any microvascular function disturbances in adult diabetic patients.

The 45 °C heating value gives information on the total capacity of the terminal vessels (Huch et al. 1981), because at this temperature the arterioles are maximally dilated. Therefore, a diminution of the 45 °C heating $PO_2$ can be regarded as a marker of organic vessel wall changes of the diabetic disease.

The rubefacient-induced $PO_2$ value represents a pharmacological stimulation test which works by transcutaneous resorption and subsequent diffusion to the small vessels of the skin. Nicotinic acid as well as its various esters are widely used vasodilating agents. The mechanism of vasodilation probably is based on the release of prostaglandins, especially prostacyclin (Andersson et al. 1977). This is a typical endothelial-cell function. Nonylic acid vanillylamide has vasodilating properties, too. It probably works by stimulation of acetylcholine receptors of the endothelial-cells

(Crossland 1970). In a second step, the endothelium-derived relaxing factor is released leading to vasodilation (Furchgott and Zawadzki 1980). So nicotinic acid butyl ester as well as nonylic acid vanillylamide induce actions which are mainly dependent on the endothelial-cell function. Therefore, the rubefacient-induced $PO_2$ value can be regarded as an indicator of endothelial-cell function disturbances.

In conclusion, the $tcPO_2$ method using the two stimulation tests does not only allow a quantitative assessment of microangiopathy, but also gives a differentiated picture of microvascular functions.

## SUMMARY

Transcutaneous $PO_2$ measurements at 37 °C can recognize and quantify diabetic microangiopathy if suitable stimulation tests are used. The following parameters are of practical importance: resting $PO_2$, postocclusive $PO_2$, rubefacient-induced $PO_2$ and 45 °C heating $PO_2$. Patients with diabetic microangiopathy show diminished $PO_2$ values in rubefacient-induced $PO_2$ and 45 °C heating $PO_2$. The former ist probably related to endothelial-cell function disturbances and the letter to organic vessel wall changes. Thus, a quantitative and differentiated picture of microangiopathy can be obtained in each individual patient.

## REFERENCES

Alexander, K., Teusen, R., and Mitzkat, H. J., 1968, Vergleichende Messungen der Extremitätendurchblutung bei Diabetikern und Stoffwechselgesunden, Klin. Wschr. 46: 234

Andersson, R. G. G., Aberg, G., Brattsand, R., Ericsson, E., and Lundholm, L., 1977, Studies on the mechanism of flush induced by nicotinic acid, Acta pharmacol. et toxicol., 41: 1

Crossland, J., 1970, Lewis's Pharmacology, Livingstone, Edingburgh, London (1970)

Ewald, U., Tuvemo, T., and Rooth, G., 1981, Early reduction of vascular reactivity in diabetic children detected by transcutaneous oxygen electrode, Lancet: 1287

Furchgott, R. F., and Zawadzki, J. V., 1980, The obligatory role of endothelial cells in the relaxation of arterial smooth muscle by acetylcholine, Nature, 288: 373

Huch, R., Huch, A., and Lübbers, D. W., 1981, Transcutaneous $PO_2$, Thieme-Stratton Inc., Stuttgart, New York

Railton, R., Newman, P., Hislop, J., and Harrower, A. D. B., 1983, Reduced transcutaneous oxygen tension and impaired vascular response in type I, (insulin-dependent) diabetes, diabetologia, 25: 340

# APPLICATION IN PERINATOLOGY

TRANSCUTANEOUS BLOOD GASES AND SLEEP APNEA PROFILE

IN HEALTHY PRETERM INFANTS DURING EARLY INFANCY

Karl H. P. Bentele, Ulrike Ancker, and Michael Albani

Department of Pediatrics
University of Hamburg, Germany

Supported by Deutsche Forschungsgemeinschaft - BE 921/1-2

SUMMARY

Studying the development of transcutaneous blood gas levels ($tcpO_2$ and $tcpCO_2$) and the sleep apnea profile in relation to sleep states in normal preterm infants between 36 and 52 weeks postconceptual age we found a dynamic increase in $tcpO_2$ during regular breathing (without apnea) and a steady decrease in $tcpCO_2$ during both regular and periodic breathing. The mean $tcpO_2$ of periodic breathing, however, persistently remained well below the corresponding level found during regular breathing. It is suggested that in normal preterm infants there is a continued maturational adjustment of autonomic respiratory control up to 3 months post term and, furthermore, that periodic breathing may persistently be associated with a relative hypoxemia.

INTRODUCTION

Compared to fullterm controls healthy preterm infants as a group were reported to have significantly more apnea and periodic breathing when arriving at term. This was followed by a decrease in all apnea variables towards 52 wks postconc. age, even below the corresponding values of the fullterm controls (1). It was assumed that preterm infants even beyond term might either have less respiratory drive or recurrent hypoxemic episodes leading to a prolonged ventilatory depression as proposed by others (2). The present study therefore attempted to test these hypotheses by investigating the development of $tcpO_2$ and $tcpCO_2$ levels during regular and periodic breathing in healthy preterm infants in relation to the amount and type of sleep apnea between 36 and 52 wks postconc. age. Since the developmental state of the autonomic chemoreceptor control of breathing is particularly revealed during NON-REM sleep all data were assessed in relation to sleep states.

METHODS

Seven healthy preterm infants with a mean gestational age of 33.4 (32-35) wks and a mean birth weight of 1975 (1160-2750) g were studied in intervals of 4 wks from 36 to 52 wks postconc. age. All infants, 3 males and 4 females, had an uneventful postnatal course, a normal ultrasound scan of the head and had never received methylxanthines or supplemental oxygen. One infant was breast-, the others were bottle fed. The polygraphic recordings covered at least 3 hrs of spontaneous sleep and were obtained during the afternoon at 36, 40, 44 and 48 wks and during the evening or early night time at 52 wks. EEG, eye movements, airflow at the nostrils (thermistor), thoracic and abdominal respiratory efforts were continuously recorded by a Schwarzer polygraph. Transcutaneous $pO_2$ and $pCO_2$ assessed with the Hellige system with electrodes heated at 44°C as well as skin and rectal temperature were marked every 20 sec on the paper write out as were the body movements according to a code system. Sleep states - NON-REM and REM - were determined in accordance with established criteria. Duration of inactive, obstructive and mixed apnea of 3 sec and more as well as of periodic breathing was calculated on the basis of 100 min each of both NON-REM and REM sleep ($AD_3$ %, PB %). The definition of periodic breathing was as follows: 3 or more successive inactive apnea lasting 3 sec or longer, interrupted by less than 20 sec of regular breathing. The mean values of $tcpO_2$ and $tcpO_2$ and $tcpCO_2$ were calculated on the basis of all successive 20 sec epochs of both regular and periodic breathing in relation to sleep states.

RESULTS

Between 36 and 52 wks postconc. age there was a divergent course of $tcpO_2$ and $tcpCO_2$ during regular breathing (without apnea) in both NON-REM and REM sleep (see fig. 1 and fig. 2). After an initial decrease of the mean $tcpO_2$ in 5 of 7 infants in NON-REM and in 1 of 7 in REM sleep at 40 wks all infants showed a continuous increase of $tcpO_2$ up to 48 wks followed by a slight but obvious decline at 52 wks in both sleep states. In contrast to the $tcpO_2$ the mean $tcpCO_2$ during regular breathing was found to decrease considerably between 36 and 52 wks in either sleep state as shown at the lower parts of fig. 1 and 2. This decrease of the mean $tcpCO_2$ was apparently biphasic in course with a substantial and significant fall up to 44 wks and a slight decline thereafter.

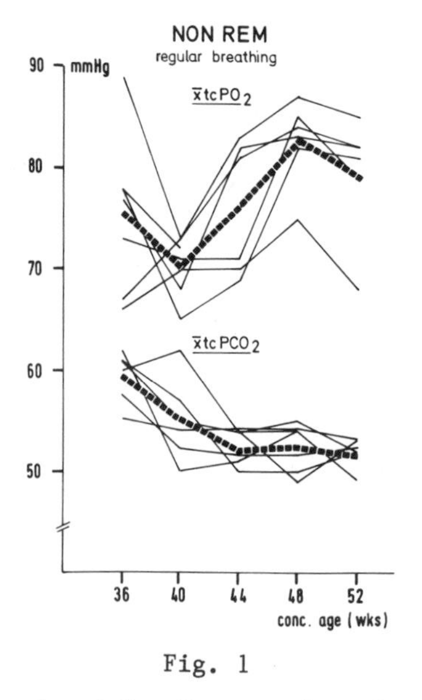


Fig. 1

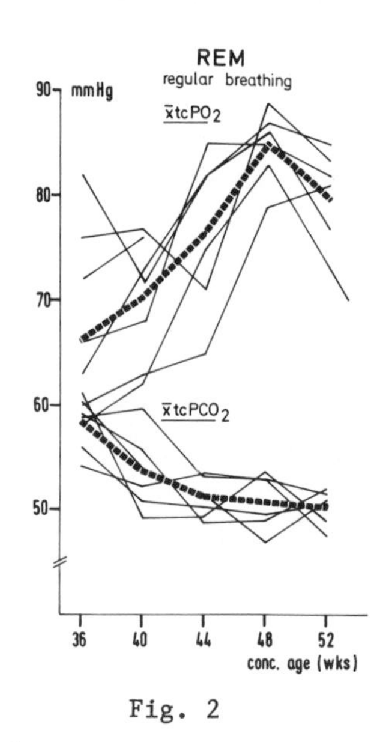


Fig. 2

Fig. 1 and 2. Course of $tcpO_2$ (upper part) and $tcpCO_2$ (lower part) during regular breathing (without any apnea) in relation to postconceptual age and sleep states given for individual infants and as the group mean (dotted line).

During periodic breathing the mean $tcpO_2$ also increased from 36 to 48 wks in both sleep states (see fig. 3 for NON-REM and fig. 4 for REM sleep). Although running a parallel course the $tcpO_2$ in periodic breathing persistently remained considerably lower (5 - 10 mmHG) than the corresponding values observed in regular breathing. From 48 to 52 wks the mean $tcpO_2$ in periodic breathing during REM sleep also decreased again whereas in NON-REM sleep periodic breathing was no longer observed at this age.

The course and range of the mean $tcpCO_2$ in periodic breathing was found to be identical with that seen in regular breathing with respect to both NON-REM (fig. 3) and REM (fig. 4) sleep. Again, there was an apparently biphasic developmental pattern with a first, more rapid and a second, more gradual decrease changing at 44 wks.

Along with these changes of the levels of transcutaneous blood gases between 36 and 52 wks there was a steady decrease of all sleep apnea variables, in particular of the amount of periodic breathing (PB %). This decrease in PB % occurred in both NON-REM (fig. 3) and REM sleep (fig. 4) and showed a somewhat biphasic pattern too. In NON-REM sleep periodic breathing disappeared at 52 wks.

The amount of obstructive and mixed apnea ($AD_3$ %) also decreased between 36 and 52 wks with respect to both sleep states. Again, the rate of decrease was found to be faster before 44 wks and more gradual thereafter as demonstrated by fig. 5 und 6. Comparable to PB % in fig. 3 and 4, $AD_3$ % is given as the mean $\pm$ 1 SD.

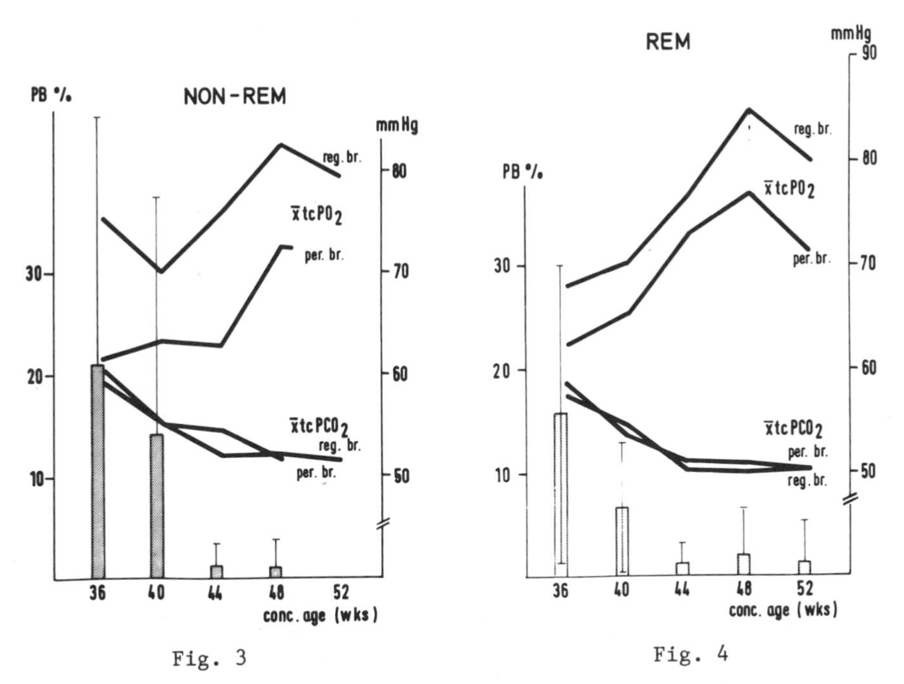


Fig. 3

Fig. 4

Fig. 3 and 4. Course of $tcpO_2$ and $tcpCO_2$ during periodic breathing in relation to post- conceptual age and sleep states given as the group mean. For comparison the corresponding data (group means) of regular breathing are given again. The amount of periodic breathing (PB %) in relation to sleep state and postconc. age is shown by the bars at the bottom (group mean $\pm$ 1 SD).

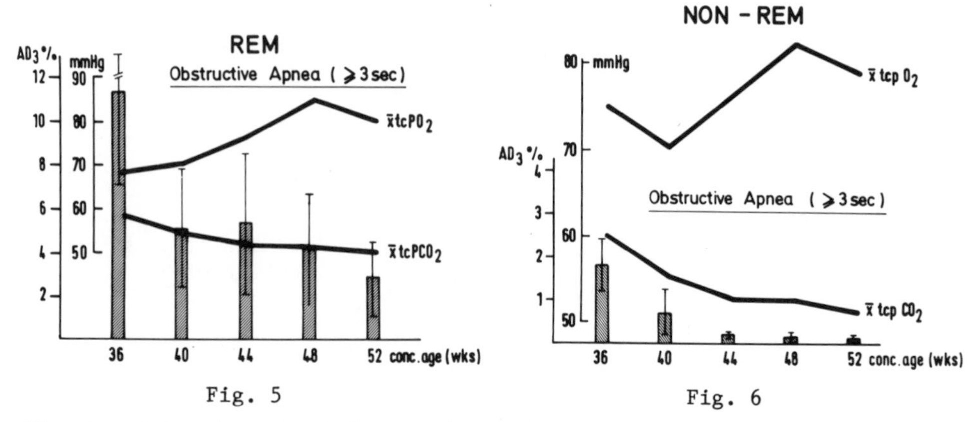


Fig. 5 Fig. 6

Fig. 5 and 6. Amount of obstructive and mixed apnea ($AD_3$ %) in relation to sleep states and postconceptual age. Bars represent group mean ± 1 SD. For synopsis the course of transcutaneous gas levels (group means) during regular breathing is given too.

As shown by fig. 7 the amount of non-periodic inactive apnea in NON-REM-sleep also decreased almost continuously between 36 and 52 wks with the exception of the higher amount observed at 44 wks that was contributed by one particular infant. In REM-sleep (fig. 8), however, there was only an initial decrease in the amount of inactive apnea up to 44 wks and no significant change thereafter.

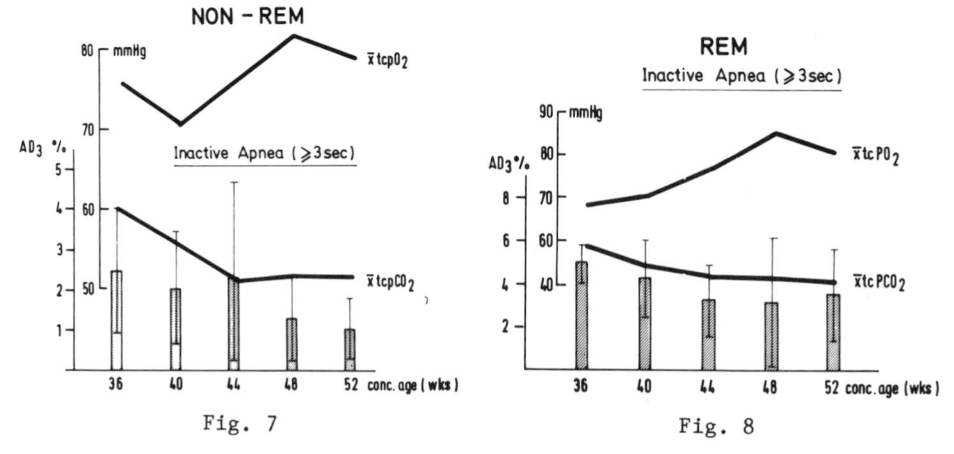


Fig. 7 Fig. 8

Fig. 7 and 8. Amount of non-periodic inactive apnea (AD3%) in relation to sleep states and postconceptual age. Bars represent group mean ±1SD. For synopsis the course of transcutaneous gas levels of regular breathing is given again.

Table 1: Synopsis of original data on transcutaneous blood gases and sleep apnea variables in relation to sleep states and postconceptual age in healthy preterm infants.

| | | NON-REM | | | | |
|---|---|---|---|---|---|---|
| Postconceptual age (wks) | | 36 | 40 | 44 | 48 | 52 |
| $tcpO_2$ | regular breathing: | 75<br>(60-89) | 70<br>(65-73) | 76<br>(69-83) | 83<br>(75-87) | 79<br>(68-85) |
| | periodic breathing: | 62<br>(52-69) | 63<br>(59-70) | 62<br>(62-64) | 73<br>(69-76) | --<br>-- |
| $tcpCO_2$ | regular breathing: | 59<br>(55-62) | 56<br>(50-62) | 52<br>(50-54) | 52<br>(49-55) | 52<br>(49-53) |
| | periodic breathing: | 60<br>(57-64) | 55<br>(50-61) | 55<br>(54-55) | 52<br>(48-55) | --<br>-- |
| % of | periodic breathing: | 21<br>( 0-39) | 14<br>( 0-65) | 1,1<br>( 0- 5) | 1,0<br>( 0- 5) | --<br>-- |
| % of | obstr. & mixed: | 1,8<br>(0,6-2,4) | 0,7<br>(0,3-1,5) | 0,2<br>(0,1-1,5) | 0,2<br>( 0-0,6) | 0,1<br>( 0-0,3) |
| % of | inactive apnea: | 2,5<br>(0,3-5,4) | 2,0<br>(0,5-4,1) | 2,3<br>(1,0-7,0) | 1,3<br>(0,6-3,3) | 1,0<br>(0,3-1,3) |
| | | REM | | | | |
| $tcpO_2$ | regular breathing: | 68<br>(58-82) | 70<br>(62-77) | 77<br>(79-89) | 85<br>(65-85) | 80<br>(70-85) |
| | periodic breathing: | 62<br>(53-68) | 65<br>(60-77) | 73<br>(64-84) | 77<br>(72-82) | 71<br>(70-72) |
| $tcpCO_2$ | regular breathing: | 57<br>(51-61) | 55<br>(48-62) | 51<br>(48-53) | 51<br>(47-53) | 50<br>(46-51) |
| | periodic breathing: | 58<br>(52-61) | 55<br>(48-62) | 51<br>(48-53) | 51<br>(47-53) | 49<br>(46-51) |
| % of | periodic breathing: | 15,7<br>( 2-39) | 6,7<br>(2,4-18,6) | 1,1<br>( 0-2,6) | 1,9<br>( 0-6,9) | 1,1<br>( 0-5,9) |
| % of | obstr. & mixed: | 11,4<br>(6,2-18,5) | 5,1<br>(3,2-9,5) | 5,4<br>(1,6-9,8) | 4,1<br>(1,3-5,7) | 2,9<br>(1,0-6,4) |
| % of | inactive apnea: | 5,0<br>(4,0-6,6) | 4,5<br>(0,9-5,7) | 3,3<br>(1,0-5,6) | 3,1<br>(0,4-8,3) | 3,5<br>(0,7-5,8) |

Figures given in table 1 represent mean and (range), blood gases are given in mmHG.

## DISCUSSION

Our data demonstrate that in healthy preterm infants there is an inverse correlation between an increase of $tcpO_2$ during both regular and periodic breathing and a decrease of the amount of sleep apnea variables, in particular of periodic breathing and obstructive/mixed apnea in both sleep states between 36 and 52 wks postconc. age. Concomitant with these changes the $tcpCO_2$ level decreased up to 52 wks without any difference as to the type of breathing or sleep state. In contrast to the sinusoidal development of the $tcpO_2$ with an overshot around 48 wks the $tcpCO_2$ decreased in a biphasic way as did the apnea variables. Whereas from earlier studies it was suggested that chemoreceptor reflexes are functionally active very early in preterm infants (3) our data seem to indicate, that a maturational process of respiratory control goes on for a much longer time along with postconc. age and results in an increasingly accurate adjustment of nominal values for $pO_2$ and $pCO_2$. In particular, the biphasic course of the decrease in $tcpCO_2$ with a significant fall up to 44 weeks and only a slight reduction thereafter which is paralleled by the also biphasic decline in the amount of periodic breathing and obstructive and mixed apnea suggest that in healthy preterm infants primarily the central chemoreceptors are immature at term and still maturing up to 44 wks post conception. Therefore it might be concluded that the sleep apnea profile between 36 and 52 wks postconc. age is related rather to a continuous developmental adjustment than to a forced respiratory drive as a consequence of recurrant hypoxemic episodes which were not observed in our infants. A recent study of Hoppenbrouwers et al. (4) supports our results and seems to confirm the conclusions drawn. A lowered $tcpO_2$ persistently recorded during periodic breathing cannot be explained by central hypoventilation since there was no difference in $tcpCO_2$ between regular and periodic breathing. It might however be an effect of altered pulmonary and/or airway mechanical properties as well as of the reduced respiratory rate in periodic breathing as suggested by others in very young prematures (5). Since it appears from our data that periodic breathing in normal preterms during the first 3 months post term persists in being associated with a relative hypoxemia which might be both an effect and a source (2) of this pattern of breathing, an increased percentage of periodic breathing seems likely to be indicative for a perhaps harmful condition in some preterm infants. Whether these infants or those whose developmental profile of transcutaneous blood gases and sleep apnea deviates from the range reported here will be at increased risk of developing prolonged apnea or even sudden infant death seems worthwhile to be studied prospectively.

## REFERENCES


1. Albani M, Bentele KHP, Budde C, and Schulte FJ, 1985, Infant sleep apnea profile: preterm vs. term infants. Eur. J. Pediatric 143: 261-268

2. Rigatto H, Brady JP, 1972, Periodic breathing in preterm infants: II. Hypoxia as a primary event. Pediatrics 50: 219-228

3. Rigatto H, Brady JP, Chir B, and de la Torre Verduzco R, 1975, Chemoreceptor reflexes in preterm infants: II. The effect of gestational and postnatal age on the ventilatory response to inhaled carbon dioxide. Pediatrics 55: 614-620

4. Hoppenbrouwers T, Hodgman JE, Arakawa K, Cabal L, Durand M, 1986, Transcutaneous gases in early infancy. Pediatric Research 20: 413 A

5. Hodgman JE, Hoppenbrouwers T, Shirazi M, Cabal LA, Durand M, 1984, Effect of breathing pattern on oxygenation during sleep in prematures. Pediatric Research 18: 1784 A

# TRANSCUTANEOUS MONITORING AS TRIGGER FOR THERAPY OF HYPOXEMIA DURING SLEEP

Marianne E. Schlaefke, Thorsten Schaefer, Harald Kronberg*, Georg J. Ullrich* and Joachim Hopmeier*

Abteilung fuer Angewandte Physiologie, Ruhr-Universitaet, D-4630 Bochum, and Frauenklinik des Ev. Krankenhauses, D-4330 Muelheim/Ruhr, *Hellige GmbH, D-7800 Freiburg i.Br.

## SUMMARY

Based on results on central chemosensitivity in cats, paired stimuli were applied for therapy to infants with central respiratory insufficiency of various degrees. An unspecific respiratory stimulus, e.g. light for 1 s, was followed by a jet of either $O_2$ or 2% $CO_2$ in $O_2$ for 1.5 s. The unspecific and the chemical stimuli were interspaced by 0.5 s. The combined stimulation was repeated every 10 s. The program was triggered by using threshold values of transcutaneous $pO_2$. In infants with intratrachial tubes or tracheostoma we used the end tidal $pCO_2$ for triggering the stimulation. The method could prevent hypoxemia during sleep in non-ventilated subjects with sleep apnea syndromes or in infants with severe hypoxemia during sleep after being rescued from Sudden Infant Death Syndrome (SIDS). In patients with Ondine's Curse Syndrome (OCS) with its $CO_2$ insensitivity, paired stimuli were used in order to condition the chemical function of the respiratory system. Polysomnograms from 310 clinically healthy infants including healthy siblings of SIDS victims revealed instability of arterial $pO_2$ and low $CO_2$ sensitivity during sleep within the second month and the fourth to ninth month of life, respectively. These data challenge the described method as a potential preventive or therapeutic measure to defeat SIDS and sleep apnea syndromes in conjunction with disturbed chemical regulation of respiration.

## INTRODUCTION

In cats, after elimination of the intermediate area on the ventral medullary surface, the central chemosensitive drive is completely abolished (Schlaefke et al., 1979). Using paired stimuli in this preparation, e.g. an air jet as unspecific stimulus followed by a short increase of $FICO_2$, respiration regains its sensitivity to $CO_2$ within 30 to 90 minutes of stimulation; not so in controls without the specified stimulation (Burghardt and Schlaefke, 1986). There are recent observations indicating that these findings may be applicable to human conditions, and further that the possibility of conditioning the central chemosensitive function may be of interest for therapeutic purposes (Schlaefke and Burghardt, 1983), e.g. in connection with the Sudden Infant Death Syndrome and with the Ondine's Curse Syndrome: (1) Neuropathological studies on the brain stem of victims of SIDS and OCS (in cooperation with Messer, Strasbourg) revealed that neurones within the ventral medullary surface medial and rostromedial of the hypoglossal nerve are missing uni- or bilaterally (Kille and Schlaefke, 1986). (2) Unilateral coagulation of the intermediate area in chronic experiments with cats shows similar breathing and heart-rate patterns as in infants at risk of SIDS (Schaefer and Schlaefke, 1986). (3) Bilateral elimination of the intermediate area in chronic experiments with cats reflects the symptoms of Ondine's Curse Syndrome (Schlaefke et al., 1974; 1979). In this study by polysomnographical means, we investigated breathing of infants between two days and two years of age in order to defeat disturbances during sleep. Artificially ventilated and non ventilated infants with severe hypoxemia during sleep and with lack of central chemosensitive drive were exposed to paired stimulation with light and short increases of $O_2$ or $CO_2/O_2$ in the inspired air.

## METHODS

213 polysomnograms of clinically healthy infants, 123 of them from siblings of SIDS victims were studied during night sleep within the first 12 months of life, in order to detect potential disorders in the respiratory system related to a higher risk of SIDS (group I). In the same manner, we observed two infants continuously during 7 days after being rescued from SIDS (group II). Furthermore, we measured periods of spontaneous respiration during night sleep in 7 infants with OCS or other severe sleep apnea syndromes being dependent upon artificial ventilation (group III). We recorded continuously and simultaneously EEG, EOG, ECG, ventilation by thoracal and abdominal induction plethysmography and the voltage sum of both (RESPITRACE), transcutaneous $pO_2$ and $pCO_2$ as well as $O_2$ saturation by pulse oximetry. For respiratory responses to $CO_2$ during NREM sleep, infants inhaled 2% $CO_2$ in air or 2% $CO_2$ in $O_2$. Respiratory responses were calculated per Torr increased transcutaneous $pCO_2$. Acute drops in $O_2$ were defined as a decrease in transcutaneous $pO_2$ by at least 10 Torr within 60 s. The patients of group II and III received paired stimuli during sleep in hypoxic phases.

In group III we started recording before the patients were disconnected from the respirator during sleep or immediately after sleep onset before starting the artificial ventilation. Spontaneous breathing during sleep was enabled when paired stimulation was applied. For unspecific arousal stimulation (fig. 1) we used light, a sound, or an air jet blown into the face for 1 s, followed by insufflation or inhalation of either 2% $CO_2$ in $O_2$ or $O_2$ alone for 1.5 s as the chemical breathing stimulus. The unspecific and the chemical stimuli were interspaced by 0.5 s and were repeated every 10 s. The stimulation was triggered by a given threshold value of transcutaneous $pO_2$, usually 45 to 50 Torr. When hypoxemia correlated with the heart rate, a heart-rate threshold was also used as a trigger. In intubated or tracheotomized patients end tidal $pCO_2$ served as a trigger. In such cases patients were allowed to breathe spontaneously within a range of an arterial $pCO_2$ of 40 to 60 Torr. $O_2$ or $CO_2/O_2$ mixtures were gently blown towards the nostrils, either by a funnel (fig. 2), or by small tubes fixed to the comforter, or were blown into the tracheal tube. In cases of decreased $CO_2$ sensitivity, $CO_2/O_2$ served as a second stimulus as long as the end tidal $CO_2$ did not exceed 60 Torr. In cases of predominant hypoxemia, we selected $O_2$ for the second stimulus.

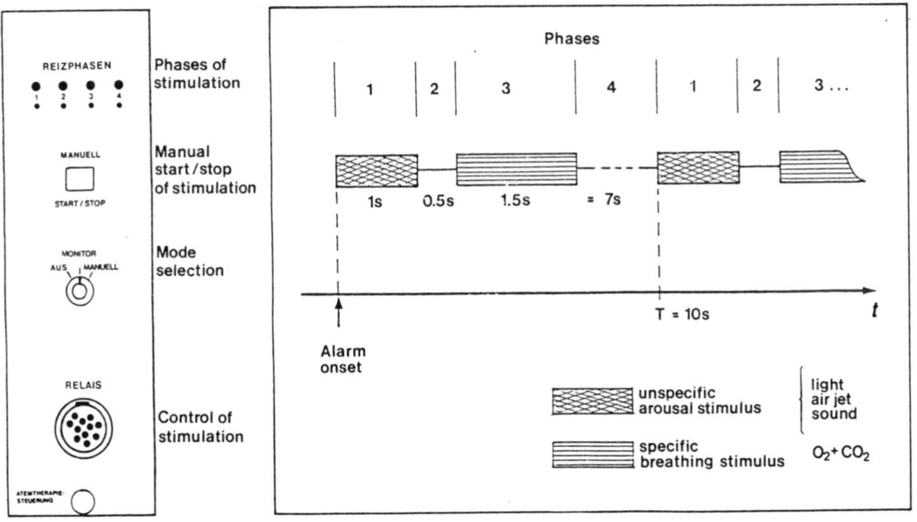


Fig. 1 Time pattern of paired stimuli for conditioning the respiratory chemoreflex

RESULTS

In group I in siblings of SIDS victims as well as in 'controls' a peak of 7.11 (6.99 in 'controls') acute drops in $pO_2$ per hour was found at an age of two months; the number decreased to 1.39 (1.88 in 'controls') at an age of more than 9 months (fig. 3). We found low $CO_2$ sensitivity (ventilatory ratio per Torr transcutaneous $pCO_2$ less than 1.25) during NREM sleep in siblings of SIDS victims between the fifth and ninth month of life, not so in 'controls'. In single cases of siblings $CO_2$ sensitivity remained low until the end of the first year of life, in these cases one and more near misses for SIDS events did occur.

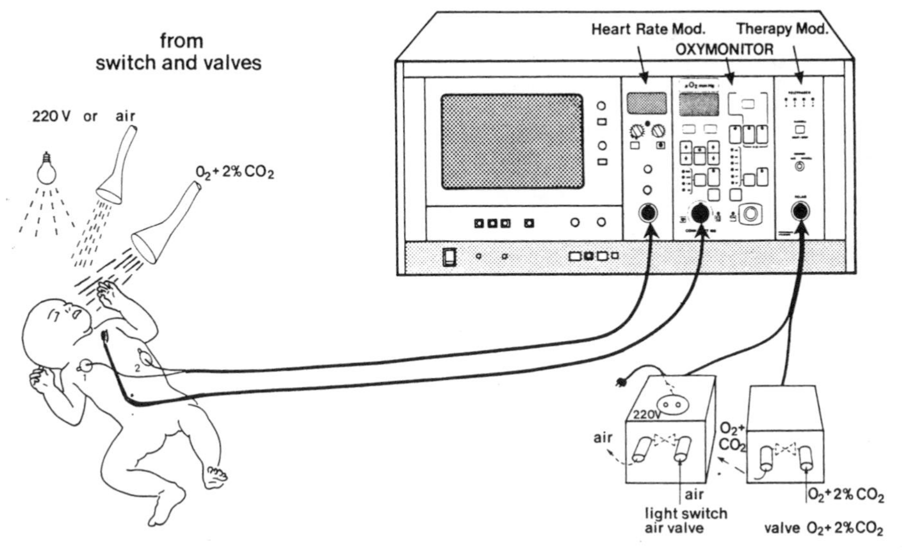


Fig. 2 Breathing therapy system for conditioning chemical regulation of respiration demanded by hypoxemia or brady-/ tachycardia

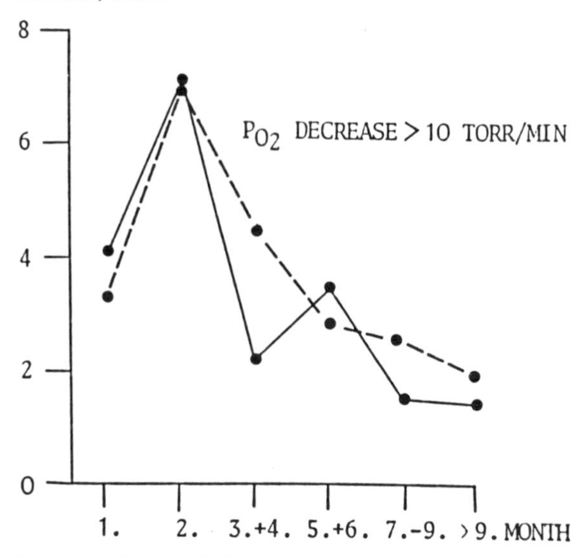


Fig. 3 Mean number of drops in transcutaneous $pO_2$ in 123 siblings of SIDS victims and 90 non siblings (broken line)

$pO_2$ values of group II infants during NREM sleep were between 15 and 20 Torr for 8 to 24 hours after a near miss for an SIDS event. Two cases were followed up polysomnographically to the 7th day after the event. Paired stimuli (light and $CO_2/O_2$) were applied according to the transcutaneous $pO_2$ values. After one week of intermittent stimulation, the $pO_2$ values became stable and were higher than 45 Torr during sleep. No repetition of a near miss for a SIDS event was observed. The babies were controlled polysomnographically 4 to 6 times during the following year and developed normally.

The infants of group III were unable to breathe during sleep and were therefore artificially ventilated. We started paired stimulation immediately upon disconnection from the respirator. We usually started with oxygen as the second stimulus, since the arterial $pCO_2$ values initially were 60 Torr or more (fig. 4). After training for a week or more with light and $O_2$, conditioning $CO_2$ sensitivity by using $CO_2$ as a second stimulus became possible and successful in patients who originally did not respond to $CO_2$ at all. The training periods lasted about two hours per night. The results were favorable in that spontaneous activity increased, provided that the length and the frequency of therapy were adequate in relation to the severity of the respiratory insufficiency. E.g. in a three months old male baby, the phases of spontaneous breathing during sleep were prolonged from 60 s to three hours after gradual increase of the time in which the baby was stimulated or disconnected from the respirator. Not to overtax the respiratory system, we supported the spontaneous respiration by repetitive ventilation. Under comparable conditions in a 5 months old girl with complete loss of central chemosensitivity, the infant, after two weeks of training, was able to breathe throughout the entire night. However, subsequent occasional training by paired stimuli was necessary (Schlaefke and Burghardt, 1983). Together with the capability of spontaneous breathing during sleep in all tested babies of group III, the respiratory response to $CO_2$ was increased.

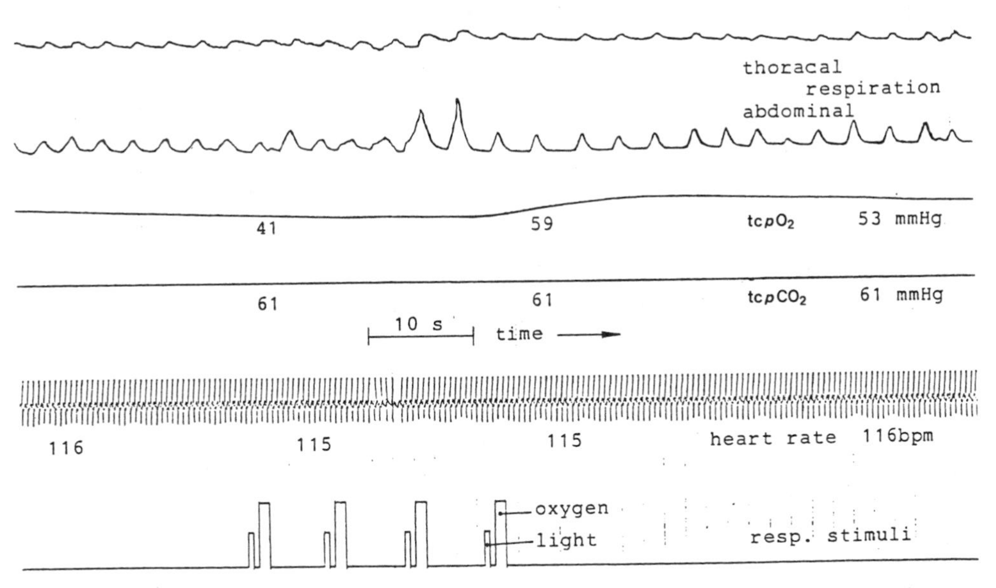


Fig. 4 Respiratory response to paired stimuli (light and oxygen) triggered by transcutaneously detected hypoxemia

## DISCUSSION

Our approach to overcome central respiratory insufficiency during sleep is based on the role of the reticular formation for the activity of the integrating respiratory system as well as for the influence of peripheral and central chemosensitive afferents (Schlaefke, 1981). The unspecific stimulus increased the basic reticular activity, so that the chemical stimulus finds adequate conditions to contribute to a sufficient synaptic input. The respiratory system can then establish its homeostatic function. Chemosensitive neurones in the intermediate area respond not only to local pH-changes, but also to unspecific stimulation. Reticular neurones in the vicinity of chemosensitive neurones within the ventral medullary surface reduce their spontaneous tonic discharge frequency, when $FICO_2$ is increased in the artificially ventilated cat. Preliminary data show that unspecific electrical stimulation of the femoral nerve paired with increases of $FICO_2$, timed as described for the infants and applied over three periods of 30 minutes, changed the $CO_2$ response of the reticular neurone from an inhibited state to a much greater firing rate as compared to the control group.

In cats, after elimination of central chemosensitivity, and in patients with OCS, conditioning with $CO_2$ as a second stimulus raised the $CO_2$ response curve and shifted it to the left (Schlaefke and Burghardt, 1983). The results reported here confirm the former observations which were attributed to a conditioning of the central chemoreflex. The question arises whether or not the function of peripheral chemoreceptors may become strengthened by this method as well. When operating with $O_2$ as a second stimulus, the unspecific stimulus is useful in counteracting the potential hyperoxic depression of respiration.

The data obtained from group I showed (1) that acute decreases of oxygen partial pressure are frequent during the second month of life, (2) that weak $CO_2$ responses characterize sleep phases of siblings of SIDS victims between the fifth and the ninth month of life. These two main findings from polysomnograms of group I challenge us to prove whether or not the described demanded conditioning of chemical regulation of respiration may be useful as preventive measures in defeating SIDS in so-called clinically healthy cases, in which the hypoxemic phases and weak $CO_2$ responses are detectable by polysomnographical means.

The data from group II indicate that in phases of hypoxemia during sleep as a high risk of SIDS (Naeye, 1980), the application of paired stimuli contributes to a stabilization of respiration and normoxia.

Data from group III confirm the principle of plasticity for the chemosensitive drive of respiration. Paired stimuli may be a useful means to support spontaneous breathing during weaning from artificial respiration. Conditioning of the respiratory system for $CO_2$ may help to overcome the OCS by training the homeostatic function of respiration during sleep. Besides the transcutaneous $pO_2$ as a trigger for conditioning, heart rate, transcutaneous $pCO_2$, end tidal $pCO_2$, pulse oximetry, or any other parameter reliably indicating life-threatening events may be used. Longer periods of therapy and further investigations are necessary in order to validate the descibed method.

## REFERENCES


Burghardt, F., and Schlaefke, M. E., 1986, Loss of central chemosensitivity: an animal model to overcome respiratory insufficiency, J. Autonom. Nerv. Syst. Suppl., 105-109.

Kille, J. F., and Schlaefke, M. E., 1986, Do ventral medullary neurones control the cardiorespiratory system in man?, Pfluegers Arch. Suppl., 406: R 25.

Naeye, R. L., 1980, Sudden infant death, Sci. Am., 242:4, 52-59.

Schaefer, T., and Schlaefke, M. E., 1986, Polysomnographically detectable data on respiration in infants with regard to the Sudden Infant Death Syndrome, Pfluegers Arch. Suppl., 406: R 25.

Schlaefke, M. E., 1981, Central chemosensitivity: A respiratory drive, Rev. Physiol. Biochem. Pharmacol., 90: 171-244.

Schlaefke, M. E., Kille, J. F., and Loeschcke, H. H., 1979, Elimination of central chemosensitivity by coagulation of a bilateral area on the ventral medullary surface in awake cats, Pfluegers Arch., 379: 231-241.

# EFFECTIVENESS OF COMBINED TRANSCUTANEOUS $Po_2/Pco_2$ MONITORING OF NEWBORNS

Harald Schachinger and
Christian Kolz

Kinderklinik der Freien Universität
Heubnerweg 6
D-1000 Berlin 19 (West)/Germany

## INTRODUCTION

Non-invasive monitoring of oxygen pressure by means of a transcutaneous $Po_2$ electrode has become standard practice in newborn intensive care units (2). This also holds for transcutaneous measurements of $Pco_2$ (1). It is of clinical importance that an infant's skin surface is as available as possible, for example for X-ray and phototherapy. It is therefore appropriate to accommodate both the $Po_2$ electrode and the $Pco_2$ electrode into one sensor.

## METHODS

More than 200 $Po_2/Pco_2$ measurements were carried out on premature and term infants with temporary disorders of adaptation and some with severe cardiopulmonary illness using a combined electrode (Fig. 1) for the monitor type Oxycapnometer S (Draeger Company). The skin thickness and the underlying subcutis were measured in order to ascertain the dependency of transcutaneously measured $Pco_2$ and $Po_2$ upon the thickness of skin and subcutis. This was done by a caligraph (Holtan Company).

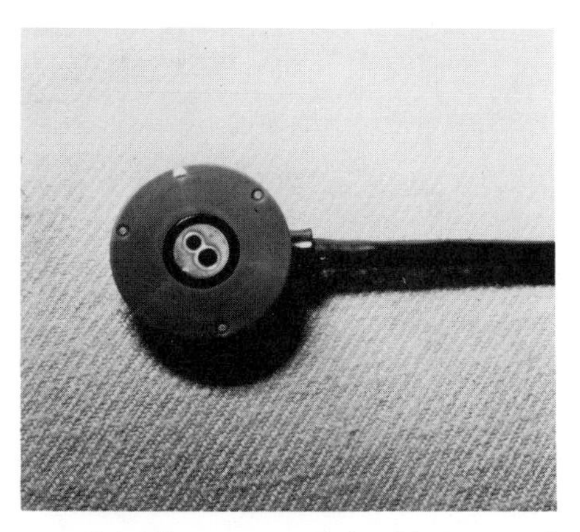

Fig. 1. The new combined sensor with the $Po_2$ and $Pco_2$ electrode (Draeger Company).

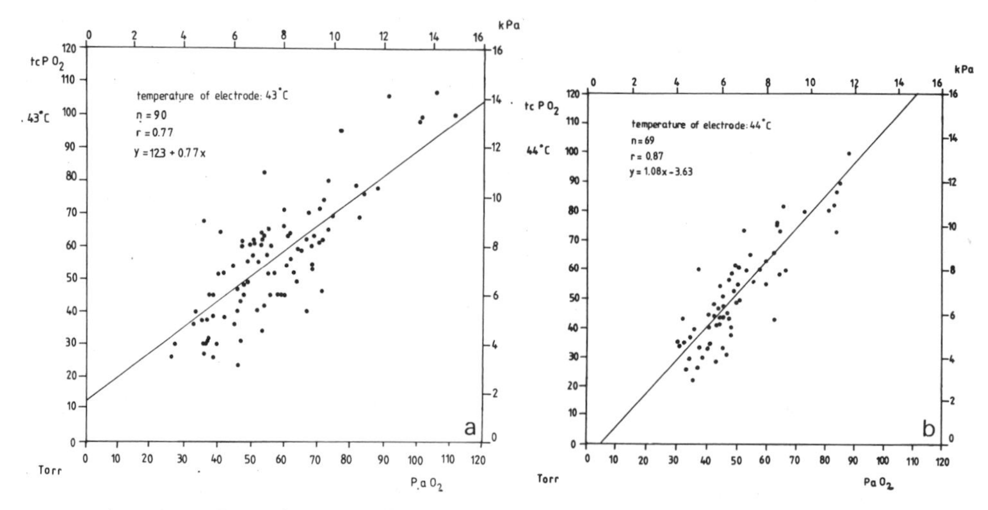


Fig. 2a + b. The correlation between $Pao_2$ and $tcPo_2$ with the combined sensor at a temperature of a) 43° C and b) 44° C.

The combielectrode was tested at 43° C and at 44° C. After 1 or 2 hours of measurement at these temperatures the patients' skin revealed only minor red blemishes.

RESULTS AND DISCUSSION

The correlation between transcutaneous $Po_2$ and arterial $Po_2$ was r = 0.87 at 44° C and r = 0.77 at 43° C (Fig. 2a + b). The $Po_2$ values from the combined electrode correlates well with the values from a single $Po_2$ electrode (5).

The changes in transcutaneous $Pco_2$ observed at 44° C or 43° C, using the combined electrode, were similar. Sudden variations in breathing were promptly registered. As expected, the niveau of transcutaneous $Pco_2$ values was significantly higher than that of arterial $Pco_2$ values (4, 6).

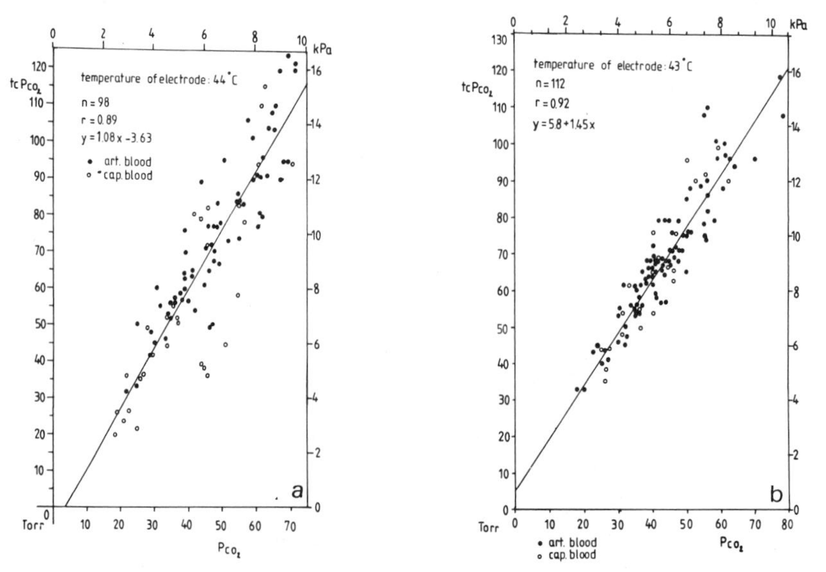


Fig. 3a + b. The correlation between $Paco_2$ and $tcPco_2$ with the combined sensor at a temperature of a) 44° C and b) 43° C.

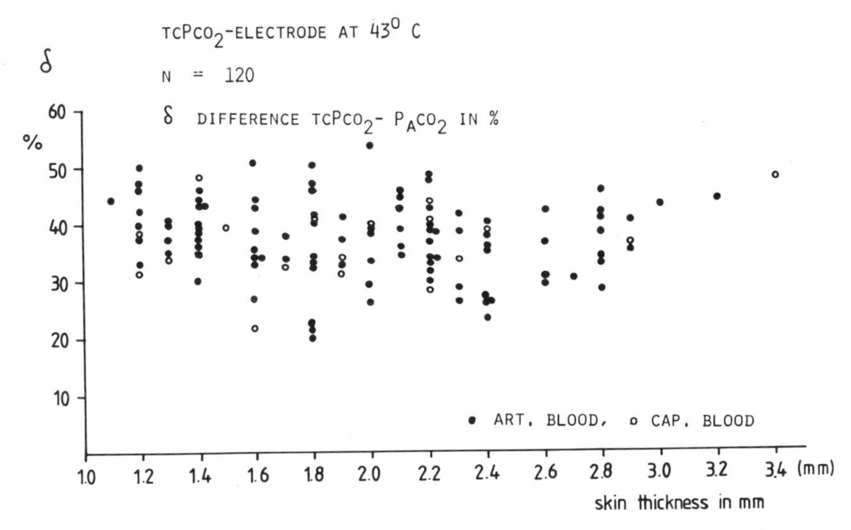


Fig. 4. The thickness of the skin and subcutis against the relative difference between $tcPco_2$ and $Paco_2$.

The correlation coefficients were r = 0.89 at 44° C and r = 0.92 at 43° C. At a temperature of 44° C the average deviation was 39.5% and 37.5% at 43° C (Fig. 3a + b). The correlation coefficients calculated at hypo-, normo- and hypercapnia were 0.79, 0.64 and 0.64, respectively.

Taking the corrective factors during calibration into account, a good approximation of the arterial $Pco_2$ was obtained.

The thickness of the skin of the newborn infants had no significant influence on the $tcPco_2$-$Paco_2$ difference as seen in Fig. 4. As Fig. 5 shows, the same applies to $tcPo_2$. Similar results were obtained at 43° C and at 44° C for both $tcPo_2$ and $tcPco_2$.

A further advantage of the combined $Po_2/Pco_2$ sensor is its ability to monitor also the infant's heart rate, thereby replacing one of the three ECG electrodes (3). This is possible by means of a special silver/silver chloride-plated membrane.

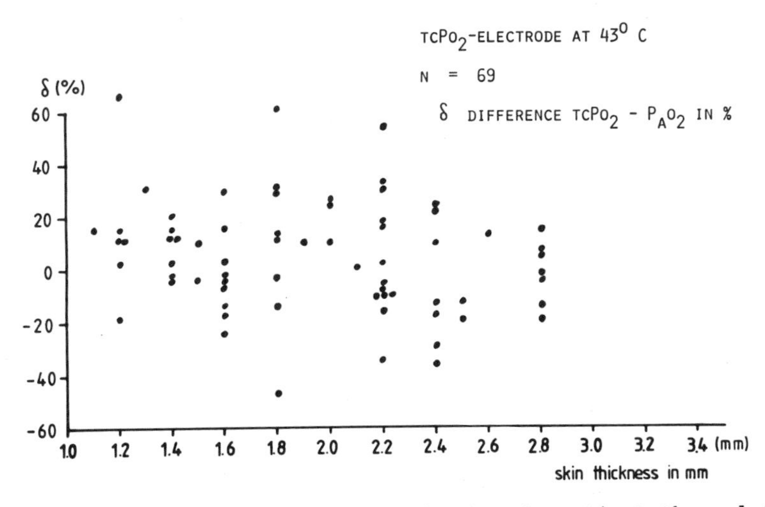


Fig. 5. The thickness of the skin and subcutis against the relative difference between $tcPo_2$ and $Pao_2$.

## CONCLUSIONS

The application of the combined transcutaneous $Po_2/Pco_2$ electrode is an improvement to neonatal monitoring. If one is familiar with the characteristics of the sensor, adequate monitoring of the levels and changes of blood gases may be obtained.

## REFERENCES


1. R. Bhat, A. Shukla, W. D. Kim and D. Vidyasagar, Continuous tissue pH and transcutaneous carbon dioxide ($tcPco_2$) monitoring in critically ill neonates, in: "Continuous transcutaneous blood gas monitoring", R. Huch and A. Huch (eds.), Marcel Dekker, Inc., New York and Basel, 361 (1983).
2. R. Huch, A. Huch, M. Albani, M. Gabriel, F. J. Schulte, H. Wolf, G. Rupparth, P. Emmrich, U. Stechele, G. Duc and H. Bucher, Transcutaneous $Po_2$ monitoring in routine management of infants and children with cardiorespiratory problems, Pediatrics 57: 681 (1976).
3. U. Hölscher, A novel approach for an ECG electrode integrated into a transcutaneous sensor, this volume.
4. R. Lemke, D. Klaus and D. W. Lübbers, The effect of carbonic anhyase inhibition on the $tcPco_2$, 2nd International Conference, Fetal and Neonatal Physiological Measurements, Oxford, 2nd-4th April 1984.
5. K. T. M. Schneider, F. Kraehenmann, A. Huch and R. Huch, Erste Erfahrungen mit der $tcPo_2/tcPco_2$-Kombinationselektrode in der Perinatalmedizin, in: "Perinatale Medizin", Bd. 10, J. W. Dudenhausen und E. Saling (Hrsg.), Thieme Verlag, Stuttgart, 368 (1984).
6. P. D. Wimberly, P. S. Frederiksen, J. Witt-Hansen, S. G. Melberg and B. Friis-Hansen, Evaluation of a transcutaneous $Pco_2$ monitoring in a neonatal intensive care department, in: "Perinatale Medizin", Bd. 10, J. W. Dudenhausen und E. Saling (Hrsg.), Thieme Verlag, Stuttgart, 140 (1984).

# INCIDENCE AND SEVERITY OF RETINOPATHY IN LOW BIRTH WEIGHT INFANTS MONITORED BY $TCPO_2$

Itsuro Yamanouchi, Ikuko Igarashi and Entaro Ouchi

Okayama National Hospital
Children's Medical Center
Okayama, Japan

SUMMARY

The incidence and severity of retinopathy of prematurity(ROP) were studied in 561 survivors of low birth weight infants(LBWIs) weighing under 2,000g. They were the survivors out of 604 LBWIs admitted to Okayama National Hospital during the five years from January 1981 to December 1985. The $tcPO_2$ of all the LBWIs on the respirator or under oxygen administration was monitored using an Oxymonitor SM 361. The upper limit of $tcPO_2$ was maintained between 50 to 80 torr. The ophthalmologic funduscopy was continued until 12 months of age. The early active stages of ROP were recorded by "the International Classification 1984". The cicatricial change of retina was classified by the criteria of the Joint Committee for the Study of Retrolental Fibroplasia in Japan. In the survivors, 55 cases of active ROP were found. (Stage 1:29 cases, 2:17, 3:9, and 4:0) In the survivors, 27 cases of cicatricial ROP were found, (Grade I:27 cases, II:0, III:0, IV:0) None of them developed blinding retinopathy. Neither the artificial ventilation nor the duration of oxygen administration influenced the incidence of ROP with statistical significance. The incidence and severity of ROP in this study seems to be one of the lowest among those hitherto published. Though our study is not a controlled one, it can be concluded that $tcPO_2$ monitoring eliminates the blinding ROP.

## INTRODUCTION

Inspite of numerous human and animal studies, the etiology of ROP remains still obscure. A critical review of the literature of ROP by Lucey et al[3] indicates that the cause of this disease is not yet known. This means that ROP especially in the extremely LBWI is a particularly difficult disease to prevent. The preventive means with judicious use and careful monitoring of supplemental oxygen, while decreasing the incidence, have not eradicated this complication of prematurity. The introduction of Huch's transcutaneous oxygen monitoring in the middle of 1970s offered the opportunity for continuous noninvasive measurements of $PO_2$. The present authors expected that the $tcPO_2$ technique would allow virtual elimination of hyperoxia, with consequent prevention of ROP, hopefully of blinding ROP. In 1980, these authors demonstrated a 7% incidence of cicatricial ROP among LBWIs weighing under 1,500g who had $tcPO_2$ monitoring for just 3 hours per day; the control group, with intermittent blood sampling, had a 26% incidence. Furthermore, they[2] also indicated very low incidence of cicatricial ROP among 111 survivors weighing under 1,500g out of 144 LBWIs. Grade I cicatricial ROP occurred in 20 LBWIs weighing under 1,500g. No patients had more than Grade I cicatricial disease. No case of blindness was found. In 1984 a new international classification of active ROP was established. This time, the following study was designed to investigate the incidence and severity of ACTIVE ROP by the new classification.

METHOD

The subject of this study was 561 LBWIs weighing under 2,000g. They were the survivors out of 604 LBWIs admitted to Okayama National Hospital from January 1981 to December 1985. The $tcPO_2$ was monitored for a whole day continuously with the Hellige Oxymonitor SM 361, during the more acute period of RDS with or without artificial ventilation. After the acute period, the $tcPO_2$ was monitored intermittently for 3 hrs a day if oxygen was administered. When the LBWIs needed extra oxygen, the upper limit of $tcPO_2$ was maintained between 50 and 80 torr during oxygen administration and/or artificial ventilation. Capillary blood specimens were collected intermittently for the monitoring of $PCO_2$ and pH. They were all cared for in the conventional closed type incubator with more than 70% humidity. The mean volume of daily total fluid intake was gradually increased according to the postnatal day, not exceeding 50 ml/kg/day on day 1, and it reached 110ml /kg/day in very LBWIs and 150ml/kg/day in the larger infants weighing over 1,500g, respectively one week after birth. The feeding policy was, initially freshly expressed colostrum from the infant's own mother. This was followed by freshly expressed or freeze-stored own mother's milk. Vitamin E was not administered to the LBWIs. Indirect ophthalmoscopy of all LBWIs and follow-up of the retinal changes after discharge until 12 months of age were routine. One of the present authors(E.O.) examined the eyes of all LBWIs who received or did not receive oxygen therapy. This examination was done regulary at intervals of 1, or 2 weeks from 3 weeks of age to the time of discharge. The eye changes of ACTIVE retinopathy were recorded by the International Classification[4]. The CICATRICIAL changes of retina was classified by the criteria of the Joint Committee for the Study of Retrolental Fibroplasia in Japan. Six cases with stage 3b were treated by cryocoagulation.

RESULTS

Fifty-five cases of active stage ROP were observed in the study population. No cases of retinal detachment were found. The distribution of incidence and severity of the active state both in different birth weight groups and in different gestational age groups were shown in Tables 1 and 2. All of the stage 3 ROP were observed in the birth weight group under 1,250g and in the gestational age group under 28 weeks. The incidences of the acute stage ROP according to the duration of $O_2$ administration were shown in Table 3. No association was found between them. The incidences of acute stage ROP according to artificial ventilation were shown in Table 4 and 5. No association was found between them. The distribution of incidence and severity of cicatricial grade both in different birth weight groups and in different gestational age groups were shown also in Tables 1 and 2. No ROP with Grade 2-4 was found. That is to say, none of the LBWIs developed blinding ROP.

DISCUSSION

The incidence and severity of active stage and cicatricial grade in this study seem to be one of the lowest among those hitherto published. The reasons associated with the low level of incidence and severity are still the subject of continuing investigation. Factors other than oxygen which may influence the occurrence of ROP, blood perfusion and partial pressure of carbon dioxide have been studied by some of the investigators. However, any other factor which influences retinal blood flow and/or retinal architecture must be potentially a cause of ROP. Therefore the effects of overhydration,which has been carefully controlled in the present study, may also be causally related to retinal pathology.

CONCLUSION

Though our study was not a controlled one, it is reasonable to say even the intermittent use of $tcPO_2$monitoring eliminates the blinding ROP.

Table 1. INCIDENCE OF ROP BY BIRTH WEIGHT (1981~85)

| Birth Weight Group | Survivors | Active Stage* | | | | | | Cicatricial Grade | | |
|---|---|---|---|---|---|---|---|---|---|---|
| | | St. 1 | St. 2 | St. 3 | St. 4 | St. 1, 2, 3 | Inci-dence | Gr. I | Gr. II, III, IV | Inci-dence |
| 500~ 749 | 6 | 3 | 1 | 2 | 0 | 6 | 100% | 4 | 0 | 67% |
| 750~ 999 | 30 | 9 | 7 | 5 | 0 | 21 | 70% | 14 | 0 | 47% |
| 1,000~1,249 | 56 | 10 | 4 | 2 | 0 | 16 | 29% | 6 | 0 | 11% |
| 1,250~1,499 | 95 | 6 | 4 | 0 | 0 | 10 | 11% | 3 | 0 | 3% |
| 1,500~1,749 | 150 | 1 | 1 | 0 | 0 | 2 | 1% | 0 | 0 | 0% |
| 1,750~1,999 | 224 | 0 | 0 | 0 | 0 | 0 | 0% | 0 | 0 | 0% |
| | 561 | 29 | 17 | 9 | 0 | 55 | | 27 | 0 | |

*International Classification 1984

Table 2. INCIDENCE OF ROP BY GESTATIONAL AGE (1981~85)

| Gestational Age | Survivors | Active Stage* | | | | | | Cicatricial Grade | | |
|---|---|---|---|---|---|---|---|---|---|---|
| | | St. 1 | St. 2 | St. 3 | St. 4 | St. 1, 2, 3 | Inci-dence | Gr. I | Gr. II, III, IV | Inci-dence |
| 23 | 1 | 0 | 0 | 1 | 0 | 1 | 100% | 1 | 0 | 100% |
| 24 | 1 | 1 | 0 | 0 | 0 | 1 | 100% | 0 | 0 | 0% |
| 25 | 11 | 1 | 3 | 4 | 0 | 8 | 73% | 7 | 0 | 64% |
| 26 | 13 | 2 | 3 | 1 | 0 | 6 | 46% | 5 | 0 | 38% |
| 27 | 24 | 4 | 4 | 3 | 0 | 11 | 46% | 7 | 0 | 29% |
| 28 | 31 | 8 | 2 | 0 | 0 | 10 | 32% | 3 | 0 | 10% |
| 29 | 34 | 9 | 2 | 0 | 0 | 11 | 32% | 2 | 0 | 6% |
| 30 | 49 | 4 | 2 | 0 | 0 | 6 | 12% | 2 | 0 | 4% |
| 31 | 50 | 0 | 1 | 0 | 0 | 1 | 2% | 0 | 0 | 0 |
| 32 | 53 | 0 | 0 | 0 | 0 | 0 | 0 | 0 | 0 | 0 |
| 33 | 43 | 0 | 0 | 0 | 0 | 0 | 0 | 0 | 0 | 0 |
| | 310 | 29 | 17 | 9 | 0 | 55 | | 27 | 0 | |

*International Classification 1984

Table 3. INCIDENCE OF ROP BY DURATION OF $O_2$ ADMINISTRATION (1981~85)

| Birth Weight | Duration (days) | Sur-vivor | Birth Weight Mean | Birth Weight SD | Gestation Mean | Gestation SD | ROP | Active Stage St.1 | St.2 | St.3 | St.4 |
|---|---|---|---|---|---|---|---|---|---|---|---|
| 1,000~1,249 | <30 | 13 | 1,154 | 60 | 29.5 | 1.2 | 4 | 3 | 1 | 0 | 0 |
| | ≧30 | 36 | 1,116 | 60 | 27.5 | 1.1 | 12 | 7 | 3 | 2 | 0 |
| 1,250~1,499 | <30 | 18 | 1,394 | 68 | 29.9 | 1.2 | 3 | 2 | 1 | 0 | 0 |
| | ≧30 | 37 | 1,377 | 64 | 29.1 | 1.3 | 7 | 4 | 3 | 0 | 0 |

Table 4. INCIDENCE OF ROP BY ARTIFICIAL VENTILATION

(1981~85)

I. CLASSIFIED BY BIRTH WEIGHT

| Birth Weight | Venti-lation | Sur-vivor | Birth Weight Mean | Birth Weight SD | Gestation Mean | Gestation SD | ROP | Active Stage St.1 | St.2 | St.3 | St.4 |
|---|---|---|---|---|---|---|---|---|---|---|---|
| 600~ 999 | (+) | 21 | 826 | 95 | 25.9 | 1.1 | 16 | 7 | 4 | 5 | 0 |
| | (−) | 14 | 896 | 82 | 27.4 | 1.3 | 11 | 7 | 3 | 1 | 0 |
| 1,000~1,249 | (+) | 20 | 1,129 | 74 | 28.1 | 1.8 | 8 | 5 | 3 | 0 | 0 |
| | (−) | 32 | 1,126 | 56 | 28.3 | 1.6 | 8 | 5 | 1 | 2 | 0 |
| 1,250~1,499 | (+) | 28 | 1,394 | 67 | 29.3 | 1.1 | 6 | 4 | 2 | 0 | 0 |
| | (−) | 45 | 1,379 | 65 | 30.0 | 1.5 | 4 | 2 | 2 | 0 | 0 |

Table 5. INCIDENCE OF ROP BY ARTIFICIAL VENTILATION

(1981~85)

II. CLASSIFIED BY GESTATIONAL AGE

| Gestational Age | Venti-lation | Sur-vivon | Birth Weight Mean | Birth Weight SD | Gestation Mean | Gestation SD | ROP | Active Stage St.1 | St.2 | St.3 | St.4 |
|---|---|---|---|---|---|---|---|---|---|---|---|
| 26~28 | (+) | 31 | 27.1 | 0.8 | 1,077 | 218 | 16 | 9 | 5 | 2 | 0 |
| | (−) | 36 | 27.4 | 0.7 | 1,124 | 156 | 11 | 5 | 4 | 2 | 0 |
| 29~31 | (+) | 25 | 29.5 | 0.5 | 1,337 | 109 | 7 | 5 | 2 | 0 | 0 |
| | (−) | 46 | 29.9 | 0.9 | 1,279 | 160 | 10 | 8 | 2 | 0 | 0 |

REFERENCES


1. Yamanouchi I: The effect of continuous oxygen pressure monitoring on the incidents of retrolental fibroplasia. Abstracts Symposia and Colloquia, XVI International Congress of Pediatrics 1980; 219
2. Yamanouchi I, Igarashi I, Ouchi E: Successful Prevention of Retinopathy of Prematurity via transcutaneous oxygen partial pressures monitoring in Huch R (ed): Continuous transcutaneous blood gas monitoring. New York and Basel, Marcel Dekker, 1983, pp 333-339
3. Lucey JF, Dangman B: A reexamination of the role of oxygen in retrolental fibroplasia. Pediatrics 1984; 73: 83-96
4. The Committee for the Classification of Retinopathy of Prematurity. An international classification of retinopathy of prematurity. Arch Opthalmol 1984; 102: 1130-1134

# TRANSCUTANEOUS OXYGEN MONITORING AND RETINOPATHY OF PREMATURITY

E. Bancalari, J. Flynn, R.N. Goldberg, R. Bawol, J. Cassady, J. Schiffman, W. Feuer, J. Roberts, D. Gillings and E. Sim

Div. of Neonatology, Depts. of Pediatrics and Ophthalmology
University of Miami/School of Medicine, Miami, Florida
Dept. of Pediatrics (R131), University of Miami/School of Medicine, P.O. Box 016960, Miami, FL 33101 U.S.A.

## SUMMARY

This study was performed to determine whether the use of continuous $tcPO_2$ monitoring could reduce the incidence of ROP in preterm infants receiving oxygen therapy. Two hundred and ninety-six infants with birth weights $\leq$ 1300 grams were randomly assigned to a continuous monitoring (CM) or a standard care (SC) group. CM infants had $tcPO_2$ monitored continuously as long as they required supplemental oxygen while SC infants had $tcPO_2$ monitored only during the more acute state of their illness. Management of both groups was otherwise identical. One hundred and one of 148 infants in the CM and 113 of 148 patients in the SC groups survived. Mean birth weights and gestational age were similar in both groups. Duration of mechanical ventilation and oxygen therapy was also similar. The overall incidence of ROP was 51% in the CM and 59% in the SC group. As birth weight for infants $\geq$ 1000 grams increased a higher risk for developing ROP was noted in the SC group. Four infants in the CM and 5 in the SC group developed cicatricial ROP. These results suggest that continuous $tcPO_2$ monitoring may reduce the incidence of ROP in infants with birth weights > 1000 grams, but not in the smaller infants in whom this complication occurs more frequently and is more severe.

## INTRODUCTION

For a number of years now there has been evidence relating oxygen therapy with retinopathy of prematurity.(1-4) Although it is likely that this is a multi-factorial disease(5), several studies have suggested a correlation between the duration of oxygen exposure and the development of ROP.(1,6) Attempts to relate different inspired oxygen concentrations with the disease have been so far inconclusive.(1,5,6) Although it has been mentioned for many years that ROP is probably related to the levels of arterial oxygen tension, several attempts to define the relationship between specific levels of arterial oxygen tension and ROP have been also unsuccessful.(5,6) It has been postulated that the reason why this relationship has not been established is because of the intermittent nature of the arterial oxygen measurements. It is possible that between arterial samples the infants were exposed to abnormal arterial oxygen tensions that were not detected by intermittent blood sampling.

The introduction of continuous oxygen monitoring devices opened the door for again testing the hypothesis that ROP may be related to abnormal arterial oxygen tensions. Although it has been suggested that continuous oxygen monitoring may reduce the incidence of ROP,(8) this has not been demonstrated in a prospective controlled study. In order to answer these questions and define the role of continuous oxygen monitoring on the incidence of ROP, a prospective controlled study was done at the University of Miami/Jackson Memorial Hospital between the years 1982-1984. The principal aim of the study was to investigate whether by continuous transcutaneous oxygen monitoring during oxygen therapy it is possible to reduce the incidence of ROP. The second aim was to evaluate whether continuous monitoring could shorten the exposure to oxygen and mechanical ventilation in infants with respiratory failure.

## METHODS

All infants born at Jackson Memorial Hospital during the study period with birth weights between 500 and 1300 grams and who required oxygen therapy at some time during the first 7 days of life were eligible for the study. Two hundred and ninety-six infants met the admission criteria and were enrolled in the study. Infants were randomly assigned to a continuous monitoring (CM) or standard care (SC) group during the first 12 hours of oxygen therapy.

To assure even distribution of weights in the two groups infants were randomized to the CM or SC groups according to their birth weight using two strata, one from 500 to 899 grams and the other from 900 to 1300 grams. After randomization infants in the CM group were kept on a transcutaneous oxygen monitor as long as they required any additional oxygen in the inspired gas to maintain their $PaO_2$ above 50 mmHg. Infants assigned to the SC group were monitored only during the more acute stage of their disease subject to availability of monitors in the unit.

In both groups arterial oxygen tension was measured intermittently in blood samples obtained through umbilical arterial lines or peripheral arteries according to the unit protocol, usually every four hours while the infant was on mechanical ventilation and less frequently after the infants were weaned from the ventilator. In all infants an attempt was made to keep arterial oxygen tension between 50-70 mmHg by adjusting the inspired oxygen concentration or the settings on the ventilator. The tc monitors on the CM infants had visual and audible alarms that were set at 50 mmHg for the low and 70 mmHg for the high limit.

Management was otherwise identical in both groups. Eye examinations were done in all infants by a pediatric ophthalmologist using indirect ophthalmoscopy. The examinations were begun when the infant reached 32 weeks of conceptional age and was in stable clinical condition and were repeated every 2 to 4 weeks until the baby was discharged from the hospital. Retinopathy of prematurity was diagnosed when one or more eye examinations were positive prior to discharge.

## RESULTS

Of the 296 infants enrolled into the study, 214 were discharged alive from the hospital. From these, 202 had a conclusive eye examination and could be classified as either ROP or non-ROP. In 12 infants, 4 in the CM and 8 in the SC group, ROP could not be definitely diagnosed or ruled out. The physical and clinical characteristics of the infants enrolled were similar in both groups. There were no differences in birth weight (1038 vs 1034 grams), gestational age, sex distribution or race between CM and SC infants in any of the birth weight strata. Although more infants developed sepsis and expired in the CM group than in the SC group, the differences did not reach statistical significance. The duration of hospitalization was also similar in the two groups. Table 1 shows the number of hours of oxygen therapy and mechanical ventilation in the survivors in the two groups and the median hours that the two groups spent on continuous oxygen monitor during the first 28 days of life. The median number of hours of oxygen therapy was similar in both groups. The duration of mechanical ventilation was longer in the CM group infants but the difference is not significant due to the large individual variation. As expected, the time on tcM was significantly greater in infants in the CM group than in the SC group.

The incidence of ROP in the two study groups is shown in Table 2. Four infants in the CM group developed cicatricial ROP, while 5 in the SC group had cicatricial ROP. The overall incidence of ROP was similar in the two groups, 51% in CM vs 59% in SC infants. In the smaller infants there was no difference in the incidence of ROP between the SC and CM groups. Above a birth weight of 1000 grams there was higher incidence of ROP in the SC group infants that became more striking in infants between 1200 and 1300 grams.

## DISCUSSION

The overall incidence of ROP in this study was higher than that reported for similar gestational age infants in previous publications.(9,10) This is most likely related to the fact that infants were examined repeatedly during their hospital stay, allowing the detection of mild forms of the disease that may be missed when infants are examined only once at the time of discharge from the hospital. As in previous reports, there was a clear relationship between ROP and birth weight with over 80% of infants under 1000 grams developing some form of the disease. The incidence of cicatricial ROP was similar to that reported previously and all cases except for one occurred in infants under 1000 grams. Although the results did not show clear evidence that continuous monitoring could reduce the incidence of ROP in the smaller infants, above 1000 grams there were more infants with ROP in the SC than in the CM group. This higher incidence in the SC group can be explained by the fact that several of these infants had a more protracted respiratory course and received oxygen for longer periods of time. While this longer exposure to additional oxygen may have been influenced by the lack of monitoring, this seems unlikely because tcM did not influence the duration of oxygen exposure in the smaller infants. There are several possible explanations for the lack of effect of tcM on the incidence of ROP in the smaller infants. It is possible that in the more immature infants factors other than arterial oxygen tension may be more important in the development of ROP. Another possibility is that the selected $PaO_2$ range of 50-70 mmHg may be too high for these very immature infants. Finally, in many infants, especially those with periodic breathing and frequent apneic episodes, there are fluctuations in arterial oxygen tension that takes the $PaO_2$ below 50 mmHg and over 70 mmHg and these fluctuations cannot be avoided in spite of using continuous oxygen monitoring.

**TABLE 1: RESPIRATORY THERAPY AND $O_2$ MONITORING FIRST 28 DAYS AMONG SURVIVORS MEDIAN (RANGES)**

| Birth Weight | 500-899 grams | | 900-1300 grams | |
|---|---|---|---|---|
| **Study Group** | **CM** | **SC** | **CM** | **SC** |
| Number of survivors | 22 | 24 | 79 | 89 |
| HRS $O_2$ THERAPY | 466 (85-659) | 410 (3-672) | 89 (0.8-672) | 109 (0.2-672) |
| HRS $FiO_2 \geq .40$ | 40 (1-377) | 47 (1-285) | 12 (0-384) | 8 (0-471) |
| HRS $FiO_2 \geq .70$ | 1 (0-46) | 2 (0-79) | 0.4 (0-51) | 0.2 (0-191) |
| REQUIRED IPPV (Number) | 22 | 23 | 60 | 65 |
| HRS IPPV | 236 (59-672) | 161 (11-672) | 49 (0-672) | 33 (0-473) |
| HRS $tcPO_2$ MONITORING (during 02 therapy) | 456 *** (82-644) | 56 (2-287) | 84 ** (0.7-672) | 35 (0.2-343) |
| NO. OF ARTERIAL BLOOD GASES | 69 (38-206) | 67 (13-136) | 33 (0-123) | 31 (1-200) |

*** $p \leq .001$

** $p \leq .01$

**TABLE 2: ROP INCIDENCE (SURVIVORS)**

CM

| B.W. (g) | N | Alive | ROP Total | ROP (Cicatr) | %ROP[1] Total |
|---|---|---|---|---|---|
| 500-699 | 12 | 1 | 1 | 1 | 100 |
| 700-799 | 12 | 9 | 7 | 2 | 78 |
| 800-899 | 19 | 12 | 11 | 1 | 92 |
| 900-999 | 26 | 18 | 12 | 0 | 67 |
| 1000-1099 | 28 | 21 | 12 | 0 | 57 |
| 1100-1199 | 27 | 18 | 7 | 0 | 39 |
| 1200-1300 | 24 | 22 | 2 | 0 | 9 |
| TOTAL | 148 | 101 | 52 | 4 | 51 |

SC

| B.W. (g) | N | Alive | ROP Total | ROP (Cicatr) | %ROP[1] Total |
|---|---|---|---|---|---|
| 500-699 | 12 | 3 | 3 | 0 | 100 |
| 700-799 | 19 | 12 | 11 | 3 | 92 |
| 800-899 | 14 | 9 | 8 | 0 | 89 |
| 900-999 | 25 | 18 | 11 | 1 | 61 |
| 1000-1099 | 23 | 20 | 12 | 0 | 60 |
| 1100-1199 | 26 | 23 | 12 | 0 | 52 |
| 1200-1300 | 29 | 28 | 10 | 1 | 36 |
| TOTAL | 148 | 113 | 67 | 5 | 59 |

(1) Indefinites treated as no ROP

## REFERENCES


1. Kinsey VE: Retrolental fibroplasia: Cooperative study of retrolental fibroplasia and the use of oxygen. Arch Opthalmol 56:481, 1956.
2. Gordon HH, Lubchenco L, Hix I: Observations on the etiology of retrolental fibroplasia. Bull Johns Hopkins Hosp 94:34, 1954.
3. Lanman JT, Guy LP, Dancis J: Retrolental fibroplasia and oxygen therapy. JAMA 155:223, 1954.
4. Patz A, Hoeck LE, DeLaCruz E: Studies on the effect of oxygen administration in retrolental fibroplasia: I. Nursing observations. Am J Ophthalmol 35:1248, 1952.
5. Lucey JF and Dangman B: A reexamination of the role of oxygen in retrolental fibroplasia. Pediatrics 73(1):82, 1984.
6. Kinsey VE, Arnold HJ, Kalina RE, Stern L, Stahlman M, Odell G, Driscoll JM, Elliott JH, Payne J, Patz A: $PaO_2$ levels and retrolental fibroplasia: A report of the cooperative study. Pediatrics 60(5):655, 1977
7. Yu VYH, Hookham DM, Nave JRM: Retrolental fibroplasia-controlled study of 4 years' experience in a neonatal intensive care unit. Arch Dis Childh 57:247, 1982.
8. Yamanouchi I, Igarashi I, Ouchi E: Successful prevention of retinopathy of prematurity via transcutaneous oxygen partial pressure monitoring: In continuous transcutaneous blood gas monitoring. Ed. R. Huch and A. Huch, Marcel Dekker, Inc., New York and Basel 1983.
9. Kalina RE and Karr DJ: Retrolental fibroplasia. Ophthalmology 89:91, 1982.
10. Campbell PB, Bull MJ, Ellis FD, Bryson CQ, Lemons JA, Schreiner RL: Retinopathy of prematurity in a tertiary newborn intensive care unit. Arch Ophthalmol 101:1686, 1983.

# COMBINED TRANSCUTANEOUS OXYGEN, CARBON DIOXIDE TENSIONS AND END-EXPIRED $CO_2$ LEVELS IN SEVERELY ILL NEWBORNS

W.B. Geven[1], E. Nagler[1], Th. de Boo[2] and W. Lemmens[2]

Departments of Neonatology[1] and Statistical Consultation[2]
University of Nijmegen
St. Radboud Hospital
The Netherlands

## SUMMARY

In 12 newborns combined transcutaneous $PO_2$ and $PCO_2$ levels were measured by means of a Sensor Medics Transend Transcutaneous Gas System simultaneously with the end-tidal $PCO_2$ ($PetCo_2$) values using an Angström Eliza $CO_2$ analyzer.

Although the individual correlation coefficients differed greatly, the diagnostic sensitivity of $PtcO_2$ and $PtcCO_2$ was satisfactory within specified limits. The $PetCO_2$ did not agree well with the $PaCO_2$ due to physiologic factors related to lung disorders. We conclude that $PtcO_2$ and $PtcCO_2$ values are very useful in monitoring severely ill newborns although arterial blood gas measurements remain necessary, but not as often as when the transcutaneous technique is not used.

## INTRODUCTION

Combined electrodes for the measurement of $PtcO_2$ and $PtcCO_2$ recently became available. However, the use of capnometers in neonatology is limited because of the considerable enlargement of the dead space and the inability to adequately measure high ventilator frequencies. These problems have been solved in a newly developed capnometer. The purpose of this study was to investigate the relation between $PaO_2$-$PtcO_2$, $PaCO_2$-$PtcCO_2$ and $PaCO_2$-$PetCO_2$ in routine use.

## PATIENTS AND METHODS

The $PtcO_2$, $PtcCO_2$, $PetCO_2$ and $PaO_2$, $PaCO_2$ levels were measured simultaneously in 12 newborns: 9 premature babies (28-34 wks) with severe RDS, 2 newborns with severe perinatal asphyxia (39-41 wks) and 1 with meconium aspiration syndrome. The $PtcO_2$ and $PtcCO_2$ values were determined using a SensorMedics Transend Transcutaneous Gas System with combined Stow-Severinghaus and Clark electrodes. An Angström $CO_2$ analyzer was used for end-tidal $CO_2$ measurements; the results are represented as the mean of the last 30 seconds. Umbilical artery catheterization in 10 patients and via the radial artery in 2 patients. Four to eight simultaneous measurements of all parameters were taken in 24-36 hours. The transcutaneous sensor was attached to the right side of the upper chest of all patients; in 5 a second electrode was placed on the left side of the abdomen. The electrode temperature was 43.5°C.

Hypoxemia was defined for $PaO_2$ values less than 6.5 kPa, normoxemia if

6.5 kPa $\leqslant PaO_2 \leqslant 10.5$ and hyperoxemia if $PaO_2 \geqslant 10.5$ kPa. Furthermore, hypocarbia $PaCO_2 \leqslant 4.2$ kPa, normocarbia if 4.2 kPa $\leqslant PaCO_2 \leqslant 6.2$ kPa and hypercarbia if $PaCO_2 \geqslant 6.2$ kPa.

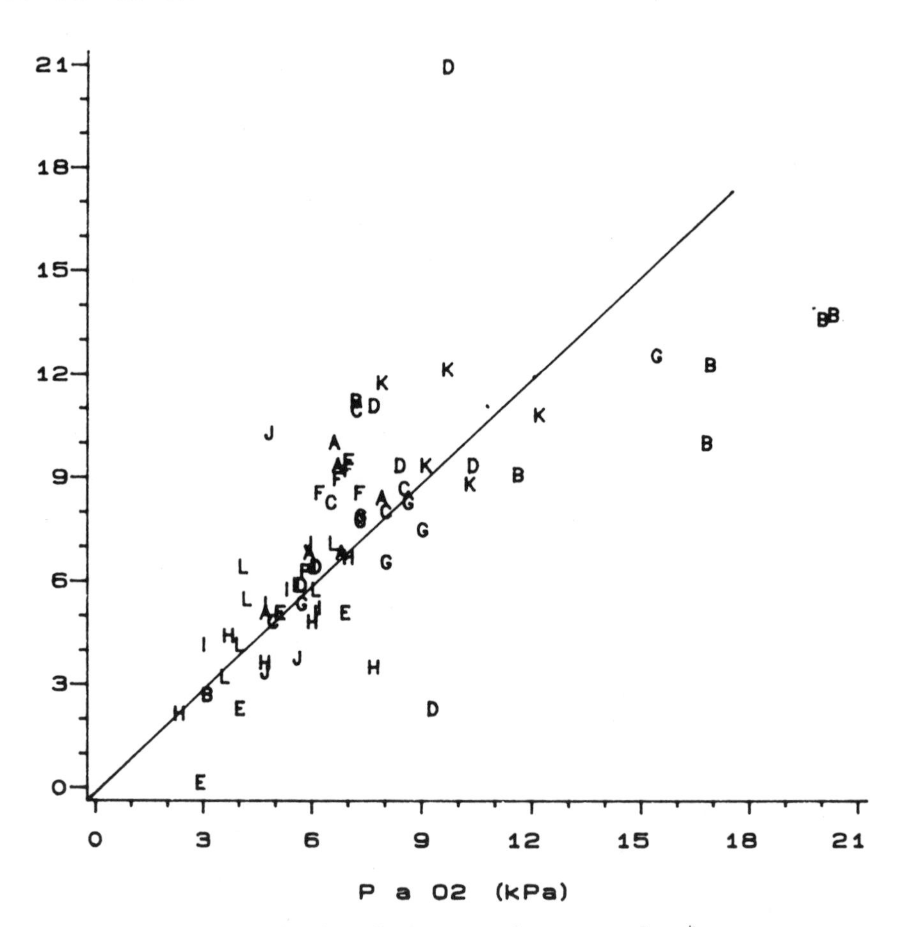


Figure 1: Analysis of $PtcO_2$ against $PaO_2$ levels in arterial blood samples of 12 newborns. The line represents identical values.

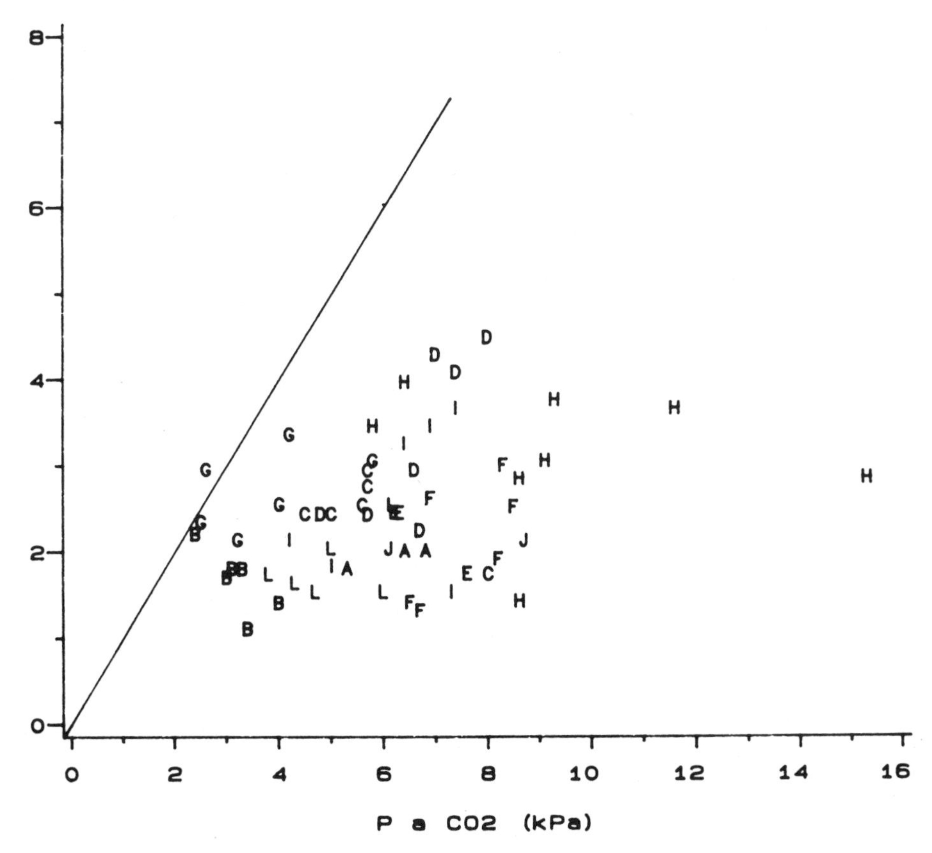


Figure 2: Analysis of $PtcCO_2$ against $PaCO_2$ levels in arterial blood samples of 12 newborns. The line represents identical values.

Table 1: Diagnostic sensitivity and agreement

| | | Right side upper chest | Left side abdomen |
|---|---|---|---|
| Diagnostic sensitivity for | hypoxemia | 28/32=0.88 | 12/13=0.93 |
| | normoxemia | 22/31=0.71 | 9/10=0.90 |
| | hyperoxemia | 5/7 =0.71 | 5/7 =0.71 |
| Agreement for hypoxemia and | hyperoxemia | 33/39=0.85 | 16/18=0.89 |
| Overall agreement | | 55/70=0.79 | 25/28=0.89 |
| Diagnostic sensitivity for | hypocarbia | 14/15=0.93 | 7/7 =1.0 |
| | normocarbia | 8/25=0.32 | 6/12=0.50 |
| | hypercarbia | 24/32=0.75 | 7/9 =0.78 |
| Agreement for hypocarbia and | hypercarbia | 38/47=0.81 | 14/16=0.88 |
| Overall agreement | | 46/72=0.64 | 20/28=0.71 |

Transcutaneous sensor position on the right side of the upper chest and on the left side of the abdomen

The sensitivity of the tc-measurement for each of these subgroups was determined in the usual way (i.e. the number of tc-measurements lying within the given kPa-interval were divided by the total number of arterial measurements in the subgroup). In addition 'agreement' in more than one of these subgroups was defined as the total of tc-measurement in the corresponding kPa-intervals, divided by the total number of arterial measurements in these intervals.

RESULTS

A total of 72 arterial blood gas samples were obtained from our 12 patients, (mean 6, range 4-8 per patient). The graph in Figure 1 of the $PtcO_2$ against the $PaO_2$ values shows the large amount of variation between the values. At very high $PaO_2$ values, the $PtcO_2$ seems to be relatively low; however, most of these values belong to one patient (B).

Table 1 shows the sensitivity and agreement of $PtcO_2$ in detecting hypoxemia, normoxemia and hyperoxemia. $PtcO_2$ agreed with the diagnosis hypoxemia and hyperoxemia in 88% and 71% of the cases respectively. In 79% the $PtcO_2$ was in the range of $PaO_2$ with the sensor on the right side of the upper chest; this figure was 89% when the sensor was placed on the left side of the abdomen. In Figure 2, $PtcCO_2$ values show considerable variation around the line representing identical values. The variation seems to be greater when the $PaCO_2$ values are over 5 kPa. Table 1 presents the sensitivity and agreement of our method of investigatic for detecting hypocarbia, normocarbia and hypercarbia.

Hypocarbia was detected in 93% of the cases and hypercarbia in 75%. Thi percentage was lower for normocarbia (32%). The end-tidal carbon dioxide levels show very large variation and only increase slightly with higher $PaCO_2$ levels.

The correlation coefficients for each individual are presented in Tabl together with the individual agreement. It appears that a low individual correlation coefficient can correspond with a good agreement (patients K,N) and also vice versa. The same is seen for $PtcCO_2$. The range for $PetCO_2$ is very much larger.

DISCUSSION

In most studies the relation between $PaO_2$ against $PtcO_2$ values and $PaCO_2$ against $PtcCO_2$ values is expressed in overall correlation coefficient, slope of regression line and intercept.[1,2,3]

Wimberley et al[4] estimated sensitivity and specificity of transcutaneo and $PCO_2$. We used individual correlation coefficients and agreement for detecting normal and abnormal oxygen and carbon dioxide levels.

Our values of overall agreement for oxygen tensions are comparable to those of Wimberley[4], while those for carbon dioxide pressure are somewhat lower. At the same time individual correlation coefficients show very large variation. When comparing data obtained using 2 sensors simultaneously on the right side of the upper chest and the left side of the abdomen, the latter position seems to be preferable.

In contrast to Epstein et al[3] we found only slightly higher $PetCO_2$ levels with rising $PaCO_2$, which was probably due to pathophysiologic factors concerning the patients' lungs. In our opinion the measurement of $PetCO_2$ can be used for detecting (partial) tube obstruction because in these cases the capnometer immediately gives an alarm. In one patient with persistent fetal circulation we observed an immediate rise in $PetCO_2$ after starting tolazoline medication due to improved circulation in the lungs. Because of the great variation in individual correlation coefficients, it is doubtful whether the in vivo calibration, according to Epstein et al[3], really improves the predictive value of $PtcO_2$ and $PtcCO_2$ if a larger number of measurements are used.

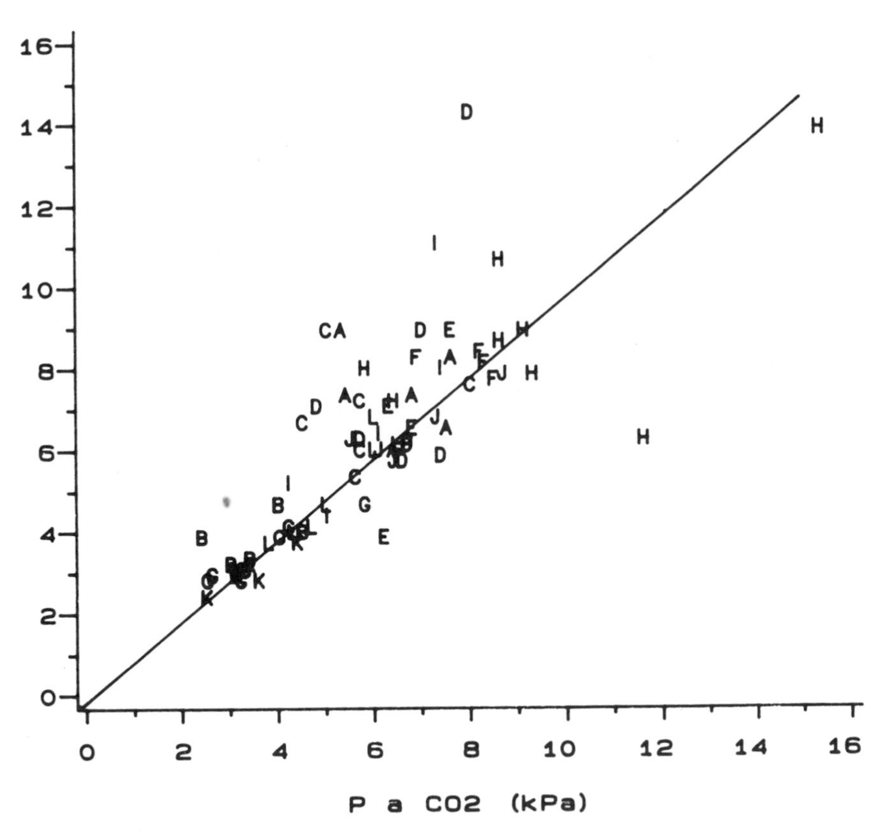


Figure 3: Analysis of $PetCO_2$ against $PaCO_2$ levels in arterial blood samples of 12 newborns. The line represents identical values.

Table 2: Individual correlation ($r$, $r_1$, $r_2$) and agreement ($a_1$, $a_2$)

| | $PtcO_2/PaO_2$ | | | | $PtcCO_2/PaCO_2$ | | | | $PetCO_2/PaCO_2$ |
|---|---|---|---|---|---|---|---|---|---|
| Patient | $r_1$ | $a_1$ | $r_2$ | $a_2$ | $r_1$ | $a_1$ | $r_2$ | $a_2$ | $r$ |
| A | 0.67 | 83 | 0.86 | 100 | -0.29 | 50 | -0.25 | 50 | 0.97 |
| B | 0.83 | 57 | 0.77 | 71 | 0.48 | 70 | 0.94 | 86 | -0.77 |
| D | 0.65 | 83 | 0.99 | 100 | 0.08 | 50 | -0.42 | 100 | -0.66 |
| E | 0.53 | 57 | -- | | 0.53 | 29 | -- | | 0.79 |
| F | 0.90 | 75 | 0.97 | 100 | 0.82 | 75 | 0.99 | 75 | -0.99 |
| G | 0.77 | 83 | -- | | 0.71 | 67 | -- | | 0.64 |
| H | 0.96 | 100 | -- | | 0.95 | 83 | -- | | 0.51 |
| J | 0.58 | 83 | -- | | 0.62 | 88 | -- | | -0.15 |
| K | 0.67 | 100 | -- | | 0.80 | 67 | -- | | 0.44 |
| L | -0.05 | 60 | -- | | 0.88 | 60 | -- | | 0.99 |
| M | -0.20 | 60 | -- | | 0.88 | 80 | -- | | - |
| N | 0.70 | 100 | 0.76 | 100 | 0.95 | 50 | 0.95 | 50 | 0.18 |

Individual correlation coefficient ($r_1$, $r_2$) and agreement ($a_1$, $a_2$) between transcutaneous and arterial values for $PO_2$ and $PCO_2$. Also individual correlation coefficient of end-expired $CO_2$ ($r$) and arterial $CO_2$. Sensor on the right side of the upper chest $r_1$, $a_1$ and on the left side of the abdomen $r_2$, $a_2$.

ACKNOWLEDGEMENTS

We thank Klees Electronics bv and Gambro bv for the loan of the instruments, Dr. L.A.A. Kollée for his valuable advice and Miss M. Collonbon for her technical assistance.

REFERENCES

1. O. Löfgren and D. Anderson, Simultaneous transcutaneous carbon dioxide and transcutaneous oxygen monitoring in neonatal intensive care, J Perinat Med, 51:56 (1983).
2. M. B. Whitehead, B. V. W. Lee, T. M. Paggdin and E. D. R. Reynolds, Estimation of arterial oxygen and carbon dioxide tensions by a single transcutaneous sensor, Arch Dis Child, 356:359 (1985).
3. M. F. Epstein, B. C. Aaron, H. A. Feldman and D. B. Raemer, Estimation of $PaCO_2$ by two non-invasive methods in the critically ill newborn infant, J Pediatr, 282:286 (1985).
4. P. D. Wimberley, P. S. Fredricksen, J. Witt-Hansen Melberg SG and B. Friis-Hansen, Evaluation of a transcutaneous oxygen and carbon dioxide monitor in a neonatal intensive care department, Acta Paediatr Scand, 352:359 (1985).

# TRANSCUTANEOUS $PO_2$ AND $PCO_2$ DURING SURFACTANT THERAPY IN NEWBORN INFANTS WITH IDIOPATHIC RESPIRATORY DISTRESS SYNDROME

S. Bambang Oetomo[1], B. Robertson[2] and A. Okken[1]

[1] Dept. Paediatrics, Div. Neonatology
University Hospital, Groningen, The Netherlands
[2] Dept. Paediatric Pathology
St. Görans Hospital, Stockholm, Sweden

## INTRODUCTION

It has been shown that surfactant treatment in infants with idiopathic respiratory distress syndrome (IRDS) results in an increase of arterial $PO_2$ and a decrease of arterial $PCO_2$. The instillation of surfactant, however, may result in a more or less severe temporate decrease of transcutaneous $PO_2$ as i.e. reported by Hallman et al.[1] The aim of this study was to monitor transcutaneous $PO_2$ and $PCO_2$ continuously before, during and immediately after the instillation of surfactant in newborn infants with IRDS. To obtain more precise information about the sequence of events associated with the administration of the surfactant. Infants with IRDS who did not receive surfactant but underwent the same procedures served as controls. All infants took part in a clinical trial on the efficacy of surfactant therapy.

## PATIENTS AND METHODS

In 8 patients weighing 1301 ± 106 g (mean ± SEM) with clinical and radiological evidence of IRDS and the need of respiratory support, transcutaneous $PO_2$ ($TcPO_2$) and $PCO_2$ ($TcPCO_2$) were measured. Four patients were treated with surfactant (Curosurf, manufactured by T. Curstedt, Karolinska Institute, Stockholm, Sweden). The instillation procedure was followed by 2 minutes manual ventilation ($FiO_2$ 1.0). Thereafter, the $FiO_2$ was lowered to keep the $TcPO_2$ in the normal range to prevent hyperoxemia. Four patients, controls, underwent the same procedures but did not receive surfactant or placebo.

$TcPO_2$ and $TcPCO_2$ were measured with a Tc M222 transcutaneous blood gas monitor with a combined electrode (Radiometer, Copenhagen). To visualize the rapid changes in lung function, following the administration of surfactant from minute to minute, during the whole procedure the $TcPO_2$ was standardized for the fraction of inspired oxygen by calculating the $TcPO_2/FiO_2$ ratio. In this way $TcPO_2$ values of patients that are ventilated with different $FiO_2$ can be compared. For example in a healthy newborn infant, breathing room air, with a transcutaneous $PO_2$ of 80 mm Hg the $TcPO_2/FiO_2$ ratio is 80/0.21 = 380. In contrast a baby with IRDS artificially ventilated with 100% oxygen and the same $TcPO_2$, has a $TcPO_2/FiO_2$ ratio of 80/1.0 = 80.

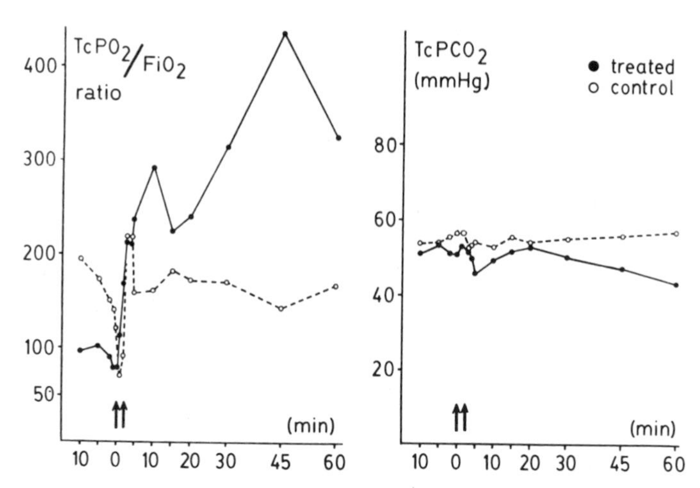


Fig. 1: Responses of $TcPCO_2$ and $TcPO_2/FiO_2$ ratio to surfactant instillation (↑↑) in surfactant treated (●) and control infants (○) with IRDS.

RESULTS

In the figure the $TcPO_2/FiO_2$ ratio and $TcPCO_2$ before, during and after surfactant treatment are presented.

Suctioning of the endotracheal tube, prior to the surfactant treatment procedure, resulted in a decrease of the $TcPO_2/FiO_2$ ratio to 74.5 ± 4.8, followed by a rise to 218 ± 33.9 one minute after the procedure in all patients. The surfactant treated patients showed a further increase of the $TcPO_2/FiO_2$ ratio to 437 ± 78 after 45 minutes. Control patients showed a decrease of the ratio to 160 ± 31 followed by maintainance of this level during the first hour after the procedure.

The $TcPCO_2$ of both groups of patients ranged from 46 to 56 mm Hg before and during the procedure. In the surfactant treated patients a gradual decrease of the $TcPCO_2$ was observed. One hour after surfactant treatment the $TcPCO_2$ was 44 ± 4.5 in the treated and 57 ± 6.3 mm Hg in control patients, respectively.

SUMMARY

We conclude that surfactant treatment in newborn infants with IRDS results in decrease in $TcPCO_2$ and an increase in $TcPO_2$ within minutes. The surfactant treatment procedure used in this study was not associated with hypoxemia. During surfactant treatment monitoring of $TcPO_2$ and $TcPCO_2$ is absolutely necessary.

REFERENCE

1. M. Hallman, T.A. Merritt, H. Schneider, B.L. Epstein, F. Mannino, and L. Gluck, Isolation of human surfactant from amniotic fluid and a pilot study of its efficacy in respiratory distress syndrome, Pediatics, 71:473 (1983).

# CONTINUOUS FETAL ACID-BASE ASSESSMENT DURING LABOUR BY tc-$pCO_2$ MONITORING

Carsten Nickelsen and Tom Weber

Department of Obstetrics and Gynaecology
Rigshospitalet, University of Copenhagen
9 Blegdamsvej, DK-2100 Copenhagen Ø, Denmark

## INTRODUCTION

Fetal well-being during labour is usually evaluated by fetal heart rate monitoring. Fetal blood sampling and analysis (FBS) of fetal acid-base status are very important supplements to this monitoring, but these procedures are invasive only giving discontinuous information. Tissue-pH monitoring is continuous (Weber,1982), but this method is also invasive and because of technical problems with the electrode not applicable for routine monitoring at present. Transcutaneous carbon dioxide (tc-$pCO_2$) monitoring is the only acid-base parameter which can be measured continuously and non-invasively. In previous studies fetal tc-$pCO_2$ has demonstrated a significant correlation to fetal capillary blood $pCO_2$ and to umbilical artery blood $pCO_2$ (Schmidt et al.,1984; Nickelsen et al.,1986).

Fetal acidosis is usually described as respiratory or metabolic, but during development of both kinds of fetal acidosis, the $pCO_2$ is usually increased - at least temporarily (Saling,1976). Consequently, development of acidosis during labour usually should be detected when fetal tc-$pCO_2$ monitoring is performed. In this paper we have studied the changes of fetal tc-$pCO_2$ during uncomplicated labour and during some cases of fetal acidosis.

## MATERIAL AND METHOD

Fetal transcutaneous carbon dioxide was measured continuously during 122 deliveries. Both high-risk and low-risk deliveries between the 36th and 42th week of pregnancy were included in the study. The Radiometer E5230 tc-$pCO_2$ electrode was in the first 80 cases calibrated by a two-step calibration ( 5% and 10% $CO_2$ gas), and in the subsequent 42 cases by a one-step calibration ( 5% $CO_2$ gas), but in these cases the electrode sensitivity was checked by exposing the electrode to the 10% gas. The electrode membrane was renewed before each monitoring and the electrode was sterilized by fluid aldehydes ( Korsolin 3% ) for 20 min. A sterile suction application ring was used for electrode application (Fig. 1). The electrode temperature was chosen to 44 °C in the first 80 monitorings and to 41 °C in the subsequent 42 monitorings. Following rupture of membranes and at a cervical dilatation of more than 3 cm, the electrode was attached to the fetal head during a vaginal examination. No application tool was

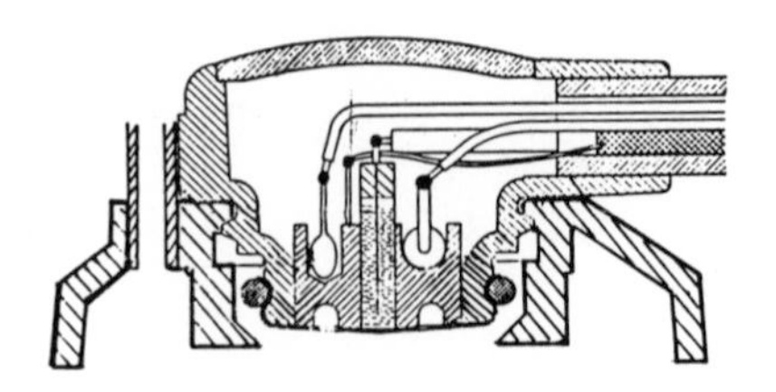

Fig. 1. Cross-section of a transcutaneous carbon dioxide electrode mounted in a suction fixation ring with a catheter to the vacuum pump. (Radiometer E5230)

used and neither cleaning nor shaving of the fetal skin was performed. An electrical pump supplied the suction ring with a continuous vacuum of 20 kPa. The tc-$pCO_2$ monitor (TCM20, Radiometer) was connected to the cardiotocograph, and the tc-$pCO_2$ value was plotted on the cardiotocogram during labour. The cervical dilatation was frequently evaluated by vaginal examination.

The fetal heart rate was monitored continuously by a direct ECG spiral electrode. The cardiotocogram was used for evaluation of fetal well-being, but the changes in tc-$pCO_2$ did not influence the management of labour. The umbilical cord was clamped immediately after delivery and blood samples from the umbilical artery and vein were analysed for pH and $pCO_2$ in a blood-gas analyser (ABL3, Radiometer). The electrode drift during monitoring was evaluated by exposing the electrode to the 5% $CO_2$ gas after each monitoring.

All tc-$pCO_2$ values were corrected for the elevated electrode temperature using a correction factor calculated from the anaerobic temperature coefficient of $CO_2$ in blood (Nickelsen et al.,1986). Consequently, the values are expressed as the tc-$pCO_2$ at 37 °C. The mean value of tc-$pCO_2$ at each cm of cervical dilatation was calculated in all cases where the newborn child had a pH of umbilical artery blood above 7.19. The cases where the umbilical artery blood pH was below 7.15 were evaluated separately.

## RESULTS

Electrode application by suction fixation was successful in all 122 cases, but in some cases the electrode dislodged during the second stage of labour. In these cases the electrode was reapplied within a few min.

Six newborns were delivered with a pH of umbilical artery blood below 7.15 and in 104 cases the pH was above 7.19. The mean values of

Table 1. Mean value of tc-$pCO_2$ (kPa at 37 °C) during labour in fetuses delivered with umbilical artery blood pH above 7.19. (n: number of observ.)

| | 4 cm | 5 cm | 6 cm | 7 cm | 8 cm | 9 cm | 10 cm | Delivery |
|---|---|---|---|---|---|---|---|---|
| Mean | 6.03 | 6.43 | 6.38 | 6.36 | 6.51 | 6.55 | 6.60 | 7.26 |
| SD | 1.11 | 1.47 | 1.40 | 1.29 | 1.33 | 1.21 | 1.16 | 1.48 |
| n | 33 | 50 | 66 | 73 | 82 | 87 | 86 | 69 |

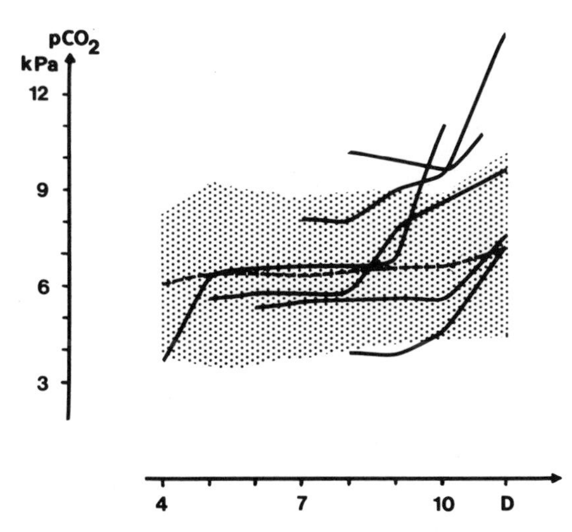


Fig. 2. Changes in tc-$pCO_2$ during six deliveries where the newborns were delivered with a pH of umbilical artery blood below 7.15.
The dotted area represents the normal range of tc-$pCO_2$ (mean $\pm$ 2SD). x-axis: cm of cervical dilatation. D: delivery.

tc-$pCO_2$ at different stages of cervical dilatation are presented in Table 1. The mean tc-$pCO_2$ value is slightly but significantly increasing from 4 cm dilatation to full dilatation, whereas the increase during the second stage is more evident (t-test). The changes of tc-$pCO_2$ during labour in the six cases where the newborn was delivered with a low pH of umbilical artery blood are presented in Fig. 2. The dotted area represents the mean ±2SD of tc-$pCO_2$ in fetuses delivered without acidosis (umbilical artery blood pH above 7.19 ). A distinct increase of tc-$pCO_2$ was found before delivery in all six cases, but only in half of the cases did the value exceed the mean + 2SD of the non-acidotic newborns.

The electrode drift during monitoring never exceeded 1 kPa, and the mean (SD) drift was -0.04 kPa (0.48).

## DISCUSSION

The suction fixation method used in this study has previously been compared to the glue fixation method, which is used by some other investigators (Schmidt et al.,1982). The fixation method does not influence the tc-$pCO_2$ value (Nickelsen and Weber,1986), but application by suction is more simple, giving less discomfort to the mother and allowing re-application following dislodgment without re-sterilization or re-calibration of the electrode.

Normal range of tc-$pCO_2$ during labour was defined as the mean $\pm$ 2SD found in newborns delivered without acidosis ( umbilical artery blood pH above 7.19 ). The mean value increased during both the first and the second stage of labour in this study. These changes were not demonstrated in previous studies of intermittent $pCO_2$ measurements (Nickelsen et al., 1986), but similar increases were found in studies of glue fixated tc-$pCO_2$ electrodes (Schmidt,S.,personal communication). In uncomplicated labour the fetal $pCO_2$ is dependent of the maternal $pCO_2$ which often is decreased during the first stage of labour because of maternal hyperventilation.

The active bearing down efforts will usually cause increasing maternal $pCO_2$ during the second stage of labour. Consequently, the fetal tc-$pCO_2$ increase during the second stage of labour may be explained by changes in maternal respiration, while the slight tc-$pCO_2$ increase during the first stage of labour must be caused by changes in fetal metabolism or in the placental exchange of $CO_2$. The mean values of tc-$pCO_2$ found in this study are close to the values found in previous studies of capillary $pCO_2$ (Nickelsen et al.,1986).

The fetal tc-$pCO_2$ was expected to be elevated at any fetal acidosis but only in three out of six cases the tc-$pCO_2$ value exceeded the normal range of tc-$pCO_2$. If all six newborns delivered with acidaemia should have been characterised by an elevated tc-$pCO_2$, the upper limit for normal tc-$pCO_2$ should be 7.3 kPa, which is the mean value of tc-$pCO_2$ at delivery of a non-acidotic child. A limit for pathological tc-$pCO_2$ of 7.3 kPa would make tc-$pCO_2$ monitoring very sensitive in predicting fetal acidosis, but the specificity would only be 50%. Although further studies might introduce other limits of tc-$pCO_2$ for detection of fetal acidosis (e.g. different limits at different stages of labour and limits of tc-$pCO_2$ increase per hour), a low specificity and low predictive value of a positive test can still be expected. Consequently, a supplementary monitoring method will often be needed. In these cases - and especially in high-risk cases - simultaneous tissue-pH and tc-$pCO_2$ monitoring offer the best continuous information of fetal acid-base, not only diagnosing the acidosis, but also distinguishing between a respiratory and metabolic fetal acidosis (Nickelsen et al., 1985). At present, further evaluation of all continuous acid-base monitoring methods is needed before routine clinical use can be recommended.

## REFERENCES


Nickelsen, C., Thomsen, S. G. and Weber, T.,1985, Continuous acid-base assessment of the human fetus during labour by tissue pH and transcutaneous carbon dioxide monitoring, Br J Obstet Gynaecol 92:220.

Nickelsen, C., Thomsen, S. G. and Weber, T.,1986, Fetal carbon dioxide tension during human labour, Eur J Obstet Gynecol Reprod Biol 22:205.

Nickelsen, C. and Weber, T.,1986, Fetal transcutaneous carbon dioxide monitoring - the effect of different fixation methods, Br J Obstet Gynaecol 93:1268.

Saling, E.,1966, Das Kind im Bereich der Geburtshilfe, Georg Thieme Verlag, Stuttgart.

Schmidt, S., Langner, K., Gesche, J., Dudenhausen, J. W. and Saling, E., 1984, Correlation between transcutaneous $pCO_2$ and the corresponding value of fetal blood - a study at a measuring temperature of 39 °C, Eur J Obstet Gynecol Reprod Biol 17:387.

Schmidt, S., Langner, K., Rothe, J. and Saling, E.,1982, A new combined non-invasive electrode for tc-$pCO_2$ measurement and fetal heart rate recording, J Perinal Med 10:297.

Weber, T.,1982, Cardiotocography supplemented with continuous fetal pH monitoring during labour, Acta Obstet Gynecol Scand 61:351.

# MICROVASCULAR DYNAMICS DURING ACUTE ASPHYXIA IN CHRONICALLY PREPARED FETAL SHEEP NEAR TERM

Arne Jensen*, Uwe Lang and Wolfgang Künzel

Department of Obstetrics and Gynecology, University of Giessen
Klinikstr. 32, D-6300 Giessen, West Germany

## INTRODUCTION

From previous studies in both human and sheep fetuses we concluded that reduced blood flow to the fetal skin after repeated episodes of asphyxia indicates circulatory centralization of the fetus, which can be detected by transcutaneous $pO_2$ measurements (Jensen and Künzel, 1980, 1982; Jensen, Künzel, and Kastendieck, 1982, 1983, 1985; Jensen, Künzel, and Hohmann, 1985a, 1985b; Jensen, Hohmann, and Künzel, 1987). This conclusion was based on two experimental findings. First, increasing catecholamine concentrations in the fetal sheep's plasma correlated with changes in blood flow to peripheral and central organs, measured by radionuclide labeled microspheres (Rudolph and Heymann, 1967). Secondly, blood flow to the skin correlated with blood flow to other peripheral and central organs. For example, the decrease in blood flow to the skin after repeated episodes of asphyxia was accompanied by a decrease in blood flow to the kidney, spleen, and small intestines, and by an increase in blood flow to the brainstem (Fig. 1).

These observations suggested that after asphyxia, reduced skin blood flow is an index of fetal circulatory centralization. However, due to both the experimental design and technical limitations of the original microsphere technique used in these previous studies, we were not able to look at the dynamics, i.e. the time course, of changes in blood flow to the skin and that to other peripheral or central organs.

To further validate skin blood flow as an index of circulatory centralization, we studied the time course of organ blood flow changes both during and after a single 2-min.episode of asphyxia.

## MATERIAL AND METHODS

We used 9 unanesthetized, chronically prepared fetal sheep between 125 and 135 days of pregnancy (term is at 147 days) to measure the time course of circulatory centralization. Blood flow measurements were made

---

*To whom correspondence should be addressed. This work was supported by the Deutsche Forschungsgemeinschaft Je 108/2-2, Je 108/4-1.

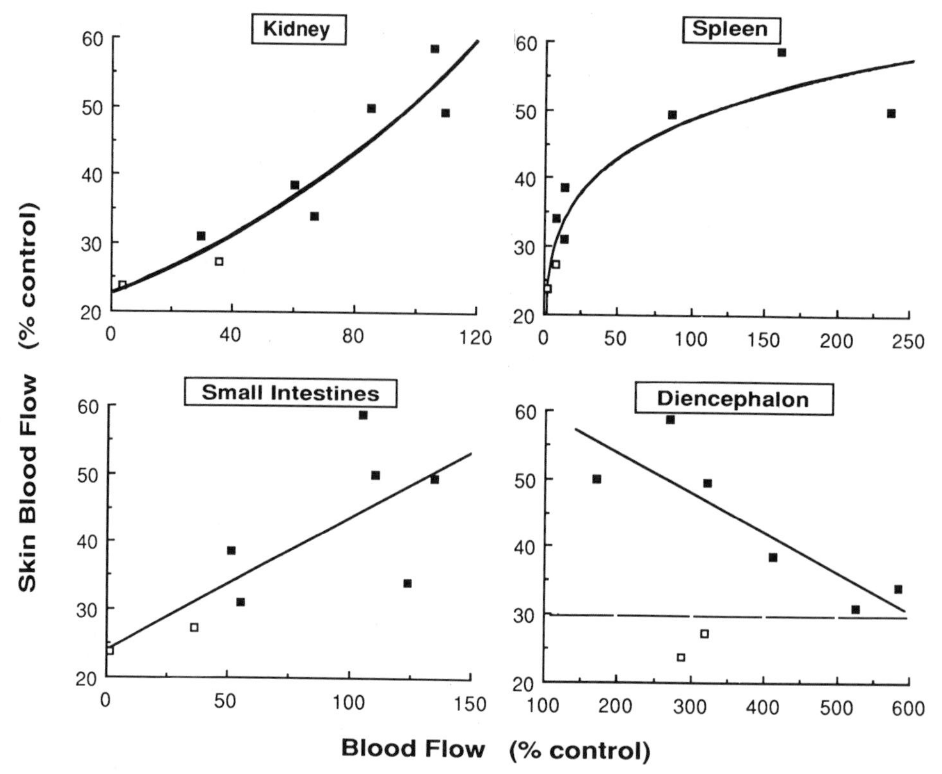


Fig. 1: Correlations between skin blood flow and blood flow in the kidney ($r = 0.95$, $n = 8$, $2a < 0.001$), spleen ($r = 0.94$, $n = 8$, $2a < 0.001$), small intestines ($r = 0.75$, $n = 8$, $2a < 0.05$), and diencephalon ($r = 0.86$, $n = 6$, $2a < 0.05$) in 8 fetal sheep near term after eleven 90-s episodes of asphyxia within 33 min caused by intermittent arrest of uterine blood flow. Note, that 2 fetuses (□), in which cutaneous, renal splenic, and intestinal blood flows were minimal, died in the recovery period. These 2 fetuses failed to increase diencephalic blood flows to the expected values (Jensen et al., J. Develop. Physiol. 9: 42, 1987).

at 1 min intervals, i.e. in the control period, after 1 and 2 minutes of asphyxia, and after 1, 2 and 28 min in the recovery period, by a modified microsphere technique (Jensen et al., 1985a). Asphyxia was induced by arrest of uterine blood flow as described previously (Jensen et al., 1985a). Before each blood flow measurement a blood sample was withdrawn from the descending aorta and analysed for blood gases, acid-base balance (Technicon BG IA, West Germany; OSM 2, Radiometer, Danmark), lactate (Roche), and catecholamine concentrations. Plasma concentrations of catecholamines were determined by reversed phase ion-pair HPLC with electrochemical detection (Jensen and Jelinek, 1986). Fetal heart rate and arterial blood pressure, amniotic fluid pressure, and maternal arterial blood pressure were recorded continuously. All measurements were made after at least 3 days recovery, in the absence of uterine contractions.

Results are given as means ± SE. Comparisons versus control were made by paired t-tests. Because more than one comparison was made in a repeated measures design, a Bonferroni correction was employed to obtain proper p-values (Miller, 1966).

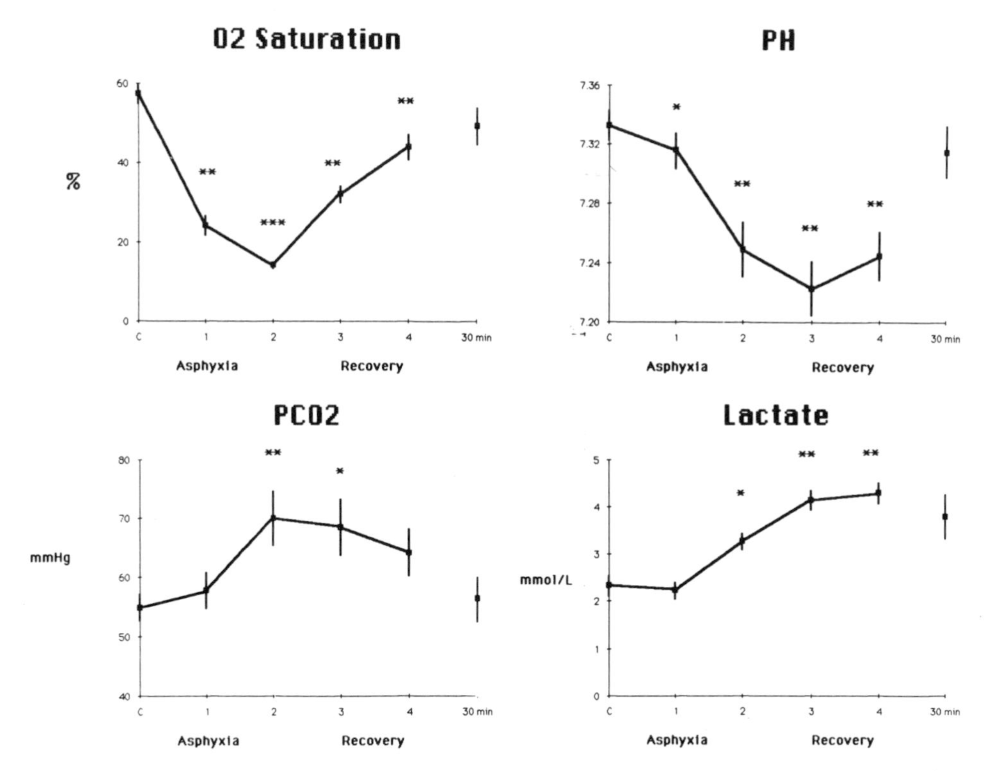


Fig. 2: Changes in arterial $O_2$ saturation of hemoglobin, pH, $pCO_2$, and lactate concentrations during and after a single 2-min episode of asphyxia, caused by arrest of uterine blood flow in fetal sheep near term.

RESULTS

In the control period the arterial blood gases, pH, base excess, lactate, norepinephrine (547 $\pm$ 58 pg/ml) and epinephrine (37 $\pm$ 11 pg/ml) concentrations were in the normal range for chronically catheterized fetal sheep near term (Fig. 2). During 2 min of asphyxia the arterial $O_2$ saturation of hemoglobin, $pO_2$, pH, and base excess decreased, and $pCO_2$, lactate, norepinephrine (26159 $\pm$ 4594 pg/ml, $p \leq 0.05$) and epinephrine (13534 $\pm$ 4239 pg/ml, $p \leq 0.05$) concentrations increased (Fig. 2).

There was a fall in heart rate (158 $\pm$ 9 vs. 97.6 $\pm$ 7 bpm, $p \leq 0.05$) after 1 min and an increase in arterial blood pressure (39 $\pm$ 4 vs. 62 $\pm$ 4 mmHg, $p \leq 0.05$; amniotic fluid pressure subtracted), which was significant only after 2 min of asphyxia. Arterial blood pressure was higher than control values ($p \leq 0.01$) until 3 min of recovery.

During asphyxia there was a tremendous increase in myocardial blood flow by 300 % ($p \leq 0.001$), which decreased rapidly in the early recovery. The blood flow to the cerebrum and cerebellum remained unchanged during asphyxia, even though arterial blood pressure rose, but increased in the immediate recovery period after asphyxia by about 100 % ($p \leq 0.01$). The adrenal blood flow did not change during, but increased by 100 % after 1 min of recovery ($p \leq 0.05$). Placental blood flow decreased transiently (-31 %, $p \leq 0.05$) at the nadir of asphyxia, in spite of an increase in arterial blood pressure, consistent with an increase in umbilical vascular resistance.

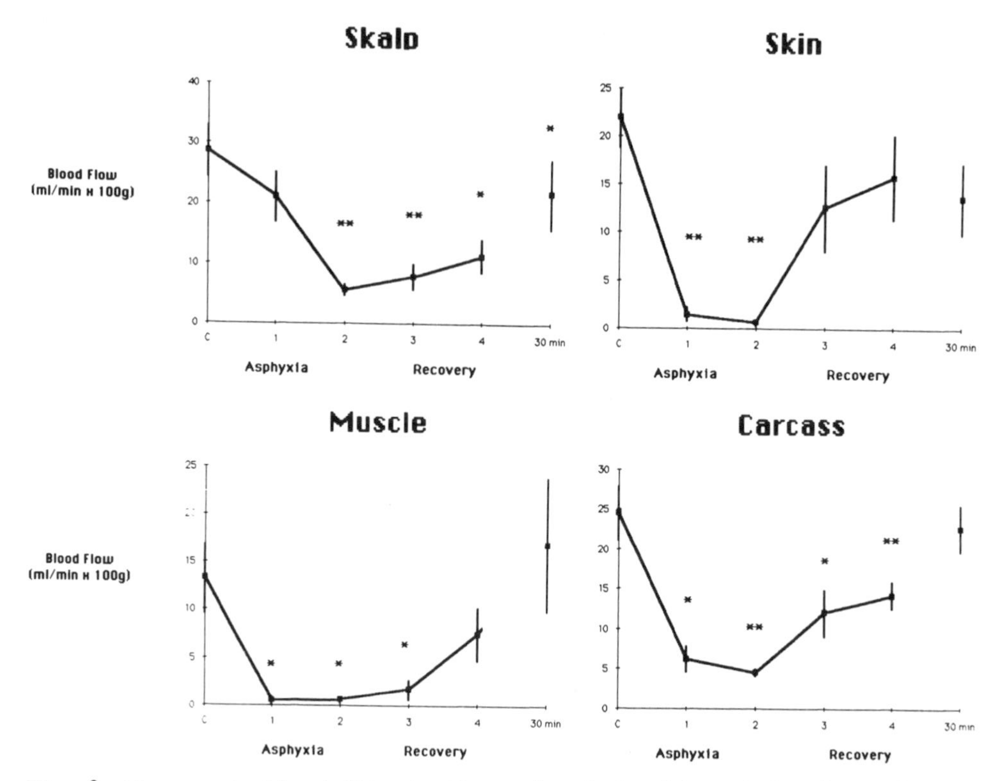


Fig. 3: Changes in blood flow to the scalp, body skin, skeletal muscle, and carcass during and after a single 2-min episode of asphyxia, caused by arrest of uterine blood flow, in fetal sheep near term. The similarity of the time courses of blood flow changes in these organs is conspicuous. Note, however, the differences in blood flow changes between scalp and body skin.

The dynamic blood flow changes within the various parts of the brain were quite different. Unlike the case with cerebrum and cerebellum, medullary blood flow almost doubled after 1 min of asphyxia ($p \leq 0.001$) and tripled after 2 min of recovery ($p \leq 0.001$). After 30 minutes of recovery the medullary blood flow returned to normal. The blood flow to the choroid plexus decreased after 1 and 2 min of asphyxia and did not recover in the first 2 minutes after asphyxia, suggesting that the blood flow regulation in the choroid plexus is much different from that in other parts of the brain.

In most of the principal peripheral organs, including muscle, carcass, small and large gut, kidney, scalp, and body skin, blood flow decreased during asphyxia and recovered gradually thereafter (Fig. 3). The pattern of changes in blood flow to these organs resembled that of scalp and body skin, although there were some differences between organs.

Notwithstanding the similarity between the time courses of blood flow changes in both cutaneous areas and those in other peripheral organs, there were some differences between blood flows to the scalp and body skin, which were consistent with previous observations (Jensen et al., 1985a). First, blood flow to the scalp did not decrease as rapidly as that to the body skin, resulting in a significant blood flow difference ($p \leq 0.01$) between both cutaneous areas after 1 min of asphyxia. Further, after 2 min of asphyxia blood flow to the body skin fell to lower values than did the scalp blood flow ($p \leq 0.001$). Finally, scalp blood flow did not recover as rapidly after asphyxia as did the blood flow to the body skin. Scalp blood flow values were significantly lower than control values after 1, 2, and 28 min of recovery (Fig. 3).

## CONCLUSION

We conclude that the decreases in blood flow to the fetal skin and scalp during asphyxia and their increases after asphyxia reflect very closely the blood flow changes in the principal peripheral organs of the fetus and hence indicate both the development of and the recovery from circulatory centralization caused by asphyxia. We therefore renew our suggestion to introduce skin blood flow measurements or measurements of variables that depend on blood flow to the skin, such as the transcutaneous $pO_2$, to detect a developing fetal circulatory centralization early during labour.

## REFERENCES


Jensen, A., Hohmann, M. and Künzel, W, 1987, Redistribution of fetal circulation during repeated asphyxia in sheep: Effects on skin blood flow, transcutaneous $PO_2$, and plasma catecholamines. J. Develop. Physiol. 4: 42.

Jensen, A. and Jelinek, J., 1986, Der Catecholamingehalt zentraler und peripherer Organe bei Normoxie und Hypoxie des Feten. Ber.Gyn. 122: 889.

Jensen, A. and Künzel, W., 1980, The difference between fetal transcutaneous $PO_2$ and arterial $PO_2$ during labour. Gynecol. Obstet. Invest, 11: 249.

Jensen, A. and Künzel, W., 1982, Transcutaneous fetal $PO_2$ under the influence of Fenoterol. In: Beta-Mimetic Drugs in Obstetrics and Perinatology (eds Jung, H. and Lamberti, G.), pp. 186 - 189, Thieme, Stuttgart, New York; Thieme-Stratton, New York.

Jensen, A., Künzel, W. and Kastendieck, E., 1982, Epinephrine and norepinephrine release in the fetus after repeated hypoxia. J. Perinat. Med. 10 (S 2): 109.

Jensen, A., Künzel, W. and Kastendieck, E., 1983, Transcutaneous $PO_2$ and norepinephrine release in the fetal sheep after repetitive reduction of uterine blood flow. In: Continuous Transcutaneous Blood Gas Monitoring (eds Huch, R. and Huch, A.), Reproductive Medicine Series, vol. 5, pp. 591 - 602, Marcel Dekker, Inc., New York, USA.

Jensen, A., Künzel, W. and Kastendieck, E., 1985, Repetitive reduction of uterine blood flow and its influence on fetal transcutaneous $PO_2$ and cardiovascular variables. J. Develop. Physiol. 7: 75.

Jensen, A., Künzel, W. and Hohmann, M., 1985a, Dynamics of fetal organ blood flow redistribution and catecholamine release during acute asphyxia. The Physiological Development of the Fetus and Newborn (eds Jones, C.T. and Nathanielsz, P.W.), pp. 405 - 410, Academic Press, London and Orlando.

Miller, R.G. jr., 1966, Simultaneous statistical interference, Mc Graw-Hill Book Company, New York, San Francisco, St. Louis, London.

Paulick, R., Kastendieck, E. and Wernze, H., 1985, Catecholamines in arterial and venous umbilical blood: placental extraction, correlation with fetal hypoxia, and transcutaneous partial oxygen pressure. J. Perinat. Med. 13: 31.

Rudolph, A.M. and Heymann, M.A., 1967, The circulation of the fetus in utero: Methods for studying distribution of blood flow, cardiac output and organ blood flow. Circ. Res. 21, 163 - 184.

# PULSE OXIMETRY

PULSE OXIMETRY: PHYSICAL PRINCIPLES, TECHNICAL REALIZATION AND PRESENT LIMITATIONS

Michael R. Neuman

Department of Reproductive Biology
Case Western Reserve University
Cleveland Metropolitan General Hospital
Cleveland, Ohio, USA

Tissue oximetry was one of the earliest applications of biomedical electronic devices to physiologic studies and patient monitoring, however, it was not until the recent development of transcutaneous pulse oximetry that the technique has enjoyed widespread application. The early work of many investigators has contributed to the modern pulse oximetry instruments. In this manuscript, the underlying principles of pulse oximetry as developed by these workers will first be examined, and the basic technical instrument that is widely used today will be described. As with any noninvasive patient monitoring technique, there are some limitations to its use, and the manuscript will conclude by pointing out some of these.

BASIC PRINCIPLES

Transmission Oximetry. The underlying concept of optically based oximetry was described in work in the 1930's and 40's. A method to spectrophotometrically measure hemoglobin oxygen saturation in tissue was described by L. Nicolai[1] in 1932. This was later applied to actual devices for measuring hemoglobin oxygen saturation in tissue by Kramer[2] and Matthes[3] in 1935. The method is based upon differences in the optical transmission spectrum of oxygenated and deoxygenated hemoglobin in the visible and near infrared portions of the spectrum as illustrated in Figure 1. Note that the spectra cross at a wavelength of 805 nm and that they are widely separated at wavelengths around 650 nm. The former point is known as an isobestic point, and the optical properties of hemoglobin solutions at this wavelength are independent of whether the hemoglobin is oxygenated or not. On the other hand at 650 nm, the optical transmission of hemoglobin solutions will be much greater for oxygenated hemoglobin than for deoxygenated hemoglobin. By making optical transmission measurements of a hemoglobin solution at both wavelengths it is possible to determine the relative amount hemoglobin present in the solution and its oxydation state. This can be done by making use of concepts from elementary optics.

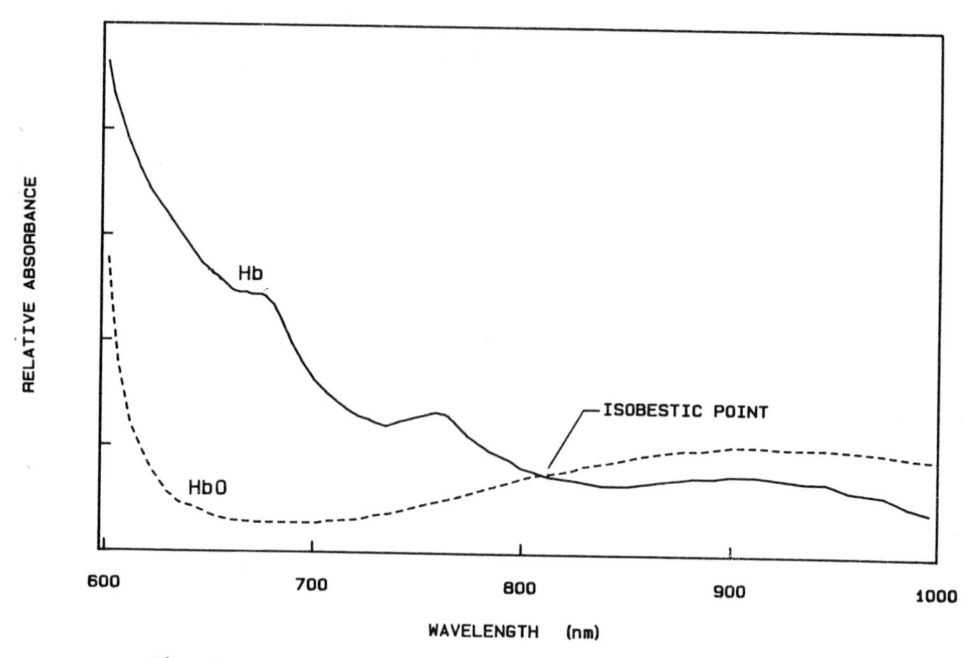


Fig. 1. Spectra for reduced and oxidized hemoglobin

The Beer-Lambert law describes the relationship between the amount of light transmitted through a solution containing a material that absorbs some of the light and the concentration of that material. For a fixed geometry between the light source and the photodectector this can be described as

$$I_t = I_o \exp(-\alpha c d) \quad (1)$$

where $I_t$ is the transmitted light intensity, $I_o$ is the light intensity at the source, $\alpha$ is the molar absorption coefficient for the absorbing material in the solution, c is the concentration of that material and d is the optical path length through the solution. Because of the exponential relationship in equation 1, it is often convenient to define relative intensities in terms of optical density.

$$OD = \log_{20} \frac{I_o}{I_t} = \alpha c d \quad (2)$$

If the measurement is carried out in a structure with fixed geometry so that the optical path, d, is constant, it is seen that the optical density will be directly proportional to the concentration of the material being investigated.

It is thus possible to determine the oxygen saturation of hemoglobin in solution, OS, by measuring the ratio of optical densities for two different wavelengths in the same cuvette. One wavelength should be in the near infrared portion of the spectrum close to the isobestic point, and the other in the visible red portion of the spectrum where one finds the greatest difference between oxygenated and deoxygenated hemoglobin.

$$OS = A - B \frac{OD_1}{OD_2} \quad (3)$$

where A and B are empirically determined constants and $OD_1$ is the optical density at wavelength $\lambda_1$ in the visible red region at approximately 660 nm and $OD_2$ is the optical density at wavelength $\lambda_2$ near 805 nm in the infrared spectrum, the isobestic point. For equation 3 to apply, measurements must be made in a fixed geometry cuvette so that the optical path for both wavelengths through the solution will be the same. The solution must also be homogenous that is the hemoglobin must be evenly dispersed throughout the cuvette. Thus, if whole blood is used, it must be thoroughly hemolyzed before the measurement can be made.

Backscatter Oximetry. Light entering a hemoglobin solution can either be transmitted through it, absorbed by it or reflected from it. In the case of oximetry, it was found that there was a similar relationship to equation 3 when backscattered light was used to look at a hemoglobin solution at two different wavelengths. An example of a backscattered light cuvette is shown in Figure 3. The oxygen saturation is once again determined from the ratio of backscattered light optical densities

$$OS = C - D \frac{R_2}{R_1} \quad (4)$$

where C and D are empirically determined constants and R refers to the backscattered optical density at each wavelength as indicated for equation 3. Although the requirement of a homogeneous solution applies for backscattered light as well as transmitted light, several investigators found that equation 4 was still approximately correct in whole blood or even tissue. Drabkin described an intravascular fiber optic sensor in 1944.[4] With it the two wavelengths of light were conducted into the blood and the backscattered light collected for photometric measurement at the proximal end of the fiber. Five years later Brinkman and Zijlstra described instrumentation to continuously measure oxygen saturation of the hemoglobin in the blood within tissue using a similar technique.[5] Good correlations between measurements made with their instruments and blood samples were obtained although the instrument did have a dependence on skin and tissue pigmentation. This work was later described in detail in a monograph.[6] Takatani et al, showed that some of this dependence could be eliminated by using more than two wavelengths and solving simultaneous equations similar to equation 4.[7]

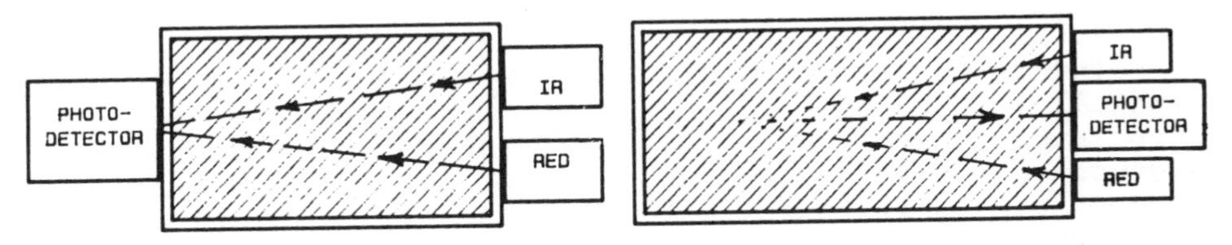


Fig. 2. Transmission oximetry

Fig. 3. Backscattered oximetry

*Ear Oximetry.* Transmission oximetry has also been applied to measurements in tissue. In 1935 Matthes and Kramer demonstrated how the transmission technique could be applied to the human ear.[2,3] Of course, the same dependence on pigmentation limited these measurements to being relative as was the case for backscattered light oximetry. In 1942 Millikan demonstrated that the blood content in the ear could be increased by heating and that this blood would more closely approximate arterial blood in terms of its oxygen content. Wood further advanced the technique of ear oximetry by developing a sensor that contained the optical components along with a bladder for applying pressure to the earlobe to squeeze out the blood from the capillaries.[9] This made it possible to determine the optical properties of the blanched tissue so that pigmentation effects could be corrected. This instrument was not only able to measure hemoglobin oxygen saturation in the ear, but it also could be used for an indirect determination of blood pressure using the bladder to apply a known external pressure to the ear.

Although there were attempts to develop commercially successful instruments based on the above principles, it was not until the Hewlett Packard Corporation developed an eight wavelength transmission ear oximeter based upon work by Shaw that the technique began to be extensively used in the research laboratory and intensive care units. This instrument made use of eight equally spaced wavelengths of light from 650-1050 nm which were sequentially conducted to the ear through a fiber optic cable and the transmitted light returned to the instrument by a similar route. A segmented rotating disk with interference filters in each segment produced the various light wavelengths for the instrument. Multicomponent Beer-Lambert equations were solved to obtain the overall relationship for oxygen saturation.

$$OS = \frac{A_o + A_1 \log T_1 + A_2 \log T_2 + \ldots\ldots + A_8 \log T_8}{B_o + B_1 \log T_1 + B_2 \log T_2 + \ldots\ldots + B_8 \log T_8} \qquad (5)$$

where $A_i$ and $B_i$ are 16 empirically determined coefficients and $T_i$ is the optical transmittance at wavelength i as defined by

$$T_i = \left. \frac{I_t}{I_o} \right|_{\lambda_i} \qquad (6)$$

The major limitations with this important research and clinical instrument were its relative high cost and bulky structure. The sensor covered the entire ear and heated it and was connected to the remainder of the instrument through a relatively stiff fiber optic cable. The instrument was also found to have increasing errors when the oxygen saturation dropped below 60%.

In 1980, Yoshiya, Shimada and Tanaka published a new technique for measuring the hemoglobin oxygen saturation of blood in a finger.[11] The idea is similar to that of the transmission ear oximeter in that two wavelengths of light are transmitted through the tissue and the ratio of their optical densities is used to determined the oxygen saturation. The unique difference of this system from previous instruments, is that it looks at the variation in optical density as a function of time resulting from capillary blood volume variations during the cardiac cycle, the photoplethysmographic pulse. At systole a fresh bolus of arterial blood enters the capillary bed, and the blood volume, and hence, the optical density will be greatest. At dyastole just before the next systole the capillary blood volume will be the least due to the venous run off. If one assumes the change in blood volume from systole to dyastole to be constant over several cardiac cycles, then the amount of hemoglobin change during the cardiac cycle will also be constant. Therefore, the optical density at the isobestic wavelength will have the same change with each arterial pulse. If the blood bolus during one pulse, however, has a different oxygen saturation from that of another pulse, the optical density at the red wavelength will differ. As the blood desaturates, the ratio of reflected optical density at the isobestic wavelength to that at the red wavelength will increase (equation 4). Thus, the pulse oximeter is able to look at arterial blood without the necessity of heating the tissue. It also is capable of monitoring the pulse rate since it must detect each arterial pulse.

Today many commercial instruments have been developed based upon this principle. Various types of sensors are used with these instruments ranging from devices for clipping on the earlobe and fingers to disposable sensors that can be taped to a finger, a toe or in the case of infants an entire hand or foot. All of these sensors make use of modern microelectronic technology to produce miniature light sources and detectors. The sources consist of light emitting diode chips that operate at nominally 665 nm for the red sensor and 955 nm for the infrared sensor. A single photodiode chip serves as the photodetector for both wavelengths. Even though the infrared wavelength is not at the isobestic point of 805 nm, the sensitivity at this wavelength is sufficiently lower than at 665 nm to allow the ratios of equation 4 to still apply. These chips are sufficiently inexpensive that disposable sensors can be manufactured at reasonable costs.

All of the instruments contain signal processing electronics to determine the pulse amplitudes for both wavelengths and to calculate the oxygen saturation based upon equation 4. Most instruments carry out this calculation by averaging several pulses together to reduce the sensitivity to random noise. All instruments also have special circuits to recognize signals that are obviously not of physiologic origin so that these can be eliminated from the calculation.

## LIMITATIONS OF OXIMETRY

Although the pulse oximeter has made it possible to noninvasively monitor hemoglobin oxygen saturation of arterial blood, there are some limitations to the measurement that are important to understand. This measurement is based upon the transmission mode of oximetry, and thus the sensor must be placed on tissues that can be easily transilluminated such as a finger, toe or ear. Unfortunately, these tissues are at the periphery of the circulation, and so for patients in shock conditions, they will receive lower perfusion. This will make the pulsatile change in capillary blood volume much smaller, so the measurements will become more susceptible to error.

The pulse oximetry sensing system is also susceptible to motion artifact. Even under the best of conditions the changes in optical density resulting from capillary blood volume changes are small compared to the overall volume of the tissue. Thus, the changes in optical density are also small, and there can be changes in optical transmission between the light sources and the detector from sources other than blood volume changes. Examples of this include optical path differences due to small position changes in the sensor and changes in tissue volume from sources other than blood volume. These changes result from motion of the sensor and tissue, and thus the reliability of the pulse oximeter generally diminishes as patient movements increase. Different manufacturers use different signal processing techniques to try to minimize the motion artifact, but none of the available instruments are uneffected by motion artifact in some way. Although the method of signal processing can help reduce the effect of motion artifact on signal quality, it can also have the detrimental effect of reducing response time and potentially making it difficult to differentiate between artifact and true signal.

Most pulse oximeters average several pulses before calculating the oxygen saturation. Thus, if one pulse is in error due to motion, its effect will be diminished by the other correct pulses that go into the calculation. Such processing also prolongs the response time of the instrument, since a number of arterial pulses must be averaged together before the response is calculated. At best, a pulse oximeter can only respond by sampling at the heart rate, and when averaging of several pulses is done the

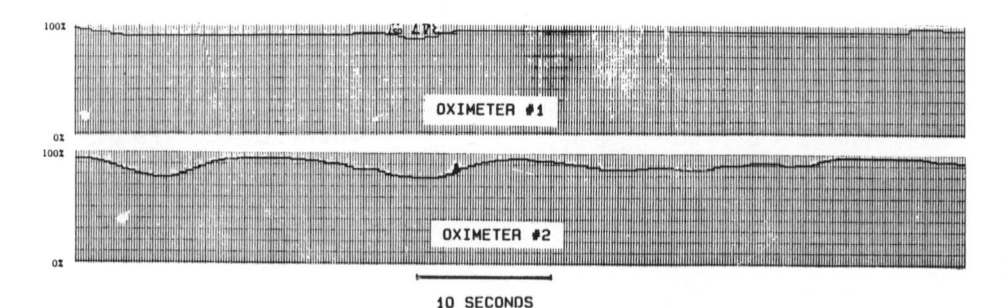


Fig. 4. Simultaneous recordings from two commercial oximeters on the same patient.

response is diminished according to the number of beats that are averaged together. When one considers a typical recording of the oxygen saturation output from a pulse oximeter as a function of time as shown in the top trace of Figure 4, it is sometimes difficult to identify whether apparent desaturations are the result of an actual decrease in blood oxygen content or of artifact being averaged over several pulses.

The fact that different pulse oximeters respond in different ways is illustrated by comparing the two traces in Figure 4. Both were recorded from the same neonate by transmission through the great toe. One sensor was attached to the left foot while the other to the right, and recordings were made simultaneously. Two things are noted from this figure: first when desaturation events occur the minimum level reached differs from one instrument to the other, and second there are some cases where one instrument records a desaturation and the other does not. This figure illustrates a clinical problem in using pulse oximetry (and most likely other instruments as well) in that one never can be certain of the reliability of the recordings obtained from the instrument. Since most clinicians do not use more than one instrument on a patient, the possibility of getting a signal corresponding to either the top or lower trace of Figure 4 is real. Needless to say, depending on which type of signal is obtained the interpretation might be different. Some clinicians and researchers who make chart recordings from pulse oximeters, therefore, record both the oxygen saturation signal and the pulse signal itself. The latter can then be used to validate the former. In situations where a stable regular pulse signal is obtained, one would expect a reliable saturation measurement to result. In circumstances where the pulse signal is irregular with rapid amplitude shifts, these are most likely due to patient movement and could introduce artifact that yielded erroneous oxygen saturation values. A comparison of three different oximeters on adult patients with sleep disorders was carried out by Strohl et al.[12] They found differences in the instruments studied that were most pronounced during periods of desaturation associated with apnea.

Equations 3 and 4 can be very sensitive to the wavelengths at which the optical densities were measured. Thus it is important to know these wavelengths. Semiconductor light and infrared emitting diodes can have small differences in wavelength from one piece to the next. Since it is not practical to determine the wavelength of the emitters in each sensor used, these wavelength differences can introduce errors in the saturation determinations.

## RESEARCH ISSUES

Although the pulse oximeter has achieved the status of routine use in many clinical settings, there are several areas where research on improvements of the technique is under way or needs to be encouraged. Noninvasive tissue pulse oximeters in use today operate in the transmission mode, and so the number of tissue sites where the instrument can be applied are limited to structures that can be easily transilluminated such as the fingers or toes. These locations are often not the most desirable due to their being at the extreme periphery of the circulatory system and therefore first to be compromised in circulatory shock. These locations are also likely to encounter more motion, and hence, more motion artifact than more central positions on the chest or abdomen. Historically it has been possible to look at in vitro oximetry from the backscatter (reflection) mode as well as the transmission. This step, however, has not been possible in practical noninvasive tissue oximeters. Research in understanding backscattered light from illuminated skin and subcutaneus tissue is necessary before practical instruments can be developed. Since light can penetrate deep in the body, one must be concerned about this depth of penetration, the effects of reflection from subdermal tissue interfaces, and the sensitivity to local anatomy, and hence, the position of the sensor in such instruments.

The availability of a practical backscatter oximeter would be an important development in that more locations would become available for clinical noninvasive oximetry. It would be possible to develop an intrapartum fetal pulse oximeter that could be used to monitor hemoglobin oxygen saturation in the fetal presenting part during labor and delivery. Although some investigators have made simultaneous recordings from transcutaneous oxygen sensors and pulse oximeters on the same subjects, the sensors had to be located at different sites due to their different location requirements. A backscatter mode pulse oximeter would make it practical to combine a transcutaneous $PO_2$ sensor and pulse oximeter into the same structure so that simultaneous measurements of oxygen tension and saturation could be made at the same site.

A second problem that needs work is the development of a reliable method for calibration and testing of currently used pulse oximeters. At the present, clinicians applying commercial devices have little choice but to accept the reading on the instrument as being correct in that the only way to check it is to take an arterial blood sample for measurement using in vitro techniques. In some clinical circumstances this is warranted, but there are others where it seems unnecessary to subject a patient to an arterial puncture to check the instrument calibration. Mendelson et al, have presented a preliminary report of an in vitro circuit consisting of a pulsatile pump, blood oxygenator and a silicone rubber "capillary bed" that they use to study oximetry devices under investigation in their laboratory.[13] While this method is still too cumbersome for clinical application, it may serve as the basis for more simplified devices.

The use of home monitoring technology as a possible means of preventing sudden infant death (cot death) has recently become quite popular especially in the United States. Monitors used are primarily infant apnea and cardiac monitors, and the relationship between the variables that they monitor and sudden infant death has not been well established. Pulse oximetry offers the potential to noninvasively monitor a third variable, arterial hemoglobin oxygen saturation, that may be an important consideration in monitoring infants at risk. At present pulse oximeters can only be used for special studies and not for routine home monitoring due to their expense. The development of smaller, less expensive instruments for home monitoring can make this additional variable available not only for infant monitoring but for the monitoring of adults as well.

Noninvasive pulse oximetry, as described in this manuscript, is limited to looking at the skin and subcutaneous tissue. It would be far more significant to look at the oxygen saturation of blood that perfuses vital organs such as the brain, myocardium, etc. Brazy et al, have shown the possibility of determining oxygen levels in the brain by transillumination of the neonatal head with lasers and measuring the scattered light at characteristic wavelengths.[14] While their preliminary report only gave qualitative results, the work in their laboratory and that of other investigators looking at the oxidation state of cytochrome $A_3$ offer future promise of new noninvasive techniques for the measurement of deep tissue oxygen levels.

SUMMARY

Oximetry for continuous patient monitoring has evolved to a mature technology that is clinically useful. Technologic limitations, however, still exist and there is need to develop more reliable, less expensive systems that will allow new applications such as fetal and home infant and adult monitoring. Although it is not possible to fully monitor patient blood gas status with a single variable, pulse oximetry provides an important view of the blood oxygen content that can be combined with other monitoring methods to more fully characterize patient status.

ACKNOWLEDGEMENT

Partially supported by NIH grant RR 00210.

REFERENCES

1. L. Nicolai, Uber Sichtbarmachung, Verlauf und Chemische Kinetik der Oxyhemoglobinreduktion im Lebenden Gewebe, Besonders in der Menschlichen Haut, Arch. Gesamte Physiol. 229: 372 (1932).

2. K. Kramer, Ein Verfahren zur Fortlaufenden, Messung des Sauerstoffgehalets im Stromenden Blute an Uneroffneten Gefassen, Zeit. f. Biologie. 96: 61 (1935).

3. K. Matthes, Untersuchungen uber die Sauerstoffsaptingung des Menschlichen Arterienblutes, Arch. f. Exper. Path. u Pharmakel. 179: 698 (1935).

4. D.L. Drabkin and C.F. Schmidt, Spectrophotometric Studies XII. Observation of Circulating Blood In Vivo and the Direct Determination of the Saturation of Hemoglobin in Arterial Blood, J. Biol. Chem. 157: 69 (1944).

5. R. Brinkman, W.G. Zijlstra and R.K. Koopmans, A Method for Continuous Observation of Percentage Oxygen Saturation in Patients Arch. Chir. Neerl. 1: 333 (1949).

6. W.G. Zijlstra and G.A. Mook, A Manual of Medical Reflection Photometry, Royal Vangorcum Ltd., Assen (1962).

7. S. Takatani, P.W. Cheung and E.A. Ernst, A Noninvasive Tissue Reflectance Oximeter, Ann. Biomed. Engrg. 8: 1 (1980).

8. G.A. Millikan, The Oximeter, An Instrument for Measuring Continuously Oxygen Saturation of Arterial Blood in Man, Rev. Sci. Instrum. 13: 434 (1942).

9. E.H. Wood and J.E. Geraci, Photoelectric Determination of Arterial Oxygen Saturation in Man, J. Lab. Clin. Med. 34: 387 (1949).

10. Hewlett Packard Corp., Operating Guide for the HP-47201A Ear Oximeter, Waltham, Massachusetts.

11. I. Yoshiya, Y. Shimada and K. Tanaka, Spectrophotometric Monitoring of Arterial Oxygen Saturation in the Fingertip, Med. Biol. Engrg. & Comput. 18: 27 (1980).

12. K.P. Strohl, P.M. House, J.F. Holic, J.M. Fouke and P.W. Cheung, Comparison of Three Transmittance Oximeters, Med. Instrum. 20: 143 (1986).

13. Y. Mendelson and J.C. Kent, An Experimental Tissue Model for Transmission Pulse Oxymetry, Proc. 39th Ann. Conference on Engrg. in Med. and Biol. No. 50.2 28: 272, (1986).

14. J.E. Brazy, D.V. Lewis, M.H. Mitnick and F.F. Jobsis vander Vleit, Noninvasive Monitoring of Cerebral Oxygenation in Preterm Infants: Preliminary Observations, Pediatrics 75: 217 (1985).

# PULSE OXIMETRY - AN ALTERNATIVE TO TRANSCUTANEOUS $PO_2$ IN SICK NEWBORNS*

Joyce L. Peabody, M. S. Jennis, and Janet R. Emery

Loma Linda University Medical Center
Loma Linda, California

## INTRODUCTION

Modern day intensive care for sick newborn infants includes continuous monitoring of oxygen as a routine. This routine stems from our concerns regarding hyperoxemia and its probable role in retinopathy of prematurity and hypoxemia and its role in CNS damage, pulmonary vasoconstriction and possibly in retinopathy of prematurity. Transcutaneous $PO_2$ ($tcPO_2$) monitoring, the most widely used method in the 1970s and early 1980s, has provided an excellent detector of fluctuations in $PO_2$ that cannot be detected by intermittent sampling of arterial blood. It has permitted rapid detection of hypoxemia associated with apnea, hypoventilation, or procedures. However, transcutaneous monitoring has limitations, including frequent calibration periods and a heated electrode, which causes first, and occasionally, second degree burns and which requires frequent site changes. Furthermore, unpredictable gradients have been reported between skin and arterial $PO_2$ values in older infants and in infants with bronchopulmonary dysplasia.

The pulse oximeter, a newly available monitor, continuously measures the arterial oxygen saturation ($SaO_2$) non-invasively. This device uses spectrophotometric principles to determine $SaO_2$ with each arterial pulsation. Its accuracy in adults and children has been reported (1). However, several characteristics of sick newborn infants raise concerns regarding the applicability of this technique to newborn medicine:

TABLE 1: Characteristics of Sick Newborns Which Could Potentially Affect Pulse Oximetry

1. Fetal Hemoglobin Concentration
2. Rapid Heart Rate
3. Measurement Site
4. Skin Pigmentation
5. Unpredictable Movement
6. Ambient Light
7. Shock States

* Modified in part from manuscript in press, Pediatrics, 1986.

Whereas some studies have included newborns, the effects of all of these characteristics have not been adequately studied. We, therefore, tested the reliability, accuracy, and practicality of the pulse oximeter for the assessment of oxygenation in sick newborn infants and compared the relative advantages of pulse oximetry to transcutaneous $PO_2$ monitoring.

## PATIENTS AND METHODS

Two groups of infants were studied. Group I was composed of 26 Caucasian infants. Group II included 5 black infants. All infants had umbilical or peripheral arterial catheters in place. For Group I, mean birth weight was 1710 g (725 to 4000 g), gestational age 30.6 weeks (24 to 40 weeks) and postnatal age at time of study ranged from 1 to 49 days. Three infants in Group I, studied early in the newborn period, were studied again when they were more than a month old. For Group II, mean birth weight was 878 g (range 780 to 930 g), gestational age 29 weeks (27 to 31 weeks), and postnatal age at time of study ranged from 4 to 48 days.

There were no selection criteria for Group I other than the presence of an arterial line and the ability to obtain informed consent. Babies in Group II were additionally selected for the presence of medium to dark skin pigmentation. The percent of fetal hemoglobin present was determined by electrophoresis at the time an infant was entered into the study and was repeated following each transfusion.

The Nellcor N-100 Pulse Oximeter was used for this study. Heart rate and $SaO_2$ are displayed digitally. Calibration is done internally by the manufacturer.

The sensor was placed around the foot or toe if an umbilical catheter was in place or on the ipsilateral hand if a radial catheter was in place. When in the correct position, the heart rate displayed on the pulse oximeter was the same as that observed on the infant's bedside monitor. A pulse oximeter reading was recorded when an arterial blood sample was obtained for clinical management. $SaO_2$ of the blood sample was measured on an Instrumentation Laboratory 282 Co-Oximeter (IL 282). In Group I no control of number of measurements per subject was attempted. In Group II, 4-6 paired samples were collected from each infant. In addition to $SaO_2$, the IL 282 determines the percentages of carboxyhemoglobin and methemoglobin. Because of the known error of $SaO_2$ measurements when fetal hemoglobin is present (2), the method of Cornelissen and co-workers (3) was used to correct the falsely elevated carboxyhemoglobin levels reported by the IL 282. Arterial $PO_2$ was measured on a Corning 170 blood gas analyzer.

In addition, heart rate and arterial blood pressure were measured. The use of pressor agents, muscle relaxants, and sedatives was recorded. The degree of dark pigmentation of each infant in Group II was assessed by a black nurse, not a member of the investigation team.

Linear regression analysis was performed to compare the pulse oximeter $SaO_2$ with the measured $SaO_2$ for all the paired measurements in Group I. Data was also analyzed according to percentages of fetal hemoglobin present, grouped as 0-24%, 25-49%, 50-74%, and 75-100%. Analysis of variance was used to compare the differences between measured $SaO_2$ and pulse oximeter $SaO_2$ among the four ranges of fetal hemoglobin in Caucasian babies. Since Group II contained a small number of babies at the time this manuscript was prepared, only mean and standard deviation for the difference between blood sample $SaO_2$ and pulse oximetry measured $SaO_2$ were calculated. These mean differences were compared to the infants from Group I who were most similar to Group II in percentage of fetal hemoglobin and postnatal age by the sign test.

RESULTS

Group 1: One hundred and seventy-seven paired arterial blood and pulse oximeter $SO_2$ measurements were obtained from the 26 Caucasian infants. The sensor was easily applied to the hand or foot in all instances, even in the smallest infant who weighed 725 g. Readings could be obtained from all infants despite heart rates as high as 220 beats per minute. It was observed that intense ambient light from heat lamps or phototherapy did interfere with obtaining a good pulse signal, so the sensor was shielded when such light sources were in use.

The linear regression equation comparing pulse oximeter $SaO_2$ to measured $SaO_2$ was $y = 0.7x + 27.2$ ($N = 177$; $r = 0.9$; standard error of the estimate = 1.87%). Measured $SaO_2$ in blood ranged from 70-100%, while pulse oximeter $SaO_2$ ranged from 71-100% (Fig. 1).

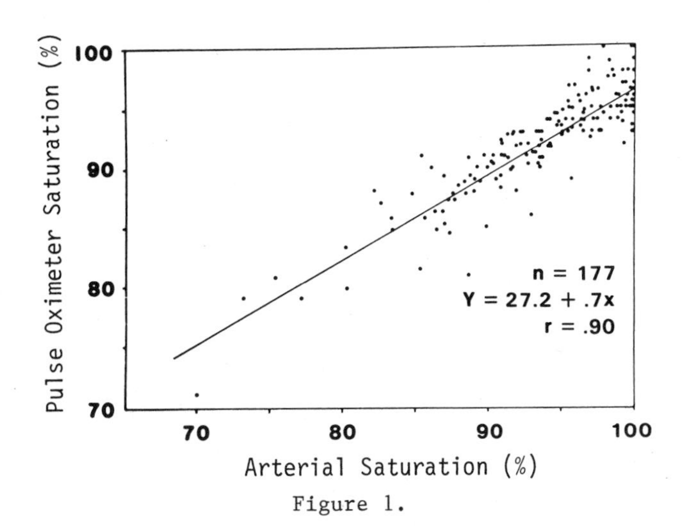


Figure 1.

Forty-nine fetal hemoglobin determinations were made in Group I. Values varied from 5 to 100%. Table 2 shows the data in the 4 ranges of the fetal hemoglobin, gives the measured $SaO_2$, the arterial $PO_2$ and the difference between $SaO_2$ and pulse oximeter $SaO_2$.

TABLE 2: Measured $SaO_2$, and $\Delta SaO_2^*$ for Ranges of Fetal Hemoglobin in Group I.

| Fetal Hgb (%) | N | Measured $SaO_2$ (IL 282) (%) | $\Delta SaO_2$* (%) |
|---|---|---|---|
| 75 - 100 | 40 | 97.4 (74.3-100)+ | 3.6 + 2.3 |
| 50 - 74 | 42 | 96.4 (70.1-100) | 2.8 + 2.4 |
| 25 - 49 | 45 | 92.8 (82.7-100) | 0.5 + 2.4 |
| 0 - 24 | 50 | 90.6 (75.7-100) | 0.3 + 1.9 |

* measured $SaO_2$ minus pulse oximeter $SaO_2^*$ expressed as mean $\pm$ SD

\+ mean and range of values
$p<.001$ compared to 0-24% fetal hemoglobin

Although $PaO_2$ values were similar among the four groups, the higher affinity of fetal hemoglobin for oxygen resulted in higher $SaO_2$ measurements when a greater percentage of fetal hemoglobin was present. The values of blood and pulse oximeter $SaO_2$ were almost identical when fetal hemoglobin was <50% but a 2.8-3.6% error was observed when it exceeded 50% ($p<0.001$). However, 68 of the 82 arterial samples were greater than 95% saturated in the two high fetal hemoglobin groups. When fetal hemoglobin was greater than 50% and $SaO_2$ was greater than 95%, pulse oximeter $SaO_2$ was always less than the measured $SaO_2$.

Heart rate, blood pressure, serum bilirubin, use of pressor agents (2 infants), muscle relaxants (9 infants), or sedatives (11 infants) were similar among the fetal hemoglobin ranges and had no obvious effect on the results.

Group II. Twenty-five paired arterial blood and pulse oximeter $SaO_2$ measurements have been obtained from 5 black infants to date. The difference between blood and pulse oximeter $SaO_2$ for the total group was -2.00 +/- 3.0 (mean +/- SD), fetal hemoglobin ranged from 3-33%. Comparing this by sign test to measurements made in Caucasian babies of similar fetal hemoglobin, a trend towards overestimation of $SaO_2$ in black babies by pulse oximetry is suggested, ($P<.05$). More measurements must be made before this trend can be definitively confirmed since these babies are of significantly lower birth weight than Group I.

## DISCUSSION

Despite a wide range of birth weights, postnatal ages, heart rates, and skin pigmentation, pulse oximetry monitoring was possible in all infants and no skin injuries were observed. Accuracy of pulse oximeter measurements varied among infants. The percentage of fetal hemoglobin present, the percent saturation, and the degree of skin pigmentation appear to affect the difference between measured $SaO_2$ and pulse oximeter $SaO_2$. The greatest accuracy was observed in Caucasian infants when less than 50% fetal hemoglobin was present. The mean difference between measured $SaO_2$ and pulse oximeter $SaO_2$ was 0.3% for 0-24% fetal hemoglobin and 0.5% for 25-49% fetal hemoglobin.

Skin pigmentation may have some effect on the pulse oximeter measurements as it may be an additional factor to affect transmission of light and may differentially transmit the light of red and infrared wavelengths, altering the internal calibration curve.

Environmental factors may affect pulse oximeter function. Heat lamps have been reported to interfere with the pulse oximeter because of the high intensity of the infrared light that they emit (4). Saturation cannot be calculated because the small changes in light transmission detected by the pulse oximeter are masked by the strong ambient light. In preliminary studies, we found similar interference with pulse oximeter measurements when infants were on open tables under radiant warmers. This problem was avoided by shielding the sensors.

The pulse oximeter requires adequate perfusion of the underlying tissues. It must detect adequate arterial pulsations to differentiate arterial from venous and capillary $SO_2$. Mihm and Halperin (5) reported signal loss when adult patients were receiving dopamine infusions. We were able to successfully monitor 2 infants who were receiving dopamine. However, 2 infants not included in this study have been monitored by our group during the dying process. Once $SaO_2$ fell below 60%, blood pressure dropped significantly and heart rate slowed, pulse detection was not possible. Further work is necessary to better define this limitation.

Movement causes a significant problem with adequate detection of pulse amplitude. We have found this to be the single greatest problem in use of this technique for monitoring sick infants. Since, currently, the pulse search alarm is identical to the alarms of hyperoxemia or hypoxemia, pulse search induced by movement causes frequent "false alarms" and results in disengagement of alarm systems even in the best of nurseries. We believe this problem must be solved before pulse oximetry will be a fully effective monitoring technique.

Despite the observed, and as yet unexplained, effect of high fetal hemoglobin, high $SaO_2$, and skin pigmentation on pulse oximeter determination, our data suggest an acceptable accuracy for clinical use in all newborn infants. Furthermore, the use of the pulse oximeter has a number of advantages over transcutaneous $PO_2$ monitoring. The pulse oximeter displays changes in oxygenation instantaneously. It is practical and easy to use, particularly during high risk periods such as transport and surgery. Calibration by the user is not required and readings are available within seconds of application.

Two groups of infants who may uniquely benefit from pulse oximeter monitoring were identified in our study. The very premature infant is particularly sensitive to skin injury caused by the heated transcutaneous

$PO_2$ electrode and fixation ring. As the pulse oximeter sensor is not heated, thermal injury does not occur. The only potential source for skin injury is the adhesive tape for mounting the sensor. However, we found no such injury in the nine infants under 1000 g. Although the manufacturer recommends daily inspection of the sensor and site, we have monitored infants for as long as three days without changing sensor location.

Older infants and infants with bronchopulmonary dysplasia are known to have increased and unpredictable gradients between arterial and transcutaneous $PO_2$. Also, because it is often difficult to obtain arterial blood from these infants, the $PaO_2$ reported may not be an accurate reflection of the infant's oxygenation. We have found the pulse oximeter $SaO_2$ to be a reliable measurement of $SaO_2$ in these infants.

SUMMARY

In summary, the pulse oximeter provides a reliable, continuous assessment of oxygenation in newborn infants. Its rapid response time and ease of use make it a practical device for use on all sick newborns. To avoid hyperoxia it should be used in conjunction with arterial blood gas measurements and we recommend a high $SaO_2$ alarm of 92% in infants with predominantly fetal hemoglobin. Finally, it is an improved way of monitoring oxygenation in very immature infants and in infants with bronchopulmonary dysplasia.

REFERENCES

1. Yelderman, M., W. New, 1983, Evaluation of Pulse Oximetry, Anes., 59:349-351.
2. Huch, R., A. Hugh, P. Tuchschmid, et al, 1983, Carboxyhemoglobin Concentration in Fetal Cord Blood, Ped., 71:461-462.
3. Cornelissen, P.J.H., C.L.M. van Woensel, W.C. van Oel, et al, 1983, Correction Factors for Hemoglobin Derivatives in Fetal Blood, as Measured With the IL 282 Co-Oximeter, Clin Chem., 29:1555-1556.
4. Brooks, T.D., D.A. Paulus, W.E. Winkle, 1984, Infrared Heat Lamps Interfere with Pulse Oximeters, Anes., 61:630.
5. Mihm, F.G., B.D. Halperin, 1985, Noninvasive Detection of Profound Arterial Desaturations Using a Pulse Oximeter Device, Anes., 62:85-87.

# APPLICATION OF THE OHMEDA BIOX 3700 PULSE OXIMETER TO NEONATAL OXYGEN MONITORING

William W. Hay, Jr., Julia Brockway and Mario Eyzaguirre

University of Colorado Health Sciences Center
Department of Pediatrics
4200 E. 9th Avenue, Denver, CO 80262

## INTRODUCTION

Recently, pulse oximeters have been introduced into nurseries to monitor clinical blood oxygenation. Pulse oximeters measure the oxygen saturation of hemoglobin in pulsed ("arterial") blood flow. Blood oxygen saturation provides a direct measure of blood oxygenation. However, this measure is variably related to the traditional measure of blood partial pressure of oxygen ($PO_2$) and is affected by factors (e.g., hemoglobin concentration and hemoglobin-oxygen affinity) that may not alter $PO_2$. Therefore, this discussion is written to review principles of blood oxygen transport and to demonstrate from data collected in newborn infants, relationships between pulse oxygen saturation ($SpO_2$) blood oxygen saturation ($SaO_2$) and blood $PO_2$ in a variety of clinical conditions.

Oxygen is transported in blood as: 1) oxygen molecules freely dissolved in plasma water, and 2) as oxygen molecules reversibly bound to hemoglobin (1). Together these two forms of blood oxygen comprise the blood oxygen content. Hemoglobin-bound oxygen accounts for about 98% of blood $O_2$ content while the dissolved oxygen accounts for only about 2% of blood $O_2$ content. Both dissolved and Hb-bound $O_2$ are directly related to blood partial pressure of oxygen ($PO_2$). The $PO_2$-$HbO_2$ relationship is called "oxygen-hemoglobin affinity" and is commonly expressed graphically as blood $O_2$ content or blood oxygen saturation [$SaO_2$, which equals $HbO_2/(HbO_2 + Hb)$] versus $PO_2$. Hemoglobin-oxygen affinity is modified by 4 main factors ([$H^+$], $PCO_2$, temperature, and [2,3-DPG]), which at increased values all act to decrease Hb-$O_2$ affinity (2). Hb-$O_2$ affinity is higher for fetal Hb than adult Hb due to a lesser effect of DPG on HbF (3,4). This higher affinity of fetal blood is important to provide adequate $O_2$ saturation and content for the fetus at the low (35-40 mmHg) uterine-placental blood $PO_2$ (5). Postnatally, Hb-$O_2$ affinity decreases as HbA gradually replaces HbF over several weeks (6,7). This is an appropriate change for the higher atmospheric $PO_2$ and for the development of postnatal "physiologic" anemia. Anemia reduces blood $O_2$ content and the lower Hb-$O_2$ affinity aids tissue oxygenation during anemia by promoting oxygen release from hemoglobin, and thus maintaining the capillary-venous $PvO_2$.

Knowledge of $O_2$ saturation in infants is important. It provides an indication of the adequacy of tissue oxygen supply. At constant [Hb], blood flow to the vital organs (brain, heart) is inversely related to $O_2$ saturation, but in other tissues (muscle, skin, gut) blood flow is directly related to $O_2$ saturation (8,9). Knowledge of $PO_2$ is also important. Pulmonary vascular tone is inversely related to $PO_2$ while ductus arteriosus tone is directly related to $PO_2$ (10). Thus, a fall in $PO_2$ will tend to reduce pulmonary blood flow and dilate the ductus, leading to a right-to-left shunt across the ductus and thus to arterial desaturation. Also very high $PO_2$ levels are associated with $O_2$ toxicity, particularly in the retinal vessels of premature infants (11,12).

Thus, both $O_2$ saturation and $PO_2$ give important information about oxygen supply. However, because the $O_2$ Saturation-$PO_2$ relationship is variable, separate measurement of each is essential.

## METHODS

Data were collected from studies of oxygenation on 142 preterm and term infants (25-42 weeks gestation, 500 to 5000 grams) using the Ohmeda Biox 3700 Pulse Oximeter. Studies were carried out in the Newborn Intensive Care Unit at University Hospital, University of Colorado Health Sciences Center, under approvals by the Pediatric Clinical Research Center and the UCHSC Human Subjects Research Committee.

All medical, surgical and nursing procedures (including blood sampling) that were studied were ordered solely on clinical grounds by the subjects' physicians. Inspired, transcutaneous and blood oxygen values were not manipulated in any way as part of the study; however, all data collected were made available to the subjects' nurses and physicians.

The pulse saturation values correlated well ($R = 0.98$, $y = 0.94x + 9.1$, $p < 0.0001$, Fig. 1) with simultaneous aortic catheter blood samples analyzed with the Radiometer OSM-2 Hemoximeter that was calibrated with placental blood (largely HbFetal) avoiding a false positive bias of about 1% for neonatal blood saturation when the Hemoximeter was calibrated with HbAdult. This discrepancy occurs because of structural differences between fetal and adult hemoglobin that lead to different extinction coefficients (constants unique to each molecular structure reflecting the fraction of light absorbed for each wavelength of light passed through that molecular substance). $TcPO_2$ values were obtained with the Hewlett Packard 78850 Oxygen Monitor.

## RESULTS

$TcPO_2$ values were more variable but trend information was nearly identical for $tcPO_2$ and pulse saturation data. Pulse saturation data also could be used to determine reliability of $tcPO_2$ readings, showing $SpO_2$ values that changed appropriately with change in $FIO_2$ when $tcPO_2$ did not change and appeared inappropriately high suggesting membrane leak. Exchange transfusion with adult blood (200 cc's/kg) resulted in apparently reduced Hb-$O_2$ affinity characterized by a "right shift" of the pulse saturation-$tcPO_2$ curve (Fig. 2). Pulse saturation values were not sufficiently accurate to predict $PO_2$ values greater than 100 mmHg (at $PO_2$ > 100 mmHg, pulse saturation = $97.1 \pm 2.4$ S.D.). A similar problem precluded accurate prediction of $PO_2$ values less than 50 mmHg (Figs. 3A and 3B). Thus, pulse saturation is a reliable indicator of hemoglobin saturation with oxygen and blood oxygenation but does not substitute for direct knowledge of $PO_2$. Also, advancing gestational age, in spite of direct transfusion of adult red blood cells totalling 100 cc's/kg, showed

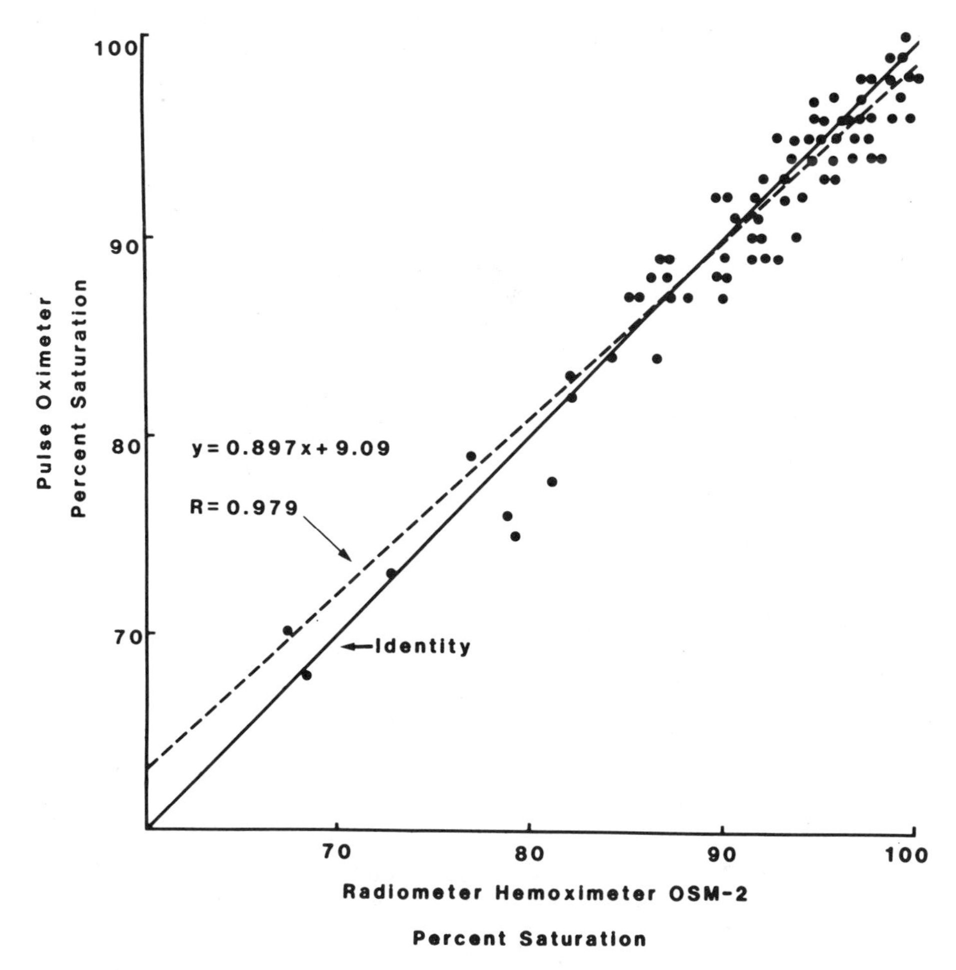


FIG. 1. Correlation of the Ohmeda Biox 3700 Pulse Oximeter with the Radiometer OSM2 Hemoximeter that was calibrated with placental blood (n = 73).

higher affinity ("left-shifted") pulse saturation-$tcPO_2$ values (Fig. 4A) but no change in pulse saturation-arterial $PO_2$ values (Fig. 4B), suggesting a separate influence of postnatal age on reducing $tcPO_2$ relative to arterial $PO_2$. Additionally, pulse saturation-$tcPO_2$ values were "left-shifted" for percutaneous-rather than for catheter-arterial saturation-$PO_2$ values (Figs. 5A and 5B), suggesting that the percutaneous stick reduced $tcPO_2$ selectively.

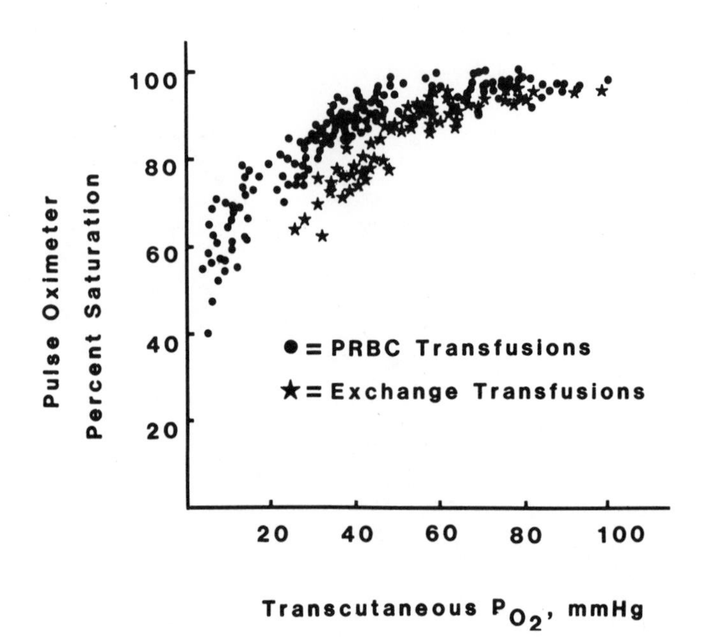


FIG. 2. Double volume (200 cc's/kg) exchange transfusion with adult blood decreased $Hb\text{-}O_2$ affinity as demonstrated by the "right shift" of the pulse saturation-$tcPO_2$ values shown here. Two infants were studied post exchange and are compared with values from 10 infants who had significant pulmonary disease.

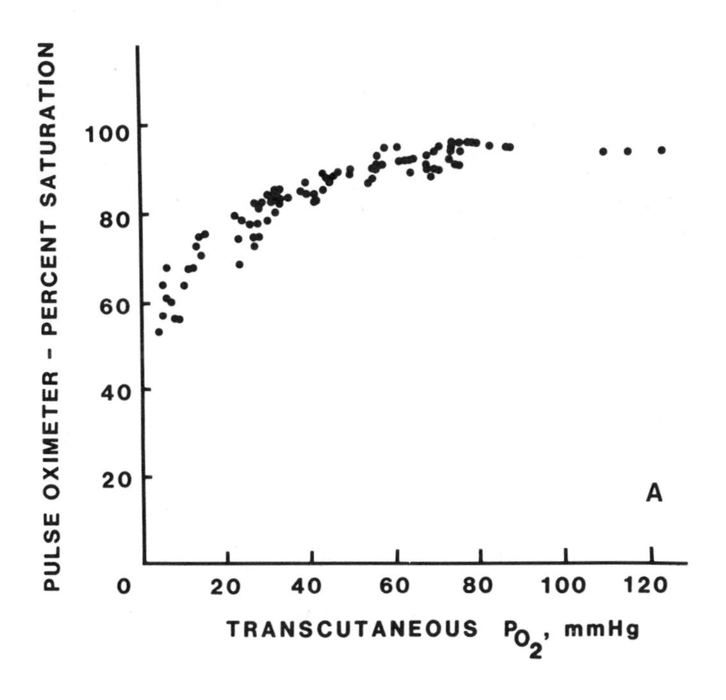


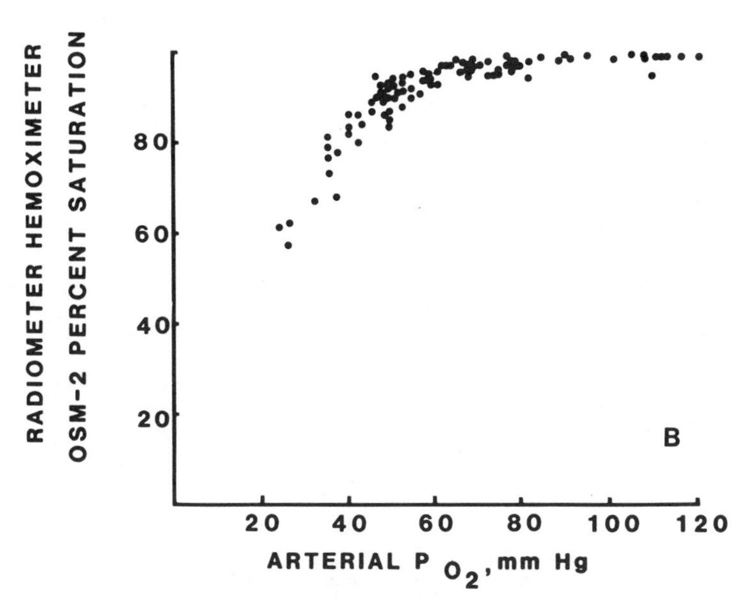


FIG. 3A and 3B. On the same group of 10 infants with significant lung disease, pulse saturation-tc$PO_2$ values below 60 mmHg $PO_2$ were "left-shifted" compared to simultaneous pulse saturation-arterial $PO_2$ values below 60 mmHg $PO_2$. This discrepancy presumably relates to decreased skin perfusion and/or oxygenation at lower $PO_2$ values and suggests that tc$PO_2$ falsely underestimates blood oxygenation in the hypoxemic range.

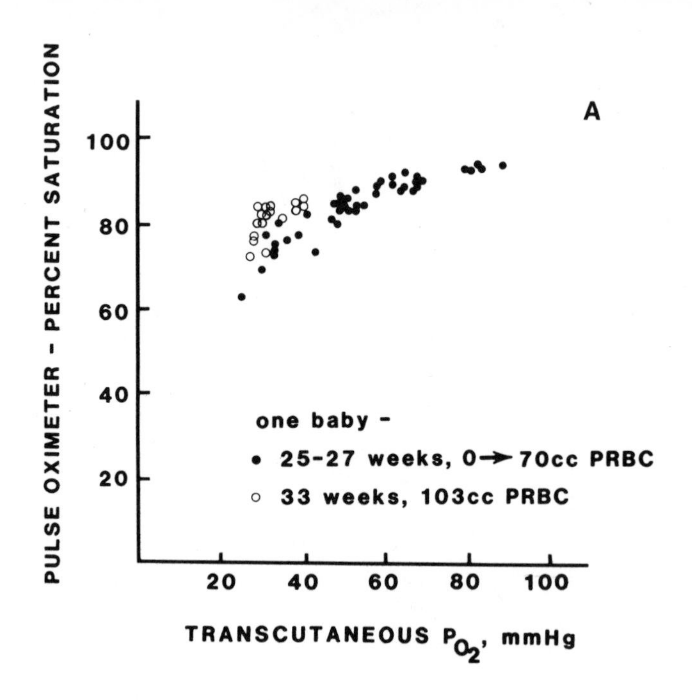


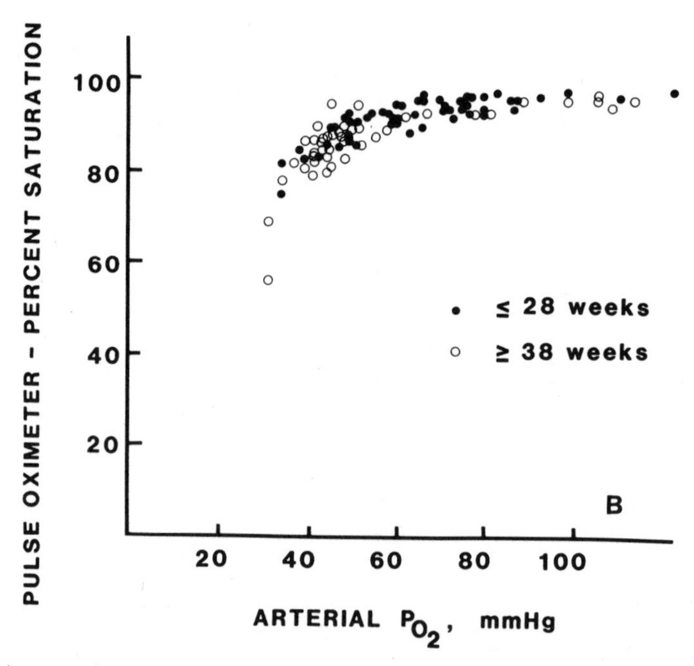


FIG. 4A and 4B. Pulse saturation-tc$PO_2$ values become "left-shifted" with increasing postnatal age in spite of increasing transfusion with adult red blood cells (4A). On the other hand (4B), there is no difference between the pulse saturation-arterial $PO_2$ values in infants at 28 weeks gestation or less compared with term infants. Together, these results suggest that with advancing gestational age, tc, but not pulse or arterial $O_2$ measurements, underestimate blood oxygenation.

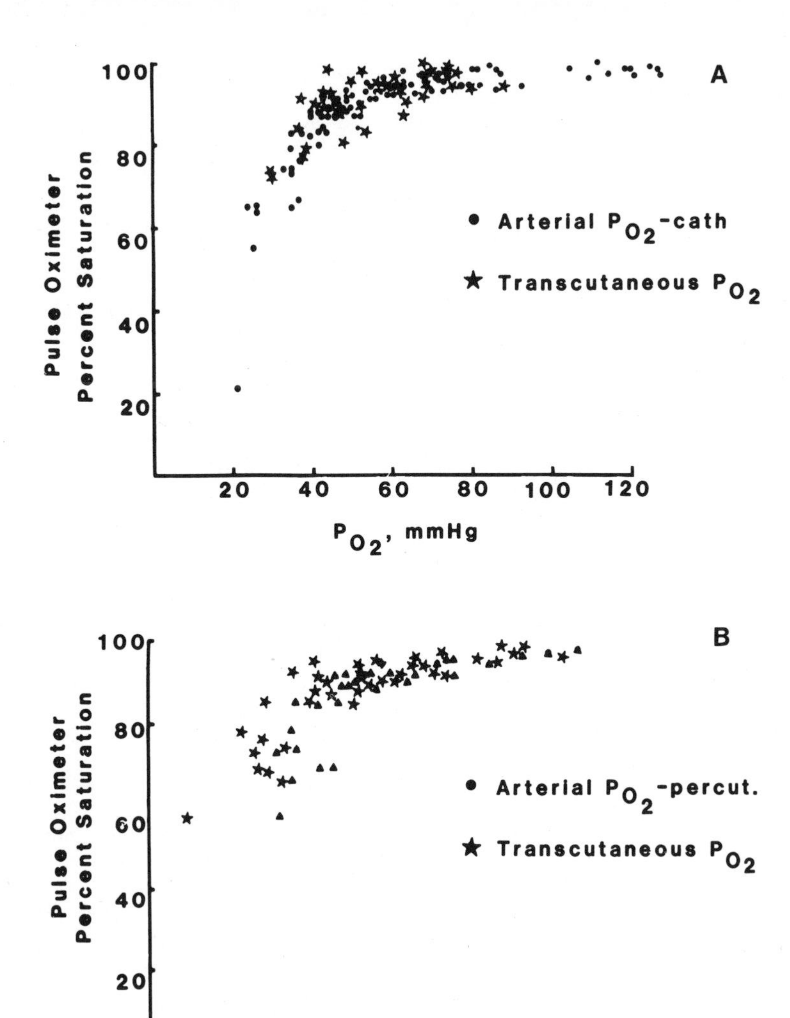


FIG. 5A (top) and 5B (bottom). Pulse saturation-$tcPO_2$ values are "left-shifted" when blood is sampled percutaneously (5A) rather than per catheter (5B), suggesting that during the percutaneous stick procedure, skin blood flow and/or oxygenation is reduced compared with blood oxygenation.

Simultaneous $tcPO_2$ and pulse saturation values were compared for absolute values and percent change before and during gavage and nipple feedings, chest percussion, endotracheal tube suctioning, lung reexpansion with bag ventilation, heelstick and percutaneous arterial blood sampling. $tcPO_2$ decreased in all cases (-6.7 mmHg $\pm$ 5.7 S.D. or -9.1% $\pm$ 9.0 S.D.). Pulse saturation also decreased but to a much lesser extent (-2.8% $\pm$ 2.5 S.D.). Marked and significant ($p < 0.01$) decreases of $tcPO_2$ were observed with nipple feeding and endotracheal tube suctioning (-21.1% $\pm$ 6.7 S.D. and -23.9% $\pm$ 18.6 S.D., respectively), while pulse saturation only decreased significantly ($p < 0.05$) for endotracheal tube suctioning (-8.4% $\pm$ 6.2 S.D.).

It is essential also to appreciate that greater changes in $PO_2$ relative to $SaO_2$ are partly to be expected, given that at $SaO_2$ values > 85%, changes in $SaO_2$ will be accompanied by increasingly larger changes in $PO_2$ due to the characteristic position of the $Hb-O_2$ affinity relationship at high levels of $SaO_2$.

## CONCLUSIONS

Pulse oxygen saturation accurately reflects arterial blood oxygen saturation. However, pulse oxygen saturation does not accurately reflect $PO_2$ values > 100 mmHg. $TcPO_2$ and pulse saturation trends are comparable but processes other than blood oxygenation (e.g., changes in skin blood flow) produce more variable $tcPO_2$ versus pulse saturation values. Pulse saturation provides a reliable measure of $SaO_2$ and blood oxygenation but, particularly in infants prone to hypoxic and hyperoxic injury, pulse saturation should not be used as a substitute for $PO_2$ measurement.

## ACKNOWLEDGEMENTS

Supported by GCRC Grant RR-69 (National Institutes of Health) and by a grant from Ohmeda.

## REFERENCES

1. Davenport, H. The ABC of Acid-Base Chemistry. Sixth Edition. University of Chicago Press, Chicago, 1974.
2. Benesch, R.E., Benesch, R. The Reaction Between Diphosphoglycerate and Hemoglobin. Fed. Proc. 29:1101-1104, 1970.
3. Orzalesi, M.M., Hay, W.W., Jr. The Relative Effects of 2,3-Diphosphoglycerate on the Oxygen Affinity of Fetal and Adult Hemoglobin in Whole Blood. Experimentia 28:1480-1481, 1972.
4. Orzalesi, M.M., Hay, W.W., Jr. The Regulation of Oxygen Affinity of Fetal Blood. I. In Vitro Experiments and Results in Normal Infants. Pediatr. 48:857-864, 1971.
5. Meschia, G., Battaglia, F.C., Makowski, E.L., Droegenmueller, W. Effect of Varying Umbilical Blood $O_2$ Affinity on Umbilical Vein $PO_2$. J. Appl. Physiol. 26:410-416, 1969.
6. Oski, F. Clinical Implications of the Oxyhemoglobin Dissociation Curve in the Neonatal Period. Crit. Care Med. 7 (9):412-418, 1979.
7. Delivoria-Papadopoulos, M., Roncevic, N.P., Oski, F.A. Postnatal Changes in Oxygen Transport of Term, Premature, and Sick Infants: The Role of Red Cell 2,3-Diphosphoglycerate and Adult Hemoglobin. Pediatr. Res. 5:235-245, 1971.
8. Peeters, L.L.H., Sheldon, R.E., Jones, M.D., Jr., Makowski, E.L., Meschia, G. Blood Flow to Fetal Organs as a Function of Arterial Content. Am. J. Obstet. Gynecol. 135:637-646, 1979.
9. Sheldon, R., Peeters, L.L.H., Jones, M.D., Jr., Makowski, E.L., Meschia, G. Redistribution of Cardiac Output and Oxygen Delivery in the Hypoxemic Fetal Lamb. Am. J. Obstet. Gynecol. 135:1071-1078, 1979.
10. Fishman, A.P. Respiratory Gases in the Regulation of the Pulmonary Circulation. Physiol. Rev. 41-241, 1961.
11. Kinsey, V.E., Jacobus, J.T., Hemphill, F.M., et al. Cooperative Study of Retrolental Fibroplasia and the Use of Oxygen. Arch. Ophthalmol. 56:481-543, 1956.
12. Lucey, J.F., Dangman, B. A reexamination of the Role of Oxygen in Retrolental Fibroplasia. Pediatr. 73:82-96, 1984.

# PULSE OXIMETRY AND TRANSCUTANEOUS OXYGEN TENSION FOR DETECTION OF HYPOXEMIA IN CRITICALLY ILL INFANTS AND CHILDREN

S. Fanconi

Intensive Care Unit
University Children's Hospital
Zürich, Switzerland

## INTRODUCTION

Transcutaneous $PO_2$ electrodes are now widely used for monitoring oxygenation[1-3]. Within the last years pulse oximetry ($tcSaO_2$) has also become available and several studies have demonstrated its reliability between 70 and 100% saturation[4-8].

The purpose of this study is to evaluate pulse oximetry and $tcPO_2$ with particular reference to the ability to predict hypoxemia in the neonatal and pediatric intensive care unit.

## PATIENTS AND METHODS

We pooled the data from two previous studies[6,7] of 54 patients with a mean age of 2.4 years (range 1 to 19 years) who had arterial saturations between 9.5 and 98.8%. Every patient had an acute life-threatening respiratory or circulatory condition, and needed respiratory support. Admission criteria to the study were: an umbilical or radial catheter in place for clinical management, blood pressure within the normal range and no severe metabolic acidosis (base deficit >6 mmol/l).

Pulse oximetry was performed with the Nellcor and with the Biox 3700 pulse oximeters. The beat-to-beat mode was used, and the value was accepted only if the pulse oximeter heart rate did not differ by >5 beats from an independent ECG monitor, and the signal strength indicator showed an effective probe position.

$TcPO_2$ was monitored (Hellige GMBH) with a sensor temperature at 44°C, and the probe was recalibrated for each patient, and replaced every 3 to 4 hours.

$TcSaO_2$ and $tcPO_2$ probes were placed according to the position of the arterial line in order to avoid discrepancies from a patent ductus arteriosus. The pulse oximeter probe was placed on the finger, toe, hand or foot, aiming for the best pulse response.

Arterial blood samples were analysed for $PO_2$ (AVL 945) and $SaO_2$ (Corning 2500 CO-Oximeter or OSM2 hemoximeter, Radiometer); $tcPO_2$, $tcSaO_2$ and heart rate were recorded simultaneously. Fetal hemoglobin was measured in 30 of the patients.

Specificity and sensitivity of $tcSaO_2$ vs $SaO_2$ and $tcPO_2$ vs $PaO_2$ were calculated for different cut-off points[9]. The cut-off point was shifted stepwise from 40 to 90% $SaO_2$ and from 25 to 85 mmHg $PaO_2$.

Statistical analysis was performed with the Minitab software.

## RESULTS

178 data sets from 54 patients (3 to 10 samples per patient) were obtained. Linear regression analysis of $tcSaO_2$ compared with measured $SaO_2$ produced an r value of 0.95 ($p<0.001$) with an equation of $y = 21.1 + 0.749x$. The mean difference between measured $SaO_2$ and $tcSaO_2$ was $-2.74 \pm 7.69\%$ (range +14 to -29%). Linear regression analysis of $SaO_2$ vs $SaO_2 - tcSaO_2$ showed an equation of $y = -21.1 + 0.251x$ ($r=0.72$, $p<0.001$)(Fig.1).

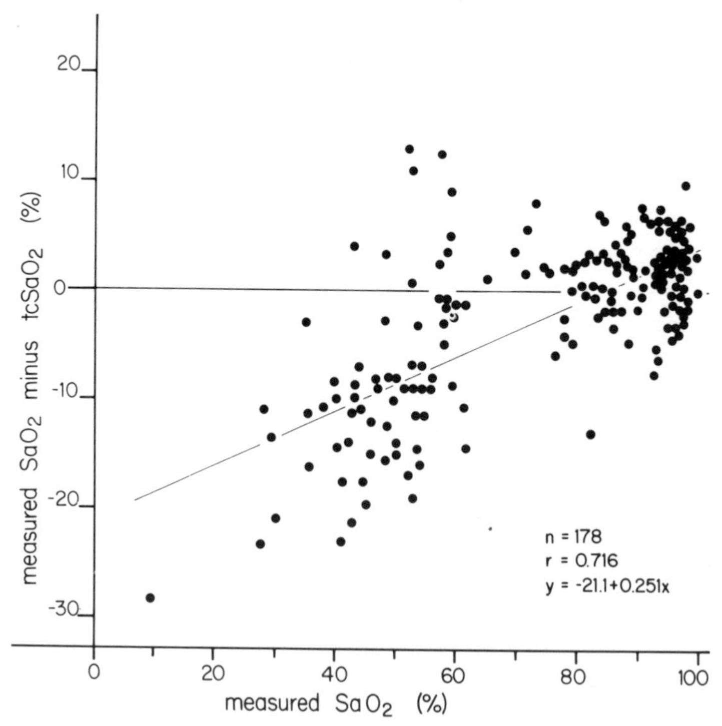


Figure 1: Relationship between measured $SaO_2$ and error of pulse oximeter $SaO_2$

Linear correlation analysis of $PaO_2$ vs $tcPO_2$ showed a correlation equation of $y = -1.04 + 0.876x$ ($r=0.95$; $p<0.001$). The mean difference between $PaO_2$ and $tcPO_2$ was $+7.43 \pm 8.57$ mmHg (range -14 to +49 mmHg) (Fig.2). Linear regression analysis of $PaO_2$ vs $PaO_2 - tcPO_2$ showed an equation of $y = 1.04 + 0.124x$ ($r=0.39$; $p<0.001$) (Fig.2).

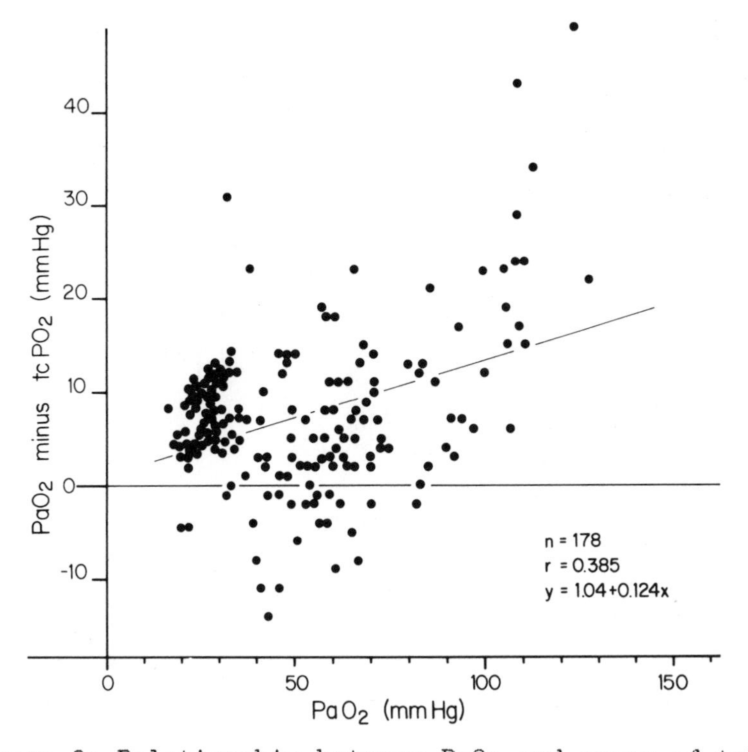


Figure 2: Relationship between $PaO_2$ and error of $tcPO_2$

We found no correlation between blood pressure, heart rate, hematocrit, fetal hemoglobin (fetal hemoglobin ranged from 1.5 to 69%) and error of $tcPO_2$ or $SaO_2$, or between error of $tcSaO_2$ and pulse oximeter device used.

The specificity of $tcSaO_2$ for hypoxemia over a wide range of cut-off points (40-80% $SaO_2$) was 93 and 100%, giving a very low false-positive ratio (3-0%); the sensitivity of $tcSaO_2$ for hypoxia was 88 to 97% for cut-off points ranging from 65 to 90% $SaO_2$, and 33 to 59% for cut-off points between 40 and 60% $SaO_2$, resulting in an increasing false-negative ratio (60-40%) at lower saturations (Fig.3).

The sensitivity of $tcPO_2$ for hypoxemia over a wide range of cut-off points (25-85 mmHg) was 89 and 100%, giving a low false-negative ratio (11-0%); the specificity of $tcPO_2$ for hypoxia was 91 to 99% for cut-off points ranging from 35 to 50 mmHg, and 76 to 79% for cut-off points between 55 and 85 mmHg, resulting in an increasing false-positive ratio with higher $PaO_2$ (Fig.4).

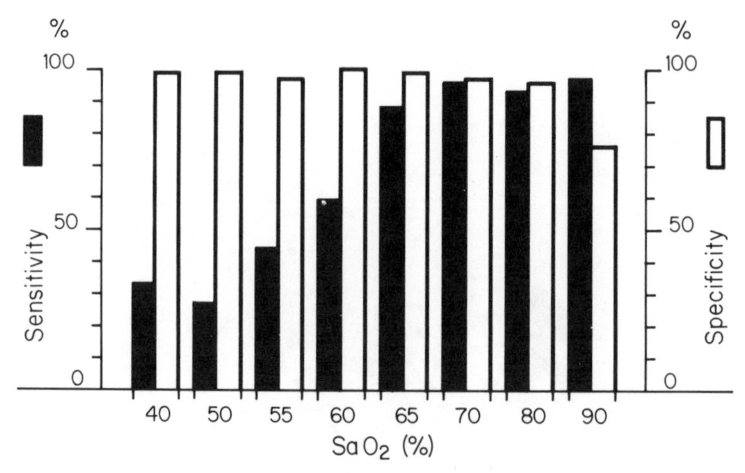


Figure 3: Sensitivity and specificity of pulse oximetry in detecting hypoxemia at different cut-off points.

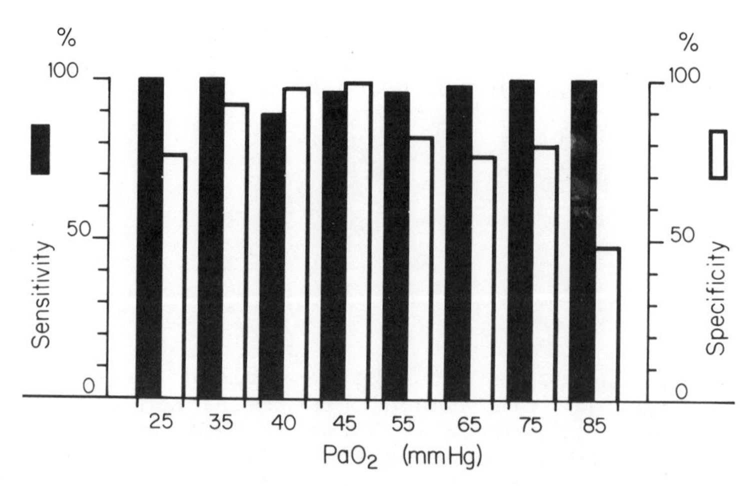


Figure 4: Sensitivity and specificity of $tcPO_2$ in detecting hypoxemia at different cut-off points.

## DISCUSSION

Our results show that both $tcPO_2$ and $tcSaO_2$ have a good correlation with the corresponding arterial values. Transcutaneous $PO_2$ tends to underestimate arterial $PO_2$, while pulse oximetry overestimates arterial saturation (Fig.1,2). The error of pulse oximetry increased with values below 65% $SaO_2$ and the one of $tcPO_2$ increased with $PaO_2$ above 70 mmHg (Fig.1,2).

The specificity of $tcSaO_2$ for hypoxemia was good (93 to 100%) over a wide range of saturations (40-80%), giving a low rate of false alarms (0 to 7%), but the sensitivity of $tcSaO_2$ was good (88-97%) only above 65% arterial saturation; below 65% $SaO_2$ there was an unacceptable rate of false-negative results (33-59%) (Fig.3).

The sensitivity of $tcPO_2$ for hypoxemia was good (89-100%) between 25 and 85 mmHg $PaO_2$, giving a low rate of false-negative results (0-11%). Between 35 and 50 mmHg $PaO_2$ there was a rate of 1-9% false alarms. The false alarms increased to 21-24% with a cut-off point between 55 and 85 mmHg $PaO_2$ (Fig.4).

According to the literature transcutaneous oxygen tension monitoring generally overestimates hypoxia and underestimates hyperoxia[3,6,8]. Some of the anesthetic gases that diffuse through the skin interfere with $tcPO_2$ values[8]. Finally $tcPO_2$ monitoring is dependent on cardiac output and peripheral bloodflow, and if this is reduced, it does not directly reflect actual measured arterial $PO_2$[1-3].

Pulse oximetry on the other side is simple and has immediately available results. It is unaffected by anesthetic gases and does not require tissue heating[4-8]. Fetal hemoglobin and other abnormal hemoglobins may interfere, but in the clinical praxis this seems to play only a minor role[8]. Reduced peripheral vascular pulsations, including hypothermia, hypotension and use of vasoconstrictive drugs interfere with pulse oximetry[6,8]. Most authors conclude that pulse oximetry is safe and accurate for monitoring arterial hemoglobin saturation in critically ill patients[4-8].

According to our results $tcPO_2$ monitoring is a nearly ideal device for detecting hypoxemia, but tends to underestimate arterial $PO_2$ with increasing $PaO_2$ (Fig.2). The flow and age dependency is a drawback, but the good sensitivity to specificity ratio in our critically ill patients with very low saturations is an important advantage and makes transcutaneous oxygen tension monitoring a good trend monitor for severely hypoxic patients (Fig.4). Pulse oximetry is reliable at arterial saturations between 65 and 90%, but unreliable below 65%, leading to a high ratio of false-negative results in severely desaturated patients (Fig.1,3). The positioning of the probe in awake children and infants is still quite cumbersome and needs improvement.

Both pulse oximetry and $tcPO_2$ are safe, continuous and noninvasive, but their reliability and precision for detection of hypoxemia depends on the cut-off point. Each of the two devices has an optimal range (65 to 90% for $tcSaO_2$ and 35 to 50 mmHg for $tcPO_2$) with a low false-positive and false-negative ratio.

SUMMARY

We tested the performance of transcutaneous oxygen monitoring ($TcPO_2$) and pulse oximetry ($tcSaO_2$) in detecting hypoxia in critically ill neonatal and pediatric patients.

In 54 patients (178 data sets) with a mean age of 2.4 years (range 1 to 19 years), arterial saturation ($SaO_2$) ranged from 9.5 to 100%, and arterial oxygen tension ($PaO_2$) from 16.4 to 128 mmHg. Linear correlation analysis of pulse oximetry vs measured $SaO_2$ revealed an r value of 0.95 ($p<0.001$) with an equation of $y = 21.1 + 0.749x$, while $PaO_2$ vs $tcPO_2$ showed a correlation coefficient of $r = 0.95$ ($p<0.001$) with an equation of $y = -1.04 + 0.876x$. The mean difference between measured $SaO_2$ and $tcSaO_2$ was $-2.74 \pm 7.69\%$ (range +14 to -29%) and the mean difference between $\bar{P}aO_2$ and $tcPO_2$ was $+7.43 \pm 8.57$ mmHg (range -14 to +49 mmHg). Pulse oximetry was reliable at values above 65%, but was inaccurate and overestimated the arterial $SaO_2$ at lower values. $TcPO_2$ tended to underestimate the arterial value with increasing $PaO_2$. Pulse oximetry had the best sensitivity to specificity ratio for hypoxia between 65 and 90% $SaO_2$; for $tcPO_2$ the best results were obtained between 35 and 55 mmHg $PaO_2$.

REFERENCES

1. R. Huch, A. Huch, M. Albani, M. Gabriel, F. J. Schulte, H. Wolf, G. Rupprat, P. Emmrich, U. Stechele, G. Duc, H.Bucher, Transcutaneous $pO_2$-monitoring in routine management of infants and children with cardiorespiratory problems, Pediatrics. 57:681 (1976).
2. K. K. Tremper, W. C. Shoemaker, Transcutaneous oxygen monitoring of critically ill adults, with and without low flow shock, Crit Care Med. 9:706 (1981).
3. P. D. Wimberley, P. S. Frederiksen, J. Witt-Hansen, S. G. Melberg, B. Friis-Hansen, Evaluation of a transcutaneous oxygen and carbon dioxide monitor in a neonatal intensive care department, Acta Paediatr Scand. 74:352 (1985).
4. M. Yelderman, W. Jr. New, Evaluation of pulse oximetry, Anesthesiology. 59:349 (1983).
5. R. Deckardt, D. J. Steward, Non-invasive arterial hemoglobin oxygen saturation versus transcutaneous oxygen monitoring in the preterm infant, Crit Care Med. 12:935 (1984).
6. S. Fanconi, P. Doherty, J. F. Edmonds, G. A. Barker, Pulse oximetery in pediatric intensive care: comparison with measured saturations and transcutaneous oxygen tension, J Peditr. 107:362 (1985).
7. S. Fanconi, Reliability of pulse oximetry and transcutaneous oxygen tension in severely desaturated pediatric patients, J Pediatr. submitted (1986).
8. C. D. Fait, R. C. Wetzel, J. M. Dean, C. L. Schleien, F. R. Gioia, Pulse oximetry in critically ill children, J Clin Monit. 1:232 (1985).
9. B. J. McNeil, D. E. Keeler, S. J. Adelstein, Primer on certain elements of medical decision making, N Engl J Med. 293:211 (1975).

# IS PULSE OXIMETRY RELIABLE IN DETECTING HYPEROXEMIA IN THE NEONATE?

P. Baeckert, H.U. Bucher, F. Fallenstein, S. Fanconi, R. Huch and G. Duc

Divisions of Neonatology and Perinatal Physiology, Department of Obstetrics and Gynecology, Department of Pediatrics University of Zürich, CH-8091 Zürich, Switzerland

## INTRODUCTION

Pulse oximetry is a new technique for continuously assessing hemoglobin oxygen saturation. It is as yet not clear whether pulse oximetry can detect hyperoxemia reliably. Because of the well-known S-shape of the oxygen hemoglobin affinity curve, a small change in oxygen saturation above 90 % may be associated with a relatively large change in $Po_2$. Detection of hyperoxemia is particularly important in premature infants where high $Po_2$ is a major risk factor in the development of retinopathy.

We tested the hypothesis that pulse oximetry using 95 % oxygen saturation as the upper limit is reliable in detecting hyperoxemia, defined as arterial $Po_2$ above 12 kPa (90 mm Hg). The level of 95 % oxygen saturation was previously suggested by other authors[1,2] as a safe threshold for detecting hyperoxemia in newborn infants. These conclusions were based on studies measuring oxygen saturation with catheter oximeters. The results of Wilkinson et al.[1] showed that an oxygen saturation below 96 % indicated a $Pao_2$ below 100 mm Hg with 95 % confidence.

## PATIENTS AND METHODS

Thirty artificially ventilated neonates with an indwelling arterial catheter were studied. The characteristics of the patients are given in Table 1. Transcutaneous $Po_2$ ($tcPo_2$) [Hellige Transoxode heated at 44°C] and oxygen saturation ($tcSo_2$) [Ohmeda Biox 3700 Pulse Oximeter] were simultaneously monitored during a 4-hour period. Arterial blood samples were drawn and analyzed for $Po_2$ ($Pao_2$) [IL 613 or AVL 945] and oxygen saturation ($Sao_2$) [Corning 2500 CO-Oximeter] at the beginning and at the end of the measuring period, every time $tcPo_2$ was above 12 kPa or $tcSo_2$ was above 95 %, and before each endotracheal suctioning. Routine care in our units includes increasing $FiO_2$ by 10 % prior to suction in order to avoid hypoxemia, although this may cause hyperoxemia of short duration.

Table 1. Characteristics of the Patients

| | Median | Range |
|---|---|---|
| Birth Weight (g) | 1560 | 890-3280 |
| Gestational Age (weeks) | 30 4/7 | 26-41 |
| Postnatal Age (days) | 3 | 1-13 |
| Hematocrit (%) | 48 | 38-70 |
| Total Serum Bilirubin (µmol/l) | 128 | 50-214 |
| Skin Temperature Near Sensor (°C) | 34.3 | 31.9-37.2 |
| Systolic Blood Pressure (mm Hg) | 58 | 35-71 |
| Diastolic Blood Pressure (mm Hg) | 37 | 20-46 |
| Fetal Hemoglobin (%) | 75 | 8-95 |

Receiver-operating characteristic (ROC) curves were depicted for graphical comparison of the performance of pulse oximetry in diagnosing arterial hyperoxemia with $tcPo_2$ and arterial $So_2$, and for determination of the optimal cut-off points. The ROC curves were constructed by shifting the cut-off point stepwise from 70 % to 100 % oxygen saturation or from 4 to 20 kPa. At every step sensibility and specificity of the method were calculated.

## RESULTS

119 blood samples (two to six per patient) were analyzed. Oxygen saturation ranged from 70 % to 100 %. Fig. 1. shows the difference between transcutaneous and arterial oxygen saturation plotted against transcutaneous oxygen saturation. The 95 % confidence interval for this difference is $\pm$ 8 % oxygen saturation. We tested to see whether or not the $tcSo_2$-$Sao_2$ difference was affected by any of the following factors: weight, systolic and diastolic blood pressure, skin temperature near the sensor, hematocrit, total serum bilirubin, percentage of fetal hemoglobin, pH and $Pco_2$ in the

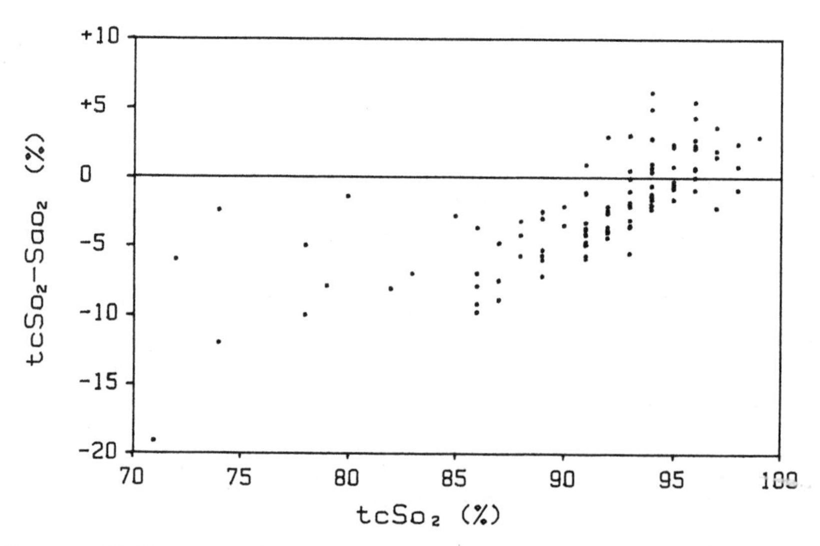


Fig. 1. Difference of transcutaneous oxygen saturation ($tcSo_2$) - arterial oxygen saturation ($Sao_2$) plotted against $tcSo_2$.

arterial blood. Using a least square regression analysis, no statistically significant correlation could be demonstrated.

As shown in Fig. 1. $tcSo_2$ underestimates oxygen saturation at lower values. In the 47 cases when $tcSo_2$ was 91 % or less, the mean value (SD) was 86.6 (5.6) for $tcSo_2$ and 91.9 (4.7) for $Sao_2$, respectively. Using the paired t-test, this difference was highly significant (P 0.001). Testing the observations where $tcSo_2$ was above 91 %, the mean values were 94.4 (1.8) and 94.6 (2.3), respectively. This difference was not statistically significant.

Fig. 2. illustrates the relationship between simultaneously measured $tcSo_2$ and arterial $Po_2$ values. One line is drawn at a cut-off point for $tcSo_2$ 95 % and the other one for $Pao_2$ 12 kPa. It will be seen that with pulse oximetry only 14 out of a total of 46 hyperoxemic episodes were identified. Consequently the sensitivity for pulse oximetry in detecting hyperoxemia was 30 %. Normoxemia was correctly recognized in 68 out of 73 cases, i.e. the specificity was 93 %. In 5 of the 19 instances when $Sao_2$ was above 95 %, $Pao_2$ was 12 kPa or less, giving a false alarm rate of 26 %.

A similar analysis, taking instead arterial oxygen saturation against arterial $Po_2$ and using the same cut-off points, gave for $Sao_2$ a sensitivity of 41 %, a specificity of 96 % and a false alarm rate of 14 %.

Finally testing transcutaneous $Po_2$ against arterial $Po_2$, using in both parameters 12 kPa as the cut-off point, the sensitivity was 78 %, the specificity 86 %, and the false alarm rate 22 %.

The effect on sensitivity, specificity and false alarm rate of lowering the cut-off point for pulse oximetry taking $Pao_2$ as the reference method is demonstrated in Table 2.

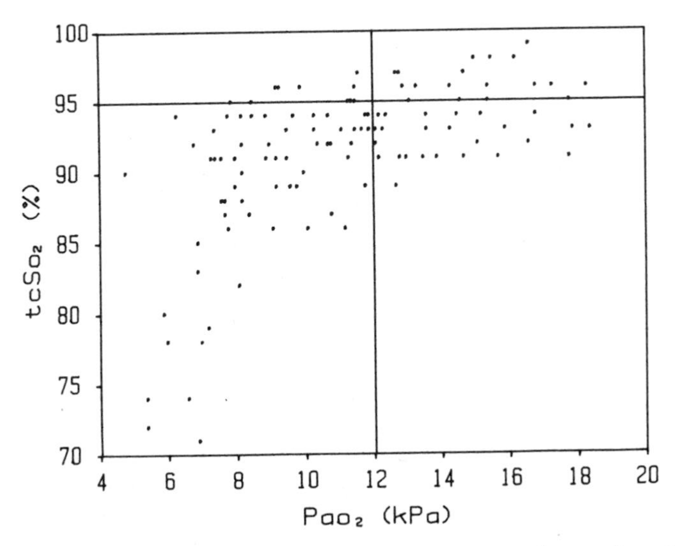


Fig. 2. Comparison between pulse oximetry ($tcSo_2$) and arterial oxygen tension ($Pao_2$).

Table 2. Sensitivity, Specificity and False Alarm Rate for Pulse Oximetry at Various Cut-off Points

| Cut-off Point for $tcSo_2$ | Sensitivity | Specificity | False Alarm Rate |
|---|---|---|---|
| 95 % | 30 % | 93 % | 26 % |
| 92 % | 70 % | 62 % | 53 % |
| 90 % | 96 % | 37 % | 51 % |

The relation between sensitivity and specificity at various cut-off points is illustrated graphically for pulse oximetry, $tcPo_2$ and $Sao_2$, using $Pao_2$ as the reference method in Fig. 3. An ideal test has a sensitivity and a specificity of 100 % and its ROC curve passes through the left upper corner. It will be seen in Fig. 3. that the ROC curves of $tcPo_2$ and $Sao_2$ rank better than that of $tcSo_2$.

As shown in Fig. 3. the best relation between sensitivity and specificity for $tcSo_2$ in separating hyperoxemia from normoxemia can be obtained by choosing a cut-off point of 92 %. However, even at this optimal cut-off point, sensitivity is only 70 % and specificity 62 %, which is not acceptable for a diagnostic test.

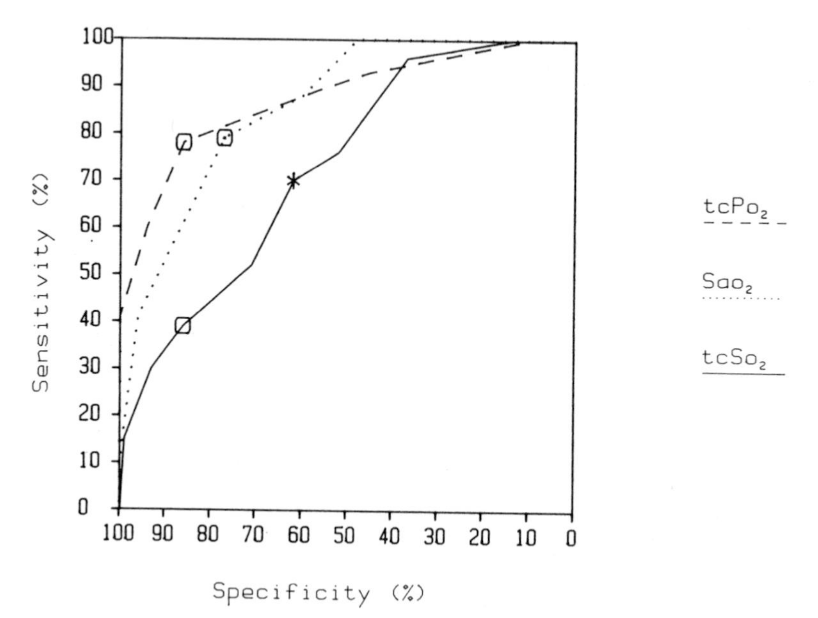


Fig. 3. Receiver-operating characteristic curves for pulse oximetry ($tcSo_2$), $tcPo_2$ and $Sao_2$. The cut-off points at 95 % and 12 kPa are marked with "▫" and that of 92 % for pulse oximetry with "*".

## DISCUSSION

Our results show that 95 % oxygen saturation as the upper limit for pulse oximetry as proposed according to measurements with catheter oximeters[1,2] is not reliable in detecting arterial hyperoxemia. Decreasing the alarm limit from 95 % to 90 % results in a better sensitivity of pulse oximetry for hyperoxemia (Table 2). However, by this procedure specificity drops considerably and the false alarm rate rises to over 50 %. This means that nurses would be called by the alarm bell twice as often as necessary, which would hardly be acceptable.

We compared the performance of pulse oximetry in diagnosing arterial hyperoxemia with $tcPo_2$ and arterial $So_2$ graphically by depicting the ROC curves. Fig. 3. shows that pulse oximetry is less reliable than the two other methods even when the cut-off point is shifted from 95 % to 92 %.

The results presented here were obtained by an Ohmeda Biox 3700 pulse oximeter. Instruments from other manufacturers may differ in design and data processing. As our $tcSo_2$ values do not differ significantly from arterial $So_2$ values in the range between 91 % and 100 % oxygen saturation, it is to be expected that the results concerning the detection of hyperoxemia are valid for instruments from other producers too.

## SUMMARY

We tested the hypothesis that hyperoxemia defined as arterial $Po_2$ above 12 kPa can be detected by pulse oximetry using 95 % oxygen saturation as the upper limit. Thirty artificially ventilated neonates with an indwelling arterial catheter were studied registrating transcutaneous oxygen saturation (Ohmeda Biox 3700 Pulse Oximeter) and transcutaneous $Po_2$ continuously during a 4-hour period and measuring arterial oxygen saturation and $Po_2$ intermittently. 46 episodes of arterial hyperoxemia were observed. Pulse oximetry had a sensitivity of 30 %, detecting 14 of these 46 hyperoxemic episodes, and a specificity of 93 %. The accuracy for separating hyperoxemia from normoxemia by pulse oximetry could be improved by shifting the cut-off point from 95 % to 92 %. With this optimal cut-off point sensitivity was 70 % and specificity 62 %. We conclude that pulse oximetry is not reliable for detection of hyperoxemia.

## REFERENCES

1. A. R. Wilkinson, R .H. Phibbs, G. A. Gregory: Continuous in vivo oxygen saturation in newborn infants with pulmonary disease. Crit Care Med 7:232-236 (1979)
2. R. W. Krouskop, E. E. Cabatu, B. P. Chelliah, F. E. McDonnell, E. G. Brown: Accuracy and clinical utility of an oxygen saturation catheter. Crit Care Med 11: 744-749 (1983)
3. B. J. McNeil, E. Keeler, S. J. Adelstein: Primer on certain elements of medical decision making. N Engl J Med 293:211-215 (1975)

# COMPARISON BETWEEN TRANSCUTANEOUS $PO_2$ AND PULSE OXIMETRY FOR MONITORING $O_2$-TREATMENT IN NEWBORNS

E. Bossi, B. Meister and J. Pfenninger

Div. of Neonatology, Div. of Pediatric Intensive Care
Dept. of Pediatrics, University of Berne, Switzerland

## SUMMARY

213 paired $tcpO_2/paO_2$-data and 186 paired $tcSO_2/SaO_2$-data measured in 25 newborns (10 term, 15 prematures) were compared. The correlation coefficient for $tcpO_2/paO_2$ was 0.796, for $tcSO_2/SaO_2$ 0.944. Sensitivity for discriminating between normo- and hypoxemia ($paO_2 < 50$ torr) was 82 % for the $tcpO_2$- and 88 % for the $tcSO_2$-method. Positive predictive values for discriminating between normo- and hypoxemia were 88 % for both methods. Sensitivity for discrimination between normo- and hyperoxemia ($paO_2 > 100$ torr) was 85 % for the $tcpO_2$- and 100 % for the $tcSO_2$-method. Positive predictive values for the discrimination between normo- and hyperoxemia were 58 % and 25 % for $tcpO_2$- and $tcSO_2$ respectively. Pulse oximetry proved to be less cumbersome than the $tcpO_2$-method. However, as $tcpO_2$, it could not be used in some very immature newborns and in those with circulatory instability. In conclusion, these preliminary results show a similar discrimination between normoxemia and hypo- hyperoxemia for both methods. A better sensitivity of pulse oximetry for hyperoxemia is counteracted by a lesser positive predictive value.

## INTRODUCTION

Transcutaneous $pO_2$ ($tcpO_2$)-monitoring of oxygen treatment in the neonate is an useful method in neonatology. In our opinion, however, the dispersion of $tcpO_2$-values at given arterial $pO_2$ ($paO_2$)-levels is too large for this methodology to be used without corroboration by repeated arterial blood gas sampling (1). Measurement of transcutaneous saturation ($tcSO_2$) by pulse oximetry is simple and harmless. The aim of this preliminary study is to find out if pulse oximetry can substitute for $tcpO_2$-measurements in newborns treated with oxygen. In order to do so, $tcSO_2$ must correlate with arterial $SO_2$ ($SaO_2$) and discriminate between normoxemia, hypoxia and hyperoxia at least as reliably as the $tcpO_2$-monitoring.

## PATIENTS AND METHODS

25 newborns (10 term, 15 prematures of gestational age 30-60 weeks) were investigated during the first 17 (mean 6, median 3) days of life. 9 had hyaline membrane disease, 8 aspiration, 2 a cyanotic cardiac malformation, 1 BPD, the other 5 suffered from various diseases. The only criterion for entering the study was the fact that the babies had an arterial line (and thus also received oxygen). Arterial blood gas sampling was performed in 13 babies through umbilical, in 11 through right, in 1 through left radial catheters. $paO_2$ was measured at 37° C with either an IL 1303 or Corning 178 blood gas analyzer. Arterial saturation was measured at 37° C, without correction for the patient's hemoglobin, with an IL 282 oximeter. $tcpO_2$ was measured either with the Hellige or with the Kontron, $tcSO_2$ with the Ohmeda (Biox 3700) device, the flex probe being used. The probes were mainly placed in the irrigation parts of the arteries catheterized; in the other cases, there was no indication for isolated right-to-left shunting through the ductus arteriosus.

### Definitions

Table I: Definitions of hypo-, normo- and hyperoxemia

| | Hypoxemia | Normoxemia | Hyperoxemia |
|---|---|---|---|
| $paO_2$ (torr) | < 50 | 50-100 | > 100 |
| $tcpO_2$ (torr) | < 40 | 40-90 | > 90 |
| $tcSO_2$ (%) | < 85 | 85-93 | > 93 |

$paO_2$-values between 50 and 100 torr were defined as mormoxemic (table I). Correlations between $tcpO_2$ and $paO_2$ having shown a regression line above the line of identity (see figure 1), $tcpO_2$-values between 40 and 90 torr were considered as representing normoxemia. The best possible combinations of positive predictive value and sensitivity for the $tcSO_2$-method at the limits of hypoxemia and hyperoxemia resulted at $tcSO_2$-values of 85% and 93 % respectively. Therefore, the range delimited by these values was chosen as the normoxemic $tcSO_2$-range.

## RESULTS

### 1. Correlations between $tcpO_2$ and $paO_2$, and $tcSO_2$ and $SaO_2$ over the whole range

Figure 1 shows the linear regression between $tcpO_2$ and $paO_2$ as well as the pertinent statistical data. The correlation coefficient was 0.796.

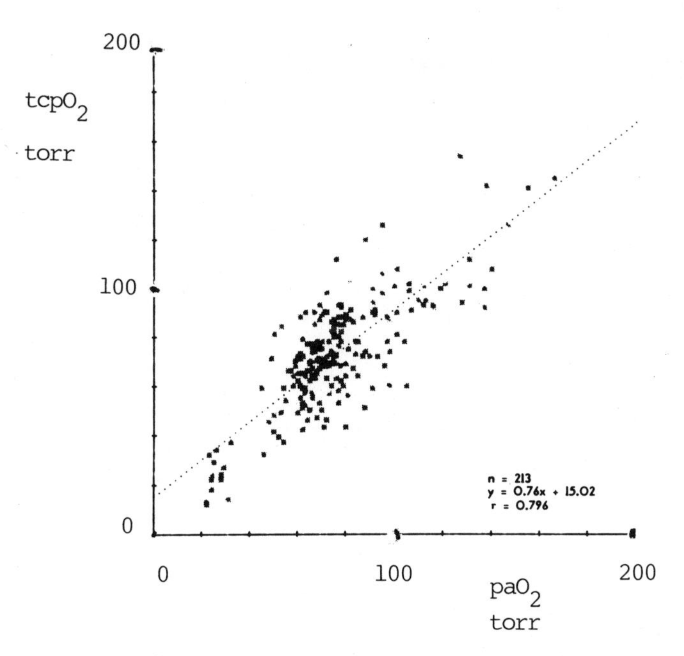


Figure 1: Correlation between $tcpO_2$ and $paO_2$

In figure 2, the same data are depicted for $tcSO_2$/art. $SO_2$. The correlation coefficient was 0.944.

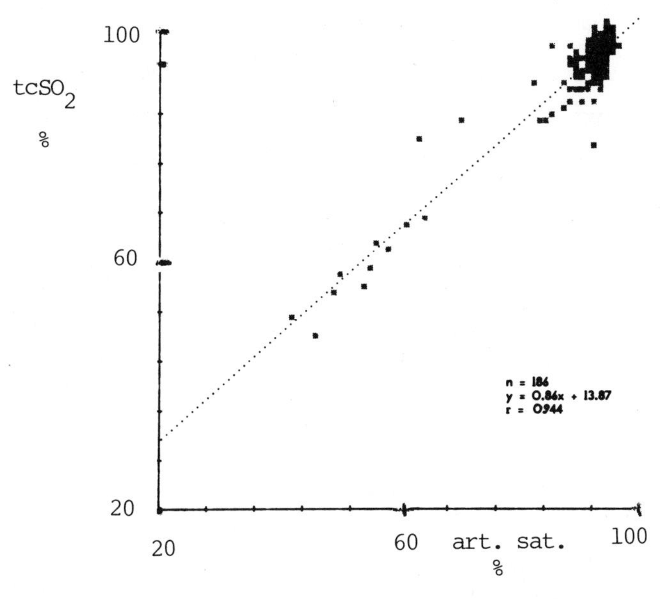


Figure 2: Correlation between $tcSO_2$ and art. $SO_2$

2. Correlations between $tcpO_2$ and $paO_2$, and $tcSO_2$ and $SaO_2$ outside the normoxic range

The data are presented in table II

Table II: Correlations in the hypoxemic and in the hyperoxemic ranges

| | Hypoxemia | | Hyperoxemia | |
|---|---|---|---|---|
| | $tcSO_2/SaO_2$ at $SaO_2$ < 85 % | $tcpO_2/paO_2$ at $paO_2$ < 50 torr | $tcSO_2/SaO_2$ at $SaO_2$ > 93 % | $tcpO_2/paO_2$ at $paO_2$ > 100 torr |
| r | 0.936 | 0.795 | 0.014 | 0.701 |
| p | 0.001 | 0.001 | n.s. | 0.001 |
| n paired values | 15 | 17 | 52 | 27 |

3. Sensitivity of $tcpO_2$ and $tcSO_2$ for discriminating normoxemia from hypo- and hyperoxemia

The data are shown in table III

Table III: Sensitivity of $tcpO_2$ and $tcSO_2$ for detecting hypo- and hyperoxemia. *absolute numbers

| $paO_2$ < 50 torr | | | | $paO_2$ > 100 torr | | | |
|---|---|---|---|---|---|---|---|
| $tcpO_2$ < 40 | 14* | $tcSO_2$ < 85 | 15 | $tcpO_2$ > 90 | 23 | $tcSO_2$ > 93 | 28 |
| $tcpO_2$ ≧ 40 | 3* | $tcSO_2$ ≧ 85 | 2 | $tcpO_2$ ≦ 90 | 4 | $tcSO_2$ ≦ 93 | 0 |
| Sensit. | 82 % | | 88 % | Sensit. | 85 % | | 100 % |

4. Positive predictive value of $tcpO_2$ < 50 or > 100 torr and of $tcSO_2$ < 85 or > 93 % for discriminating normoxemia from hypo- and hyperoxemia

The data are given in table IV

Table IV: Positive predictive value of $tcpO_2$ < 50 or > 100 torr and of $tcSO_2$ < 85 or > 93 % for detecting hypo- and hyperoxemia. *absolute numbers

| | $tcpO_2$ < 40 | $tcSO_2$ < 85 % | | $tcpO_2$ > 90 | $tcSO_2$ > 93 % |
|---|---|---|---|---|---|
| $paO_2$ < 50 | 14* | 15 | $paO_2$ > 100 | 23 | 28 |
| $paO_2$ ≧ 50 | 2* | 2 | $paO_2$ ≦ 100 | 17 | 86 |
| Pred. val. | 88 % | 88 % | Pred. val. | 58 % | 25 % |

DISCUSSION

These results show that the correlation between $tcSO_2$ and $SaO_2$ is higher than the correlation between $tcpO_2$ and $paO_2$ (figures 1 and 2). The first requirement for pulse oximetry as a substitute for $tcpO_2$-monitoring is thus fulfilled. The correlation coefficient of 0.944 is the same as the one described by Fanconi et al (0.95) with another pulse oximeter in a population of 11 newborns and 29 older children (2). Chapman et al (3) found a correlation coefficient of 0.93 in 10 adult healthy volunteers by using the BIOX II device. In Fanconi's work (2) the regression line is almost like the identity line (y = 0.973x + 0.97), in our patients, it lies above it (y = 0.86x + 13.87). The reason for this difference is not clear. It may be that the correction for the patient's temperature and hemoglobin when determining $SaO_2$, as performed by those authors but not by us, could partly explain the difference.

Correlation was also calculated for the hypoxemic and hyperoxemic range separately (table II). The number of values outside the normoxemic range is small. In hypoxemia, however, the correlations remain in the same order of magnitude as for the whole population. In hyperoxemia, correlation is low for $tcpO_2$ and non-existent for $tcSO_2$.

The second requirement for pulse oximetry, should it substitute for $tcpO_2$-monitoring in the newborn, is that it must discriminate between normoxemia and hypo- or hyperoxemia at least as well as the $tcpO_2$-method. In fact, the sensitivity of $tcSO_2$ for detecting hypoxemia is a little higher than the one of $tcSO_2$ (88 vs. 82 %). Sensitivity in the hyperoxemic range even reaches 100 % for $tcSO_2$: at $paO_2$-values above 100 torr, $tcSO_2$ is always higher than 93 %. In this hyperoxemic range, $tcpO_2$-sensitivity is 85 %.

However, when the positive predictive values are compared at hyperoxemic values, both methods are found to falsely indicate hyperoxemia very often: 42 % of the $tcpO_2$-values > 90 torr are falsely elevated. A non-existing hyperoxemia is indicated by 75 % of the $tcSO_2$-values > 93 %. (Positive predictive values 58 % and 25 % respectively, table IV). In hypoxemia, the positive predictive values of both methods are equally satisfying (88 %). Clearly, too little hypo- and hyperoxemic values have been gathered up to now for allowing definitive conclusions concerning the suitability of $tcSO_2$ for monitoring neonatal oxygen treatment when compared to $tcpO_2$. These preliminary results have to be extended by more data in the abnormal ranges.

Pulse oximetry has proven to be more practicable than $tcpO_2$-monitoring. Calibration is not necessary, the probe can be left on the same site for more than 4 hours without provoking lesions. Due to the necessity of utilizing the device in the clinical management of older children as well, we could not investigate this question further and establish time limitations. The fact that the BIOX pulse oximeter indicates if the signal it receives is sufficient or not may help to avoid insufficient correlations which otherwise would not be apparent as it is the case for $tcpO_2$-monitoring. Pulse oximetry could sometimes not be used in very immature babies, in newborns with unstable circulation and in those with important edema. In these conditions, $tcpO_2$-monitoring was also found not to yield acceptable values when compared to arterial $pO_2$. In these situations also, pulse oximetry has the advantage of indicating its non-reliability by displaying either the "low quality signal" or the "probe off patient" signal. In our experience, there are some newborns in whom the flex probe is too small, the emitter and the detector not being in a parallel position due to the thickness of the extremity of the baby. In most of these cases, the ear probe can also not be used, since it compresses the tissue too much, stopping blood flow. Again, no definitive conclusions or guidelines can be drawn from our still limited experience.

In conclusion, pulse oximetry has some advantages as compared to $tcpO_2$-monitoring, mainly concerning practicability. Correlation and discrimination data seem to indicate a similar validity of both methods, the major drawback of both, but especially of pulse oximetry, being the low positive predictive value in the hyperoxemic range. Our data are preliminary: 6 more comparisons must be made at hypo- and hyperoxemic $paO_2$-levels for allowing more definitive conclusions regarding monitoring of oxygen treatment in the neonate by pulse oximetry.

REFERENCES


1. E. Bossi et al, Cutaneous $pO_2$-measurements in newborns with respirator distress syndrome. Helv paediat Acta, 31: 335-345 (1976)
2. S. Fanconi et al, Pulse oximetry in pediatric intensive care: Comparison with measured saturations and transcutaneous oxygen tension, J Pediatr 107: 362-366 (1985)
3. K. R. Chapman et al, The accuracy and response characteristics of a simplified ear oximeter, 83: 860-864 (1983)

# THE ACCURACY OF THE PULSE OXIMETER IN NEONATES

A. Hodgson, J. Horbar, G. Sharp*, R. Soll and J. Lucey

Departments of Pediatrics and Pathology*
University of Vermont College of Medicine
Burlington, VT

## SUMMARY

The Nellcor pulse oximeter was studied in 16 newborn infants. The transcutaneous oxygen saturation (TCSaO2) was compared to both fractional saturation and functional saturation. TCSaO2 was found to be closer to fractional SaO2. 95% prediction intervals were established for the estimation of arterial oxygen saturation (SaO2) from TCSaO2. At a pulse oximeter reading of 85% to 90%, SaO2 could be as low as 83% and as high as 98%.

## INTRODUCTION

Accuracy of the Nellcor pulse oximeter has been established in adults.[1] The reading from the pulse oximeter is compared to arterial oxygen saturation as measured by a co-oximeter. Documentation of its accuracy in newborns presents difficulties because conventional co-oximeters detect a fictitious elevation of carboxyhemoglobin in the presence of fetal hemoglobin. Cornelissen et al[2] have suggested a correction to be used when samples containing fetal hemoglobin are analyzed.

The Nellcor pulse oximeter reportedly measures functional saturation[3] whereas the IL 282 measures fractional saturation. Functional saturation is the ratio of oxyhemoglobin to the total hemoglobin available for binding to oxygen. Fractional saturation is the ratio of oxyhemoglobin to all the hemoglobin present, including carboxyhemoglobin and methemoglobin. Most investigators[1,4,5] have compared the reading from the pulse oximeter, which is supposed to reflect functional saturation, to fractional saturation as measured on the co-oximeter. One of the objectives of this study was to determine which of these the Nellcor pulse oximeter more closely approximated.

$$\text{FRACTIONAL SaO2} = \frac{\text{O2Hb}}{\text{O2Hb} + \text{RHb} + \text{COHb} + \text{MetHb}} \times 100$$

$$\text{FUNCTIONAL SaO2} = \frac{\text{O2Hb}}{\text{O2Hb} + \text{RHb}} \times 100$$

Most investigators have used linear regression and correlation to establish the accuracy of the pulse oximeter. However, good correlation does not necessarily allow for precise estimation of one variable from another. Prediction intervals provide a measure of how precisely one can estimate one value from another. In this study, 95% prediction intervals have been established to estimate SaO2 from TCSaO2.

METHODS

Sixteen newborns less than 7 days of age with indwelling postductal arterial catheters were studied. Written consent was obtained from one or both parents and the study was reviewed by the Institutional Review Board. Four data points on each patient were obtained for a total sample size of 64. The neonatal oxisensor was placed postductally on the foot, wrist or hand, and was shielded from radiant warmers and phototherapy lights. The pulse oximeter saturation reading was covered until the moment of data point collection to minimize bias in the timing of the sample. When the heart rate on the pulse oximeter corresponded within 5 beats per minute to the heart rate on the Corometrics 515 neonatal monitor, the blood sample was drawn and a reading from the pulse oximeter was recorded. Fetal hemoglobin determinations were made at study entry and following a red blood cell transfusion. Quantity of fetal hemoglobin was measured using the alkali denaturation method. Fractional SaO2 was measured on the IL 282 co-oximeter and corrected for carboxyhemoglobin using the method of Cornelissen. This value was then converted to functional SaO2. Data were analyzed by repeated analysis of covariance.

RESULTS

64 data points were obtained from 16 neonates whose gestational ages ranged from 24 to 43 weeks. TCSaO2 ranged from 85% to 100%, and fetal hemoglobin ranged from 19 to 99%. The mean difference between TCSaO2 and fractional SaO2 was 0.8 (SE=1.8), while the difference between TCSaO2 and functional SaO2 was 4.3 (SE=1.8). The difference between these values was statistically significant ($p<0.0001$). There was a statistically significant correlation between TCSaO2 and fractional SaO2 ($r=0.49$). The equation for the regression line was $Y=0.56X + 41.9$. 95% prediction intervals were constructed; at TCSaO2 values of 85-95%, the fractional SaO2 could be as low as 83% and as high as 98%.

FRACTIONAL VS FUNCTIONAL SaO2

| TCSaO2 - Fractional SaO2 | TCSaO2 - Functional SaO2 |
|---|---|
| $0.8 \pm 1.8$ (SE) | $4.3 \pm 1.8$ (SE) |

$p < 0.0001$

95% PREDICTION INTERVALS

| TCSaO2 | SaO2 |
|---|---|
| 84 | 82 - 95 |
| 85 | 83 - 95 |
| 86 | 83 - 96 |
| 87 | 84 - 97 |
| 88 | 84 - 97 |
| 89 | 85 - 98 |
| 90 | 85 - 98 |

## DISCUSSION

The reading from the Nellcor pulse oximeter was found to be closer to the arterial fractional saturation rather than to the functional saturation. This is contrary to the original assumption that the pulse oximeter measures functional saturation. Reasons for this unexpected finding are not readily evident.

In spite of a significant correlation between the pulse oximeter reading and arterial oxygen saturation, the prediction intervals were very broad. To avoid hyperoxemia and desaturation, one is obliged to maintain the TCSaO2 in a very narrow range, a range that would be very difficult, if not impossible to adhere to in the clinical setting.

## ACKNOWLEDGEMENTS

Data were partially managed and analyzed using the University of Vermont GCRC CLINFO system supported by the Division of Research Resources, NIH (GCRC RR109). We would like to thank Nellcor, Inc. for use of their equipment, and Instrument Laboratories for technical assistance. We would also like to thank Nancy Moreland for her assistance in preparing this manuscript.

## REFERENCES

1. Yelderman M, New W. Evaluation of pulse oximetry. Anesthesiology 1983; 59:349-351.

2. Cornelissen PJH, van Woensel CLM, van Oel WC, et al. Correction factors for hemoglobin derivatives in fetal blood, as measured with the IL-282 Co-Oximeter. Clin Chem 1983; 29:1555-1556.

3. Nellcor Technical Note No. 2: Nellcor Incorporated, Hayward, California.

4. Jennis MS, Peabody JL. Pulse oximetry, an alternative to transcutaneous PO2. Pediatric Research 1985; 19:142A.

5. Fanconi S, Doherty P, Edmonds JF, et al. Pulse oximetry in pediatric intensive care: comparison with measured saturations and transcutaneous oxygen tension. J Pediatr 1985; 107:362-366.

# PULSE OXIMETRY AND TRANSCUTANEOUS OXYGEN TENSION IN HYPOXEMIC NEONATES AND INFANTS WITH BRONCHOPULMONARY DYSPLASIA

Harry N. Lafeber, Willem P.F. Fetter, André R. v.d. Wiel and T. Cornelis Jansen

Department of Pediatrics, subdivisions of Neonatology and Medical Electronics, Sophia Children's Hospital, Rotterdam The Netherlands

## SUMMARY

The reliability of pulse oximetry and transcutaneous oxygen tension ($tcPo_2$) was investigated in hypoxemic neonates and older infants with chronic hypoxemia due to bronchopulmonary dysplasia (BPD). It was found that during severe hypoxemia ($tcPo_2 < 40$ mmHg and saturation $< 80\%$) pulse oximetry showed a better correlation with arterial saturation than $tcPo_2$ with arterial oxygen tension. During mild hypoxemia and normoxemia ($tcPo_2$ 40-90 mmHg and saturation 80-95%) $tcPo_2$ and pulse oximetry both showed a good correlation with arterial values. Above 95% saturation and a corresponding $tcPo_2$ of 70-120 mmHg, the correlation between arterial and transcutaneous $Po_2$ was better than that between pulse oximetric and arterial saturation. Computer recording and analysis of $tcPo_2$ and pulse oximetry improves the quality of both noninvasive oxygenation parameters in older infants with BPD.

## INTRODUCTION

Recently pulse oximetry was introduced as a new clinical technique for noninvasive measurement of oxygenation [1]. We tested this technique in hypoxemic neonates and infants with congenital cyanotic heart disease and chronic lung disease (BPD). $TcPo_2$ was measured simultaneously and arterial samples were taken for $Po_2$ and saturation. Infants with BPD show a poor correlation between $tcPo_2$ and arterial $Po_2$ [2]; therefore pulse oximetry could be a better noninvasive oxygenation parameter in these infants. The two infants with congenital heart disease offered the opportunity to test $tcPo_2$ and pulse oximetry under extreme hypoxemic conditions. $TcPo_2$ and pulse oximetric recordings often show many artefacts in older infants. We therefore tested a personal computer based recording and data analysis program in older infants with BPD that were treated with low flow oxygen therapy. We evaluated whether or not a mean $tcPo_2$ or saturation calculated over a prolonged period (after artifact elimination) would add to the quality of these parameters.

## METHODS

Saturation and pulse were measured using a Nellcor N-101 pulse oximeter with special disposable sensors on the midfoot in neonates or the toe in older infants. Arterial saturation was determined using a Radiometer OSM-2 cooximeter. $TcPo_2$ was measured at 44°C using our own system that is comparable to the commercially available electrodes. Arterial $Po_2$ was measured in a IL 1302 blood gas analyzer. All neonates and infants had radial artery catheters for arterial bloodgas sampling, with the exception of the older infants with BPD, in which computer recording and analysis of $tcPo_2$ and pulse oximetry were tested. Details of the computer registrations and data analysis are described elsewhere [3]. Patient information and recordings are stored on floppy disks. The computer calculates and presents a time-amplitude histogram within optional limits. The histogram is printed with the mean and standard deviation, duration of registration and the percentage of time the signal exceeds upper and lower limits.

Four normoxemic neonates were studied at a postnatal age of 5-10 days. They had a birthweight of 1260-1580 g and a gestational age of 28-32 weeks. All has recovered from mild respiratory distress syndrome. They had not received blood transfusions and had an arterial pH of 7.28-7.36 and a $Pco_2$ of 38-54 mmHg at the time of study. Two hypoxemic infants with BPD were 1020 and 1340 g with a gestational age of 28-30 weeks. They received 3-5 blood transfusions and had an arterial pH of 7.32-7.40 and a $Pco_2$ of 52-65 mmHg. Two extremely hypoxemic neonates both suffering from tetralogy of Fallot and severe pulmonary stenosis were studied. They were born at term with birthweights of 2980 and 1040 g. The hypoxemic neonate was studied at day 5-7 before i.v. prostaglandins were given in order to reopen the ductus arteriosus. No transfusions were given and arterial pH was 7.32-7.36 and $Pco_2$ 24-36 mmHg. The second baby with 4F, was studied at 3 months shortly after the surgical introduction of a Blalock shunt that unfortunately ceased functioning. At that time the bodyweight was 2450 g, the arterial pH 7.11-7.38 and the $Pco_2$ 34-62 mmHg. Fifteen infants with BPD were registered for $tcPo_2$ and saturation every 4 weeks during 24 hrs. All were initially ventilated because of respiratory distress syndrome. Their birthweights were 730-2180 g with a gestational age of 25-33 weeks. All had severe BPD (Northway scale IV [4], Toce score 15-25 at day 21 [5]),were weaned from the ventilator (17-120 days) and received oxygen by nasal cannula (0.1-2.0 l/min 100% $O_2$) for 2-15 months. Nine infants received oxygen therapy at home and were admitted to the hospital for the $tcPo_2$ and saturation recordings.

## RESULTS

The results of the simultaneous measurement of $tcPo_2$ and arterial $Po_2$ are given in fig. 1; the line of identity is shown and a linear regression was calculated for each group of neonates or infants. A good correlation between $tcPo_2$ and arterial $Po_2$ was found in the normoxemic neonates (r=0.91). A moderate correlation was found in the infants with BPD (r=0.74) and in the infant with congenital heart disease (r=0.73). A poor correlation was found in the newborn with congenital heart disease (r=0.20). The results of the simultaneous measurements of arterial saturation and pulse oximetry with a linear regression of each group of neonates or infants are presented in fig. 2. A moderate correlation was found between arterial saturation and pulse oximetry in the normoxemic neonates (r=0.73). A good correlation between arterial saturation and pulse oximetry was found in the infants with BPD (r=0.92) and the neonate and infant with congenital heart disease (r=0.98 and r=0.90).

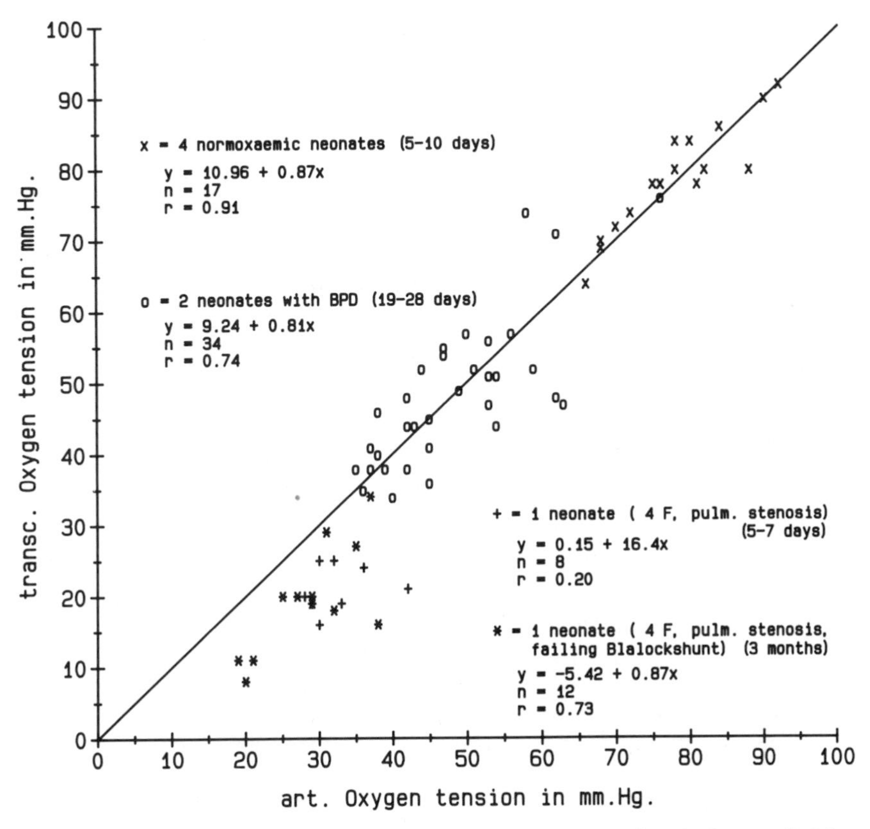


Fig. 1. Relationship between arterial $Po_2$ and $tcPo_2$ (with line of identity).

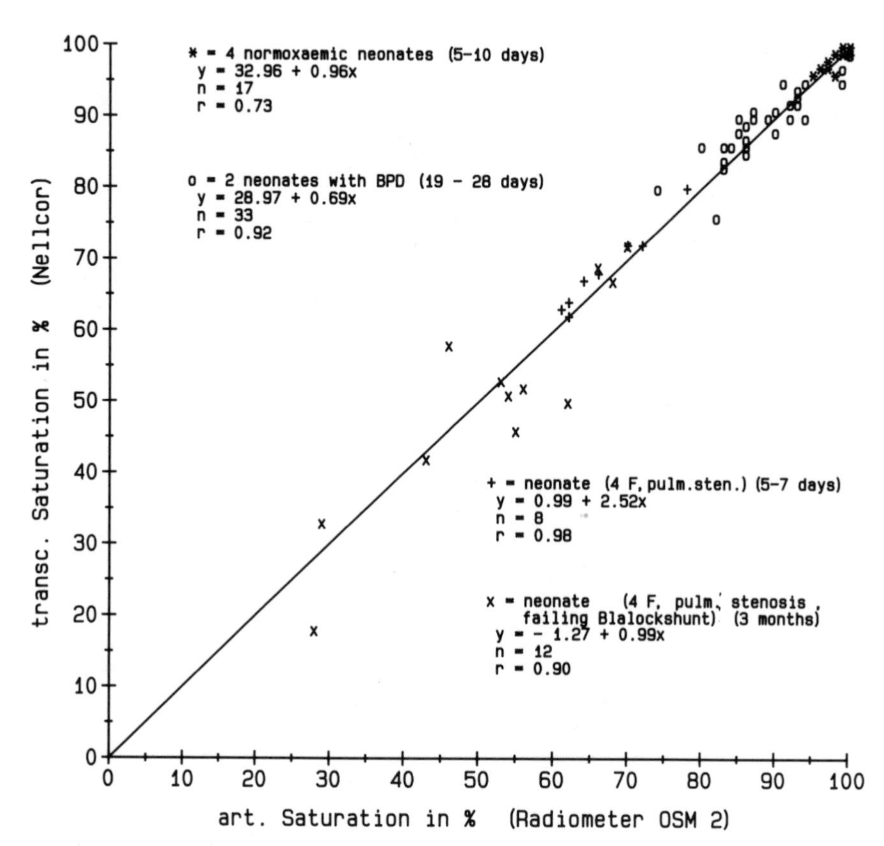


Fig. 2. Relationship between arterial saturation and Nellcor pulse oximeter (with line of identity).

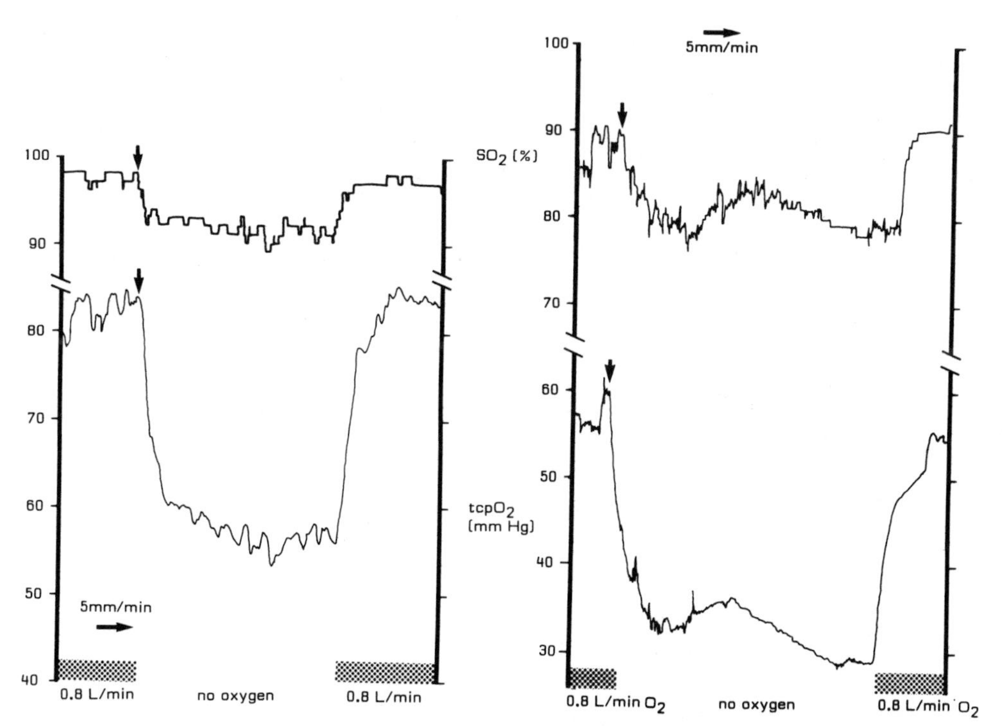


Fig. 3. Hypoxemia tests in an infant with BPD at 3 months (left) and 9 months (right). The improvement in basic $tcPo_2$ and saturation levels reflects improved lung function.

Computer recordings of $tcPo_2$ and pulse oximetry were performed at monthly intervals in infants with BPD. Calculation of the mean $tcPo_2$ and saturation with standard deviation helped to give an objective indication of the lung condition and the amount of supplementary oxygen that was needed. Growth in BPD infants was optimal if mean $tcPo_2$ was kept above 50 mmHg (at 44°C) with corresponding saturation above 90%. Each month oxygen was stopped for a short period to get an impression of lung function. As shown in fig. 3 $tcPo_2$ and saturation decreased markedly when oxygen was stopped in a 3 month old infant with BPD shortly after extubation. Six months later withdrawal of oxygen no longer resulted in hypoxemia.

DISCUSSION

In practice pulse oximetry is preferable to $tcPo_2$ monitoring since no calibration procedures are needed, the application time to the skin is not limited and probes are always available without using new membranes frequently. Since the number of patients studied is still limited, statistical analysis must be regarded with care, but the correlation coefficients show that pulse oximetry is better than $tcPo_2$ recording during severe hypoxemia ($tcPo_2 < 40$ mmHg and saturation $< 80\%$). A possible explanation is that pulse oximetry is less sensitive to diminished skin perfusion during hypoxemia than $tcPo_2$ measurements. The lower limits of pulse oximetry in clinical use must still be determined using more extended techniques.

In the normal range pulse oximetry and $tcPo_2$ are reliable parameters but as soon as saturation exceeds 95%, $tcPo_2$ registration is preferable. This is partly due to the preset microprocessing of the pulse oximeter around 100% saturation. But also from the relation between oxygen tension and saturation in the oxygen dissociation curve it seems obvious that $tcPo_2$ is a more sensitive parameter than saturation during hyperoxemia.

Computer registration and calculation of mean $tcPo_2$ and saturation over a longer period proved to be helpful parameters for the setting of the oxygen flow in older infants with BPD. Regular short term hypoxemia tests in these infants are helpful as a lung function parameter. It is remarkable to find low $Po_2$ but sufficient saturation values in these chronic hypoxemic infants. This suggests a shift to the left in the oxygen dissociation curve with an increased oxygen affinity of the erythrocytes as is seen in newborns with a high fetal hemoglobin content.

In conclusion we prefer pulse oximetry to $tcPo_2$ measurement in hypoxemic neonates and infants for practical reasons and better reliability. During mild hypoxemia and normoxemia both parameters are reliable but during mild hyperoxemia $tcPo_2$ monitoring is preferable to pulse oximetry. Computer registration and analysis of $tcPo_2$ and pulse oximetry improves the quality of both parameters especially in older infants.

REFERENCES


1. M. Yelderman and W. New, Evaluation of pulse oximetry, Anesth. 59: 349-352 (1983).
2. E. S. Rome, E. K. Stock, W. A. Carlo and R. J. Martin, Limitations of transcutaneous $Po_2$ and $Pco_2$ monitoring in infants with broncho-pulmonary dysplasia, Pediatr. 74: 217-220 (1984).
3. A. R. van der Wiel, T. C. Jansen, H. N. Lafeber and W. P. F. Fetter, Multichannel recording and analysis of physiological data using a personal computer. This volume.
4. W. H. Northway, R. C. Rosan and D. Y. Porter, Pulmonary disease following respiratory therapy of hyaline-membrane disease. Broncho-pulmonary dysplasia, N Engl J Med. 276: 357-363 (1967).
5. S. S. Toce, P. M. Farrell, L. A. Leavitt, D. P. Samuels and D. K. Edwards, Clinical and roentgenographic scoring systems for assessing bronchopulmonary dysplasia, Am J Dis Child. 138: 581-585 (1984).

# CONTROL OF PEDICLE AND MICROVASCULAR TISSUE TRANSFER BY PHOTOMETRIC REFLECTION OXIMETRY

[1]H.P. Keller and [2]D.W. Lübbers

[1]Chirurgische Universitätsklinik, 8700 Würzburg, FRG
[2]Max-Planck-Institut für Systemphysiologie
4600 Dortmund 1, FRG

Postoperative control is of great importance for the success of the transfer of axial pattern flaps, replanted fingers, and freely transferred microvascular flaps (4). Therefore, we were interested in testing methods which should allow quick, non-invasive, transcutaneous evaluation of the healing process. For this purpose we investigated whether reflection oximetry of the skin, using a method proposed by Lübbers and Hoffmann (7), was easy to apply and gave sufficiently reliable results without needing recalibration.

## METHOD

The Oxyscan (Sigma Instruments, Berlin), a reflection photometer registering the extinction spectra of the skin, was used for the measurements. The spectrum was scanned by rotating interference filters at approximately 1900 r.p.m., covering a wavelength range between ca 500 nm and 600 nm. The reflection spectrum was displayed on an oscillograph screen. The measuring head comprised a flexible light guide, the fibers of which partly illuminated the skin and partly collected the reflected light.

The extinction spectrum of oxygenated hemoglobin has two absorption peaks. During deoxygenation, the amplitude of these two peaks becomes smaller, as does the distance between the two peaks. In an $O_2$ saturation range between 80 % and 40 %, the wavelength distance between the two peaks can be used as a rough estimation of the oxygen saturation of the cutaneous blood, c-Hb.$SO_2$ (1, 2, 6). The instrument allows measurements of the distances between the peaks by positioning two dots on the corresponding peaks. However, such an evaluation is not possible at saturation values below 40 %, because the spectra then flatten and the single peak of deoxygenated hemoglobin develops.

More accurate $O_2$ saturation values in the whole saturation range were obtained by multicomponent analysis of the reflection spectra using fully oxygenated and deoxygenated hemoglobin as reference spectra and taking into account spectral distortions caused by reflection in the skin (5, 7). The evaluation is done by a microprocessor (Sigma Instruments, Berlin). The c-Hb.$SO_2$ was determined with a precision of at least 5 %. The values of hemoglobin saturation are shown in a liquid crystal display

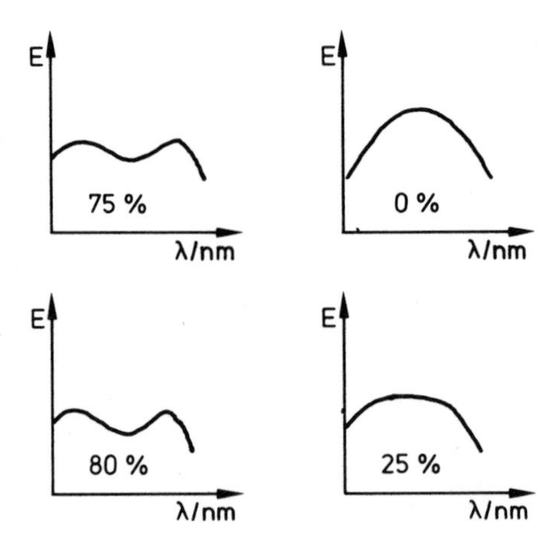


Fig. 1. Skin reflection spectra during revascularization of an axial pattern flap. Abscissa: wavelength; ordinate: extinction. Numbers: cutaneous hemoglobin oxygen saturation.

and printed on paper. The oscillograph display can be photographed with an instant camera.

RESULTS

Using the Oxyscan, the process of healing has been investigated on three groups of patients: 1.) patients with axial pattern flaps (n = 6), 2.) patients with free microvascular tissue transfer (n = 10). The cases included transfer of toes, a part of a toe, and skin or skin muscle flaps. The tissues were freely transferred by microsurgery or covering skin defects to restore function. 3.) patients with replanted extremities (n = 30) with normal healing. The c-Hb.$SO_2$ of replanted fingers was 40 - 60 %.

In all 46 cases the c-Hb.$SO_2$ measured and evaluated by the Oxyscan was found to be a reliable measurement of the healing process.

Fig. 1 demonstrates the revascularization of an axial pattern flap. A groin flap was transferred to cover the skin defect on the dorsum of the hand. Top left of Fig. 1 shows the reflection spectrum of the skin with 75 % c-Hb.$SO_2$. After the pedicle was pinched, the cutaneous hemoglobin became reduced (c-Hb.$SO_2$ = 0 %), (top right). As a sign of revascularization 14 days postoperative a spectrum with a hemoglobin oxygen saturation of 80 % was obtained (bottom right). At the same time, c-Hb.$SO_2$ in the middle of the flap was only 25 % (bottom left), clearly demonstrating the direction of vascularization.

Fig. 2 demonstrates a case in which a subtotal amputated big toe was replanted. During the first hours postoperative, c-Hb.$SO_2$ values measured on the replanted toe were 0 %. However, in the following time, the oxygenation of the toe became better and the c-Hb.$SO_2$ values increased to 30 % and later on to 52 %. The healing was complete.

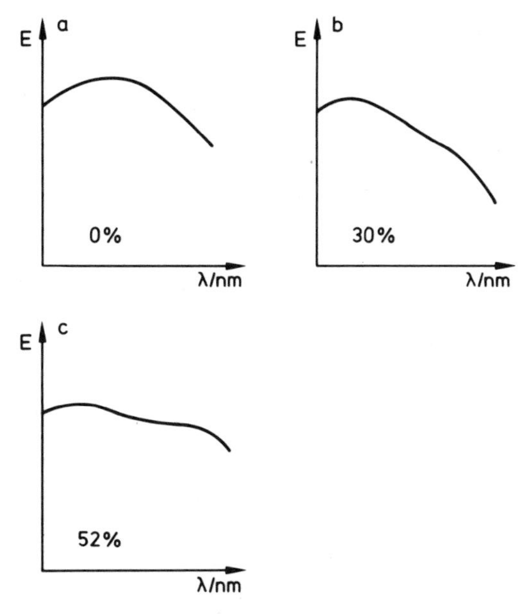


Fig. 2. Skin reflection spectra during replantation of a subtotally amputated big toe. (For explanation see Fig. 1.)

DISCUSSION

The flexible light guide permits easy application to the skin, even if the skin is moving slightly. Measurements can be quickly carried out anywhere and they are, above all, non-invasive and repeatable. They can be taken by nursing staff, if the appropriate limit reading is provided and the corresponding warning signal is determined.

The illuminated spot (4,5 mm in diameter) is sufficiently small in size to allow a regional investigation of the oxygen supply. The optical signal originates from skin capillaries as well as from deeper arterial and venous plexus. As a result the Hb saturation values on replanted fingers are in the range of 40 - 60 % and on axial pattern flaps and free transfer flaps in the range between 60 - 80 %. But, we also found values as low as 25 % c-Hb.$SO_2$ on skin flaps with perfectly normal healing. This relatively large range of values can be explained by the structure of the skin vasculature, which forms rather large venous pools (3). Consequently, c-Hb.$SO_2$ values as low as 25 % can still indicate an adequate oxygen supply. However, patients with such low c-Hb.$SO_2$ values should be carefully observed. Permanent low hemoglobin saturation during pedicle revascularization is an expression of reduced perfusion of the skin and a dangerous sign.

The intra- and postoperative results showed that with the described reflection oximetry, complications in the process of revascularization and healing could be detected fairly early, so that therapeutic measures may be taken.

REFERENCES


1. K.-H. Austermann, P. Tetsch, and D.W. Lübbers, Clinical experiences with reflection-oximetry on pedicle skin flaps, in: "Quantitative Measurement of Revascularization of Skin Transplants", D.W. Lübbers, H.P. Keller , and H.-R. Figulla, eds., Arzneim.-Forsch (Drug Res.) 29:1198 (1979)

2. H.-R. Figulla, K.-H. Austermann, and D.W. Lübbers, A new non-invasive technique for measuring $O_2$-saturations of hemoglobin using wavelength distances and its application to human and animal skin transplants, in: "Oxygen Transport to Tissue". Adv. Physiol. Sci., Vol. 25, A.G.B. Kovách, E. Dóra, M. Kessler and I.A. Silver, eds., Akadémiai Kiadó, Budapest: Pergamon Press (1981) pp. 317-318

3. R. Huch, A. Huch and D.W. Lübbers, "Transcutaneous $PO_2$", Georg Thieme, Stuttgart-New York (1981)

4. H.P. Keller and U. Lanz, Transcutaneous $PO_2$ measurements to evaluate blood flow in skin flaps, in: "Continuous Transcutaneous Blood Gas Monitoring", R. Huch and A. Huch, eds., Marcel Dekker, Inc., New-York-Basel (1983) pp. 673-679

5. G. Kortüm, "Reflexionsphotometrie", Springer-Verlag, Berlin-Heidelberg-New York (1969)

6. D.W. Lübbers, D. Piroth and R. Wodick, Bestimmung der Sauerstoffsättigung des Hämoglobins bei inhomogener Farbstoffverteilung. Naturwissenschaften, 57:42-43 (1970)

7. D.W. Lübbers and J. Hoffmann, Absolute reflection photometry at organ surfaces, in: "Cardiovascular Physiology: Heart, Perpheral Circulation and Methodology", Vol. 8, A.G.B. Kóvach, E. Monos and G. Rubányi, eds., Adadémiai Kiadó, Budapest, Pergamon Press, (1981) pp. 353-361

# COMPARISON OF IN-VIVO RESPONSE TIMES BETWEEN PULSE OXIMETRY AND TRANSCUTANEOUS $Po_2$ MONITORING

F. Fallenstein, P. Baeckert and R. Huch

Perinatal Research Unit and Department of Neonatology
University of Zürich (Switzerland)

## INTRODUCTION

Monitoring simultaneously transcutaneous $Po_2$ ($tcPo_2$) and oxygen saturation by pulse oximetry ($tcSo_2$), one sometimes has the impression that $tcSo_2$ responds slightly faster to changes in oxygenation than $tcPo_2$ does. In order to investigate this phenomenon more in detail, we analyzed simultaneous pulse oximetry and $tcPo_2$ recordings from artificially ventilated neonates. In a second step, we carried out a pilot study with healthy adult volunteers breathing a gas mixture with reduced $FiO_2$ for a short period of time.

## THE ANALYSIS OF THE NEONATAL STUDY

On 13 artificially ventilated neonates, $tcPo_2$ (HELLIGE Oxymonitor) and $tcSo_2$ (OHMEDA BIOX 3700 Pulsé Oximeter) were recorded simultaneously over 4-hour periods. Routine care in our neonatal ICU includes increasing $FiO_2$ prior to suctioning in order to avoid hypoxemia. This may cause moderate hyperoxemia of short duration.

From these recordings, we tried to analyze the hyperoxemic episodes. The onset of $tcPo_2$ increase could clearly be seen in most cases. But the $tcSo_2$ values were in general already close to full saturation under normal $FiO_2$, and a change in oxygenation - as it was introduced by increased $FiO_2$ - produced only a very small variation in $tcSo_2$. Thus it was practically impossible to determine the exact point on the pulse oximetry tracing where this parameter really began to increase. Consequently we were not able to find a systematic difference in response times.

## THE ANALYSIS OF THE ADULT STUDY

In this experimental series, again $tcPo_2$ and $tcSo_2$ were recorded simultaneously, this time on 5 healthy adult volunteers who gave their informed consent. The equipment was the same as in the neonatal study. The $tcPo_2$ electrode was applied to the forearm, the pulse oximetry probe to a finger on the same arm. At the beginning of the experiment, the test person was breathing normal room air. After the $tcPo_2$ settled to a steady state, $FiO_2$ was changed by mask breathing of a gas mixture containing 13.3% oxygen.

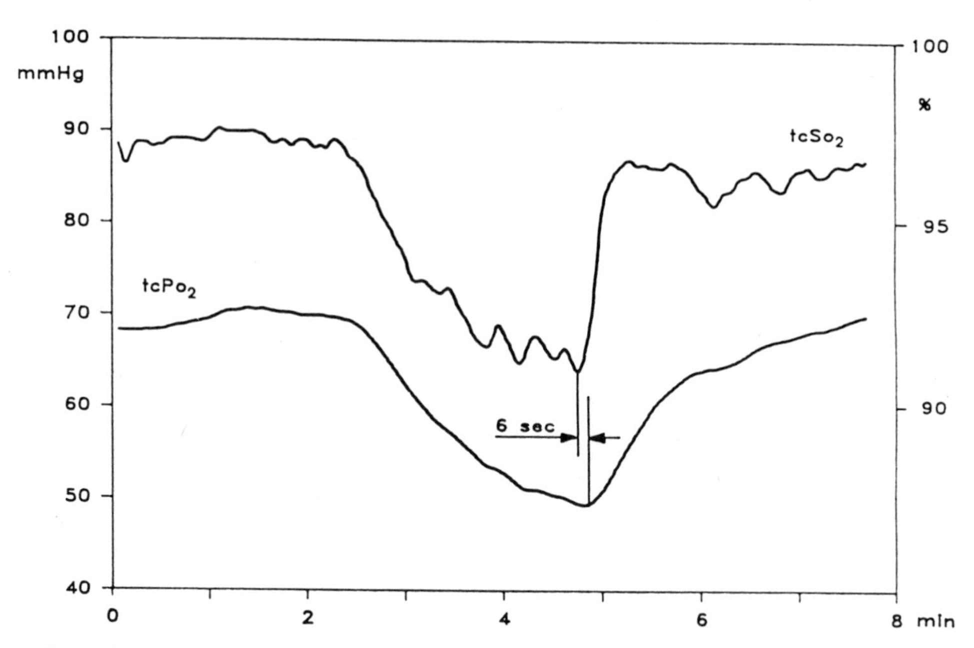


Fig. 1: Simultaneous measurement of $tcSo_2$ (pulse oximetry) and $tcPo_2$. Oxygenation decrease was effected by reducing oxygen fraction from 21% (room air) to 13.3% for 2 1/2 minutes.

This period was always 2 1/2 minutes long and $tcSo_2$ usually dropped from 97% down to 90%. Then the test person again breathed room air and the recording was continued until $tcSo_2$ and $tcPo_2$ reached the previous steady state level.

Fig. 1 shows the results of one of these experiments. This example represents what could be seen in all the recordings. Once $FiO_2$ was reduced, both pulse oximetry and $tcPo_2$ signals started to decrease almost parallel to each other until the lowest values were reached.

This has a simple physiological explanation: Coming from a normoxemic state, enough oxygen is stored in the lungs, the blood and the tissues so that the oxygenation will decrease gradually even if the oxygen supply by ventilation is suddenly reduced. The transcutaneous $Po_2$ signal, known as a relatively slow parameter, can still follow such changes.

However, after return to normal air breathing, pulse oximetry and $tcPo_2$ curves behave differently: $tcSo_2$ recovers already after a few breaths, whereas $tcPo_2$ comes back to the normal level with the usual time constant of about 2 minutes.

Furthermore, in this particular case one can see a 6-second delay between the onset of $tcSo_2$ increase and $tcPo_2$ increase. This has a physical explanation which can be demonstrated by the response function to a sudden stimulus change in a theoretical diffusion model. In Fig. 2a, with a relatively short time constant, the $tcPo_2$ signal completely reaches a steady state within the hypoxemic period and responds almost instantaneously when the oxygenation stimulus switches back to high. In Fig. 2b however, with a four-fold longer time constant, the $tcPo_2$ signal is still decreasing when the oxygenation is increased again and there is a certain time delay until the decreasing trend is stopped.

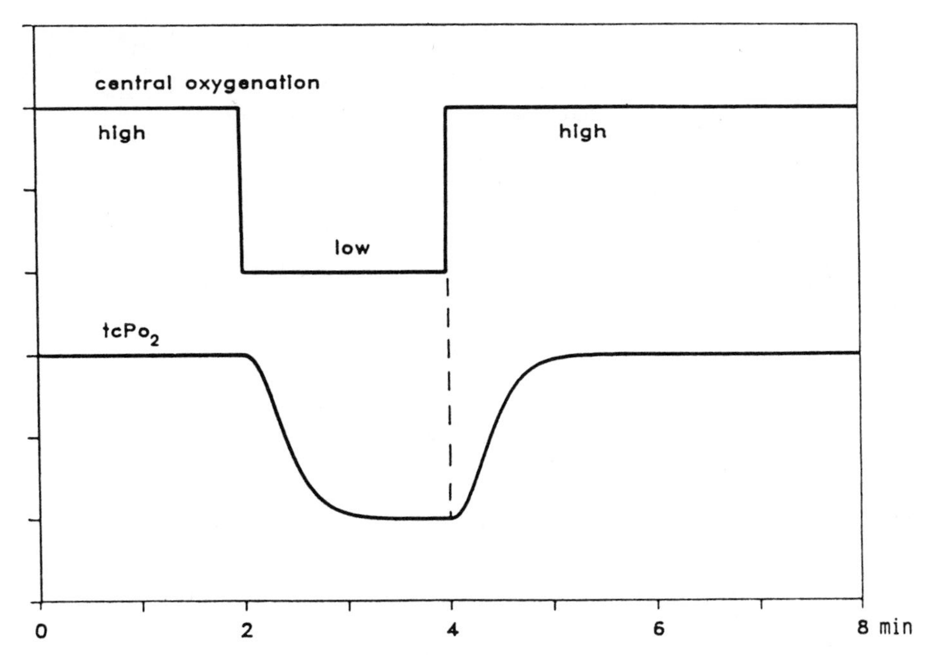


Fig. 2a: Theoretical response of $tcPo_2$ to a sudden change in central oxygenation. A short time constant results in a steady state of the $tcPo_2$ signal within the low oxygenation period.

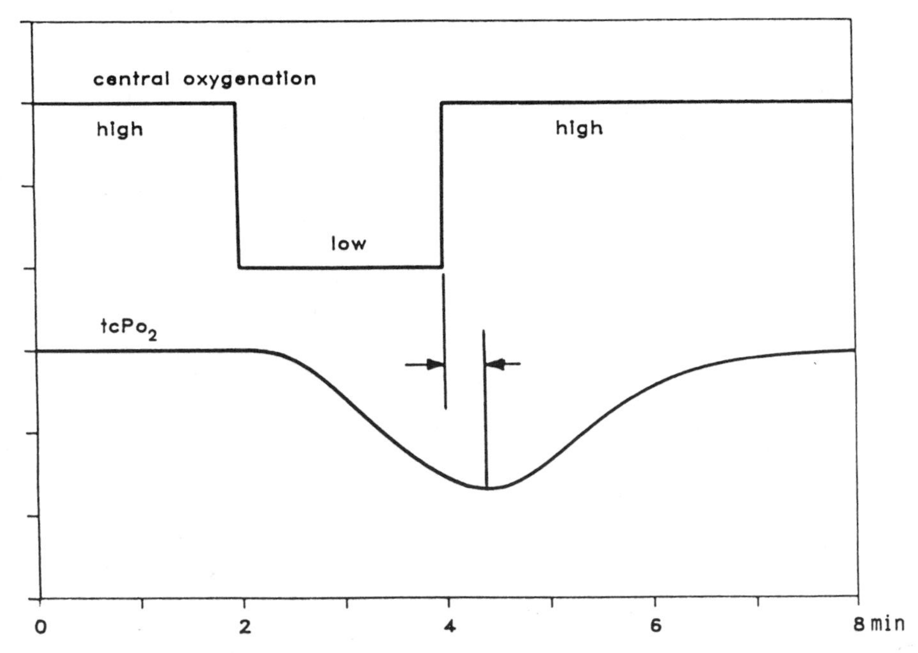


Fig. 2b: The same as in Fig. 2a, but with a four-fold increased time constant. In this case, $tcPo_2$ is still decreasing at the end of low oxygenation and responds with a time delay.

In order to describe the different response characteristics of pulse oximetry and $tcPo_2$, only the increasing sections after the low $FiO_2$ period were evaluated. As a measure for the response time, the time distances from the lowest parameter value to 50% of the total change were determined (assuming exponential curves, the 95% response time is approximately 4 times the 50% response time). In addition to this, the time delays between pulse oximetry and $tcPo_2$ were estimated and it was noted whether or not the oxygenation reached a steady state during the low $FiO_2$ period.

Table 1 gives the individual results and mean values from the 5 experiments. Only one subject (MR) had a steady state within the low $FiO_2$ period and also did not show a time delay between $tcSo_2$ and $tcPo_2$, which is in accordance with the theoretical considerations described above.

Table 1: Individual results and mean values of response times

| Subject | FL | MR | PT | NT | KR | mean |
|---|---|---|---|---|---|---|
| 50% time constant ($tcSo_2$) sec | 11.5 | 9.5 | 11.0 | 11.5 | 12.5 | 11.2 |
| 50% time constant ($tcPo_2$) sec | 34 | 42 | 33 | 133 | 46 | 57.6 |
| t ($tcPo_2$ - $tcSo_2$) sec | 6 | 0 | 11 | 12 | 12 | 8.2 |
| steady state before increase | no | yes | no | no | no | |

CONCLUSION

We found a difference in response times between pulse oximetry and transcutaneous $Po_2$. On the average, $tcSo_2$ responds about 5 times faster than $tcPo_2$. In practice, this can only be seen when the central oxygenation changes rapidly, e.g. in recovering from a moderate hypoxemia.

It should be noted that the data presented here are of course only valid for the BIOX 3700 instrument and only for the data available at the recorder output connector. (For the digital display, the user can choose between two different time constants). Pulse oximeters from other manufacturers may have other internal time constants in signal processing.

# ERRORS IN THE MEASUREMENT OF CO-HEMOGLOBIN IN FETAL BLOOD BY AUTOMATED MULTICOMPONENT ANALYSIS.

Peter Tuchschmid

Division of Neonatology
Department of Pediatrics
University of Zurich
CH-8032 Zurich

Since pulse oxymetry was introduced in clinical medicine recently, two non-invasive techniques to measure oxygen concentration in human blood are available. Each of these methods offers advantages and disadvantages which may be evaluated by direct comparison but in order to assess their precision they have to be compared against a standard method.

Whereas transcutaneous $pO_2$ is compared to blood $pO_2$-measurement, pulse oxymetry must be standardized against a technique measuring hemoglobin oxygen saturation, which requires spectroscopic evaluation of oxy- and deoxy- or total hemoglobin concentration in a blood sample. Whereas older instruments were designed to measure total hemoglobin- and oxy-hemoglobin concentrations only, a new generation of equipment allows the simultaneous evaluation of oxy-, deoxy-, carbonmonoxy- and met-hemoglobin concentrations. These instruments are based on the principle of spectroscopic multicomponent analysis which makes use of the fact that the total absorbance (A) of a mixture of a finite number of components (n) equals the sum of the absorbances of each component, the latter being the product of its concentration (c) and its extinction coefficient (E) according to Lambert-Beer's law.

$$A_{(1..n)} = c_1 \cdot E_1 + c_2 \cdot E_2 + \ldots \quad \ldots + c_n \cdot E_n$$

This law implies that for a given number of components an equal number of absorbances must be available to solve the matrix shown in Fig. 1.

$$A^{L_1} = c_1 \cdot E_1^{L_1} + c_2 \cdot E_2^{L_1} + c_3 \cdot E_3^{L_1} + c_4 \cdot E_4^{L_1}$$

$$A^{L_2} = c_1 \cdot E_1^{L_2} + c_2 \cdot E_2^{L_2} + c_3 \cdot E_3^{L_2} + c_4 \cdot E_4^{L_2}$$

$$A^{L_3} = c_1 \cdot E_1^{L_3} + c_2 \cdot E_2^{L_3} + c_3 \cdot E_3^{L_3} + c_4 \cdot E_4^{L_3}$$

$$A^{L_4} = c_1 \cdot E_1^{L_4} + c_2 \cdot E_2^{L_4} + c_3 \cdot E_3^{L_4} + c_4 \cdot E_4^{L_4}$$

Fig. 1: Matrix of Lambert-Beer's law for absorbances of a quarternary mixture:

$A^L$ = absorbance of component n at wavelength $L_i$

$c_n$ = concentration of component n

$E_n^L$ = extinction coefficient of component n at wavelength $L_i$

Fig. 2A shows absorbance spectra of hemoglobin A and the wavelengths utilized by instruments used to measure oxy-, deoxy-, CO- and methemoglobin in blood. The location of some measuring wavelengths in the steep slopes of spectra (e.g. 535 or 585 nm) is inevitable because of practical limitations (interference filters availability). The implication of this technical design is that the method becomes very sensitive to small spectral shifts of peaks towards higher or lower wavelengths.

This type of equipment was used to compare HbCO in human cord blood with maternal HbCO levels by Bureau et al. (1982). Validity of the results was discussed by Huch et al. (1983) because of arterio-venous differences of HbCO in cord blood which were physiologically difficult to explain.

It was shown by Zwart et al. (1981), Huch et al. (1983) and Cornelissen et al. (1983) that HbCO measurements were influenced by the oxygen saturation of the sample in fetal but not in adult blood and that arterio-venous differences of HbCO in cord blood were due to different oxygen saturations rather than to different HbCO contents.

It was argued that small spectral differences between fetal and adult hemoglobin derivatives (oxy, deoxy, CO and met) might be responsible for the observed influence of oxygen saturation on the measurement of CO hemoglobin concentration. We therefore measured the difference spectra of pure oxy-, deoxy-, CO- and met- derivatives of fetal versus adult hemoglobin. Results are given in Fig. 2B and show the existence of such differences in the order of 3 to 5 % of the total absorbance.

Applying the spectral differences found between HbF and HbA in Fig. 2B to the matrix shown in Fig.1, as outlined in Fig. 3, the influence of oxygen saturation on the HbCO reading for adult and fetal hemoglobin can be simulated.

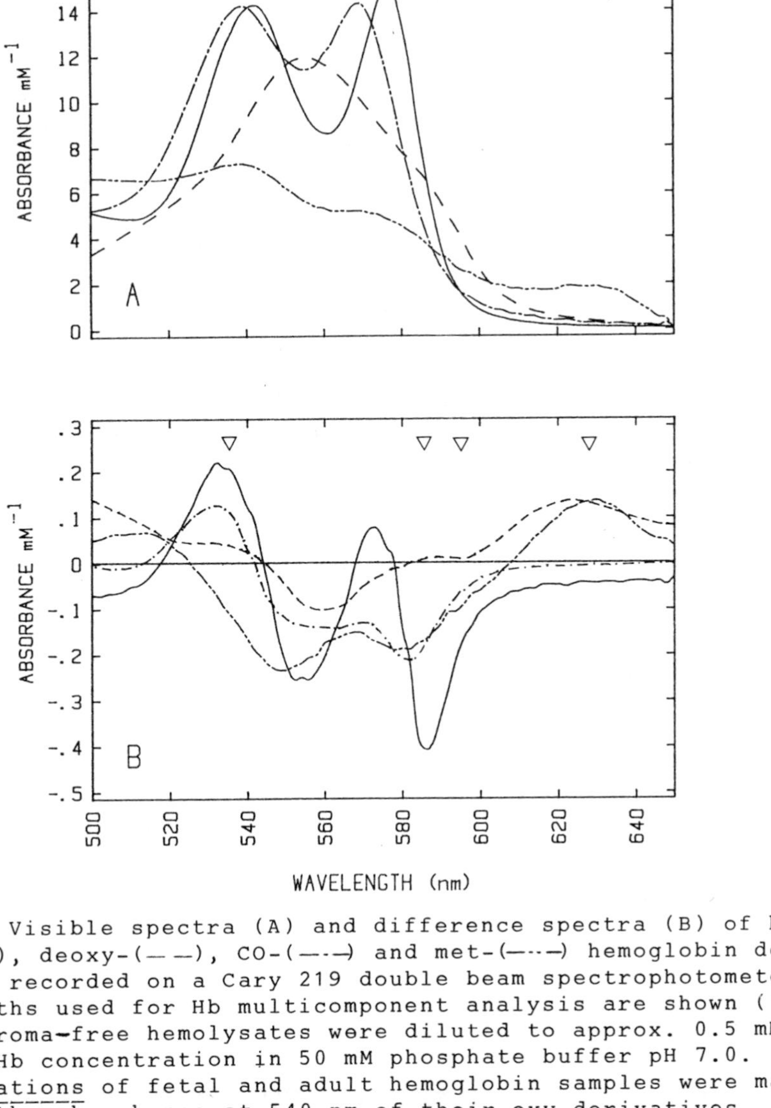


Fig. 2A: Visible spectra (A) and difference spectra (B) of human Oxy-(——), deoxy-(— —), CO-(—·—) and met-(—··—) hemoglobin derivatives, recorded on a Cary 219 double beam spectrophotometer. Wavelengths used for Hb multicomponent analysis are shown ( ▽ ). Fresh, stroma-free hemolysates were diluted to approx. 0.5 mM monomer Hb concentration in 50 mM phosphate buffer pH 7.0. Concentrations of fetal and adult hemoglobin samples were matched by the absorbance at 540 nm of their oxy derivatives.
HbF was prepared from cord blood containing more than 90% HbF.
HbA was obtained from nonsmoking individuals.
$HbO_2$ was obtained by exposure of the Hb solution to pure $O_2$.
HbDeox was prepared by treatment of $HbO_2$ with a few crystals of sodium dithionite per ml. Spectra were recorded immediately and repeated after conversion of the HbDeox sample to HbCO for comparison to HbCO obtained directly from $HbO_2$.
HbCO was prepared by exposure of $HbO_2$ and HbDeox respectively to pure carbonmonoxide for 5 to 10 minutes.
HbMet was prepared by addition of a few crystals of potassiumferricyanide per ml of dilute $HbO_2$.
All samples were finally converted to HbMetCN and checked for Hb-concentration to exclude protein loss during derivatization.

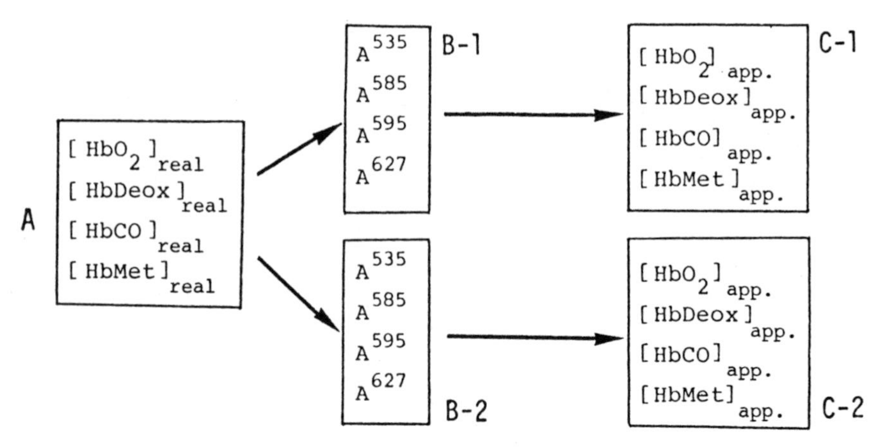


Fig. 3: Computer simulation of multicomponent analysis of a quarternary mixture of hemoglobin derivatives.

A: Concentrations of fetal- or adult hemoglobin to be simulated.

B: Absorbances calculated for 4 wavelengths using the matrix given in Fig. 1 and extinction coefficients for adult hemoglobin (B-1) and fetal hemoglobin (B-2) respectively.

C: Concentrations of hemoglobin derivatives calculated back from absorbances in B-1 and B-2 respectively using the inversion of the matrix from Fig. 1 and with adult extinction coefficients only. Results correspond to the readings expected from an instrument using multicomponent analysis based on adult extinction coefficients.

The results show that the influence of oxygen saturation on HbCO readings found by measuring cord HbCO concentrations with analyzers that use spectroscopic multicomponent analysis is fully explained by small spectral differences between adult and fetal hemoglobin. Although the Corning CO-oximeter 2500 uses a 7-wavelengths matrix, the dependence of HbCO determination on oxygen saturation in fetal blood was comparable to the one found for the IL-282 instrument. For the Corning CO-oximeter 2500 the existence of a similar error caused by HbMet or HbS was further analysed. Artefacts, such as noted for fetal CO-hemoglobin, were absent for HbS-CO as well as for the fetal HbMet derivative.

Fig. 4 compares the computer-calculated influence of oxygen saturation on HbCO using a 4-wavelength matrix with values measured on a IL-282 (Instrumentation Laboratory, Lexington, Mass. USA)and on a CO-oxymeter 2500 (Ciba-Corning, Diagnostics, Medfield, Mass. USA).

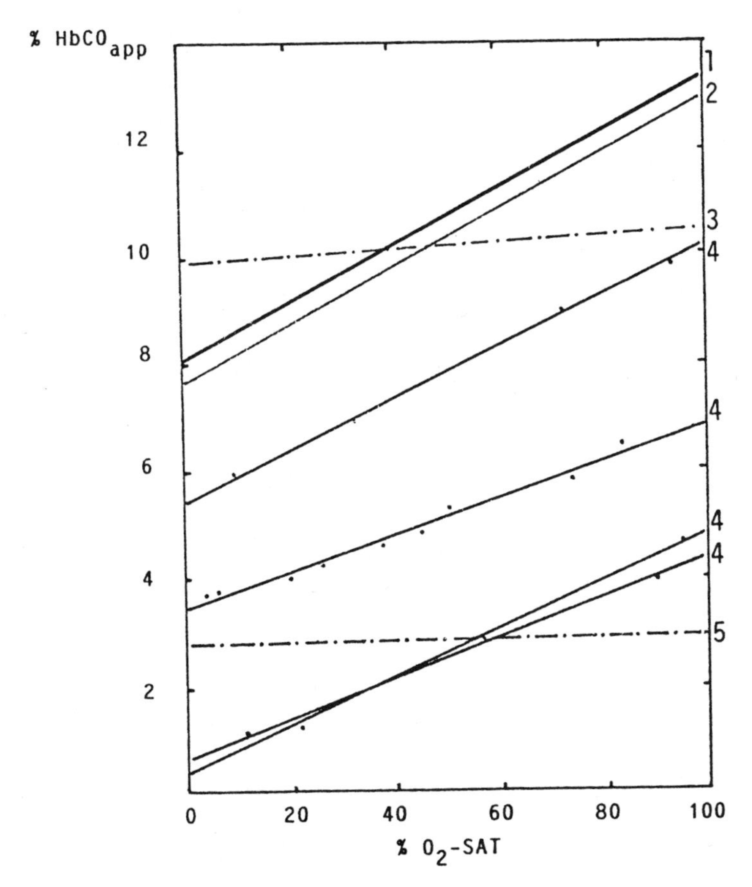


Fig. 4: Comparison of computer simulated influence of oxygen saturation on HbCO measurement (1) with data obtained by readings obtained by an IL-282 (2,3) and a Corning CO-oximet 2500 (4,5), for fetal (2,4) and adult (3,5) hemoglobin respectively.

Part of this work was supported by Ciba-Corning Diagnostics, Medfield, Mass., USA and Giessen, GFR.

Abreviations

HbA : Adult hemoglobin
HbF : Fetal hemoglobin
HbS : Sickle-cell hemoglobin
$HbO_2$ : Oxy-hemoglobin
HbMet : Met-hemoglobin
HbMetCN: Cyanmet-hemoglobin

REFERENCES

Bureau, M.A., Monette, J., Shapcott, D., Pare, C., Mathieu, J.L., Lippe, J., Blovin, D., Berthiaume, Y., and Begin, R., 1982, Carboxyhemoglobin concentration in fetal cord blood and in blood of mothers who smoked during labor, Pediatrics, 69: 371.

Cornelissen, P.J.H., van Woensel, C.L.M., van Oel, W.C., and de Jong, P.A, 1983, Corrective Factors for Hemoglobin Derivatives in Fetal Blood, as Measured by the IL 282 CO-Oxymeter, Clin. Chem., 29: 1555.

Huch, R., Huch, A., Tuchschmid, P., Zijlstra, W.G., and Zwart, A., 1983 Carboxyhemoglobin Concentration in Fetal Cord Blood. Pediatrics, 71: 461

Zwart, A., Buursma, A., Oeseburg, B., and Zijlstra, W.G., 1981, Determination of hemoglobin derivatives with the IL 282 CO-Oxymeter as compared with a manual spectrophotometric five-wavelength method., Clin. Chem., 27, 1903.

## A SIMPLE METHOD FOR HbF ANALYSIS

Ursula von Mandach, Peter Tuchschmid,
Albert Huch and Renate Huch

Perinatal Physiology Unit
Department of Obstetrics
University of Zurich
CH-8091 Zurich

### INTRODUCTION

It has been discussed by several authors[1-4] that oxygen saturation ($SO_2$) determined by CO-oxymeters are affected by fetal hemoglobin (HbF). This artefact depends on the wavelengths used to collect absorbance data and/ or the extinction coefficients used to calculate the hemoglobin derivatives. Increased HbF and oxyhemoglobin ($HbO_2$) in human blood elevate the carboxyhemoglobin (HbCO) percentage and cause falsely underestimated $SO_2$ values.

With the introduction of pulse oxymetry, i.e. transcutaneous oxygen saturation, in clinical medicine values obtained by this technique are compared with simultaneously drawn blood samples and analyzed by CO-oxymetry. In order to compensate for the effect of fetal hemoglobin, this must be determined before applying a correction such as that of Tuchschmid[5].

It is the aim of the present paper to indicate a new simple method for the analysis of HbF.

### PRINCIPLES

The oxygen saturation is defined as the ratio between the concentration of oxyhemoglobin and the concentration of deoxy- plus oxyhemoglobin (see Fig 1).

With the CO-oxymeter, it is in fact not the oxygen saturation that is determined, but the fraction of oxyhemoglobin. This fraction is only equal to $SO_2$ if the hemoglobin derivatives carboxy-, met- and sulfhemoglobin are zero. If carboxyhemoglobin increases, $SO_2$ is underestimated.

In cases where carboxyhemoglobin occurs naturally as in the blood of newborns of smoking mothers, this method for $SO_2$ determination is reliable. If HbCO is apparently elevated due to a systematic error in the

$$\text{❶ } SO_2 = \frac{[HbO_2]}{[Hb] + [HbO_2]} \qquad SO_2 = \text{oxygen saturation}$$

$$\text{❷ } F\,HbO_2 = \frac{[HbO_2]}{[Hbtot]} \qquad F\,HbO_2 = HbO_2 \text{ Fraction}$$

$$[Hbtot] = [Hb] + [HbO_2] + [HbCO] + [MetHb] + [SHb]$$

❷ = ❶ if HbCO, HbMet, HbS = 0

Hbtot = total hemoglobin
Hb = deoxyhemoglobin
$HbO_2$ = Oxyhemoglobin
HbCO = Carboxyhemoglobin
MetHb = Methemoglobin
SHb = Sulfhemoglobin

Fig 1
Definitions of oxygen saturation ($SO_2$) and fraction of oxyhemoglobin ($FHbO_2$).

presence of HbF, a correction must be involved. As Tuchschmid[5] shows, the analysis of carboxyhemoglobin in the presence of fetal hemoglobin can be corrected either by substitution of adult extinction coefficients with fetal ones or by a nomographic technique. Both types of correction require the percentage of fetal hemoglobin present in the blood specimen which is normally unknown.

We have developed a simple method for the estimation of HbF in human blood which utilizes the described systematic error.

HbCO apparently increases with a rise in $HbO_2$ and the degree of this increase is related to the actual HbF percentage. The present method instead measures the slope of the increase and deduces from this the HbF percentage. As there is a linear relationship between the oxyhemoglobin and the measured carboxy derivative, it is enough to establish the slope by two points.

## METHOD AND RESULTS

To test the reliability of the following described method, five experiments were performed on five different days each with fresh adult and fetal blood. All measurements were made in duplicate.

Adult and fetal blood were mixed to obtain samples with zero to hundred percent fetal hemoglobin. All samples were deoxygenated in an open

open tonometer by flushing with nitrogen and 5% $CO_2$. After 30 minutes the first measurements of HbCO and $HbO_2$ were made with the CO-oxymeter. Oxygen was then added stepwise and HbCO and $HbO_2$ measurements were repeated. In the presence of HbF there was an apparent increase in HbCO with increasing $HbO_2$, as an example representative for our five cases demonstrates in Fig 2.

As will be seen there is no influence of increased oxygen saturation on the percentage of HbCO in adult blood.

The slopes of the five lines were then correlated with the corresponding fetal hemoglobin percentages and a linear relationship was found. Fig 3 shows this linear correlation (r=0.995) where the slope values are means of 5 slopes calculated for each percentage of fetal hemoglobin.

Once the relation between HbCO and HbO is established by measuring at two different $HbO_2$ percentages, the slope is entered in Fig 3 and the corresponding HbF is read off.

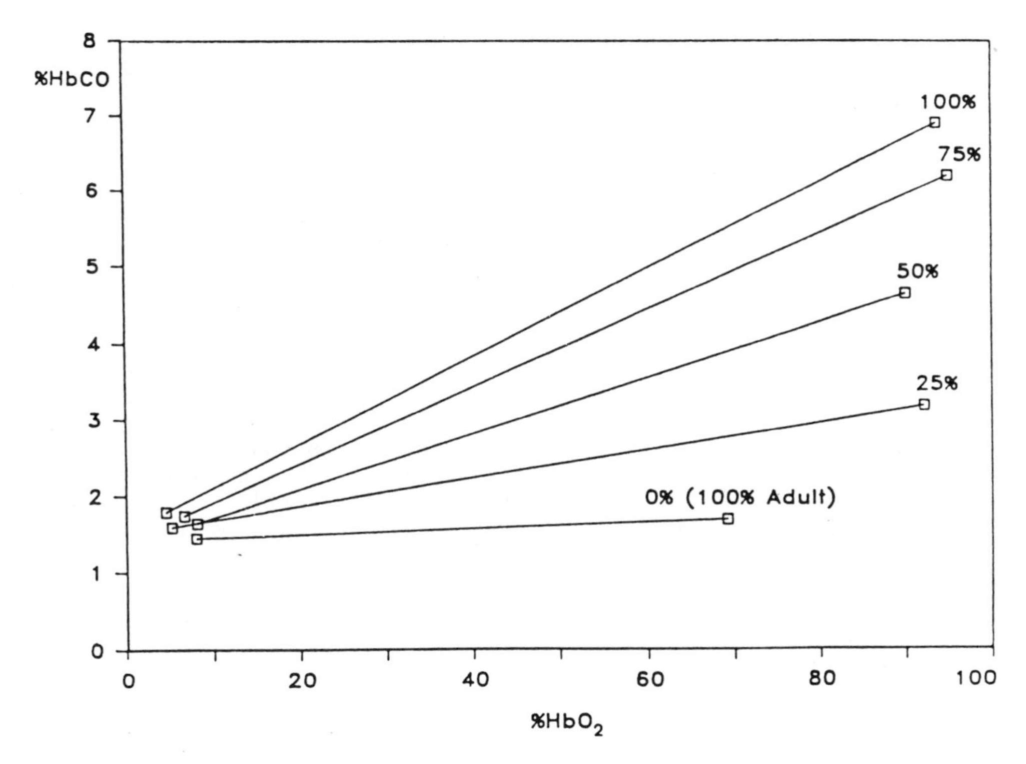


Fig 2
Carboxyhemoglobin (HbCO,%) versus oxyhemoglobin ($HbO_2$,%) measured in blood samples with different percentages of fetal hemoglobin (0-100% HbF).

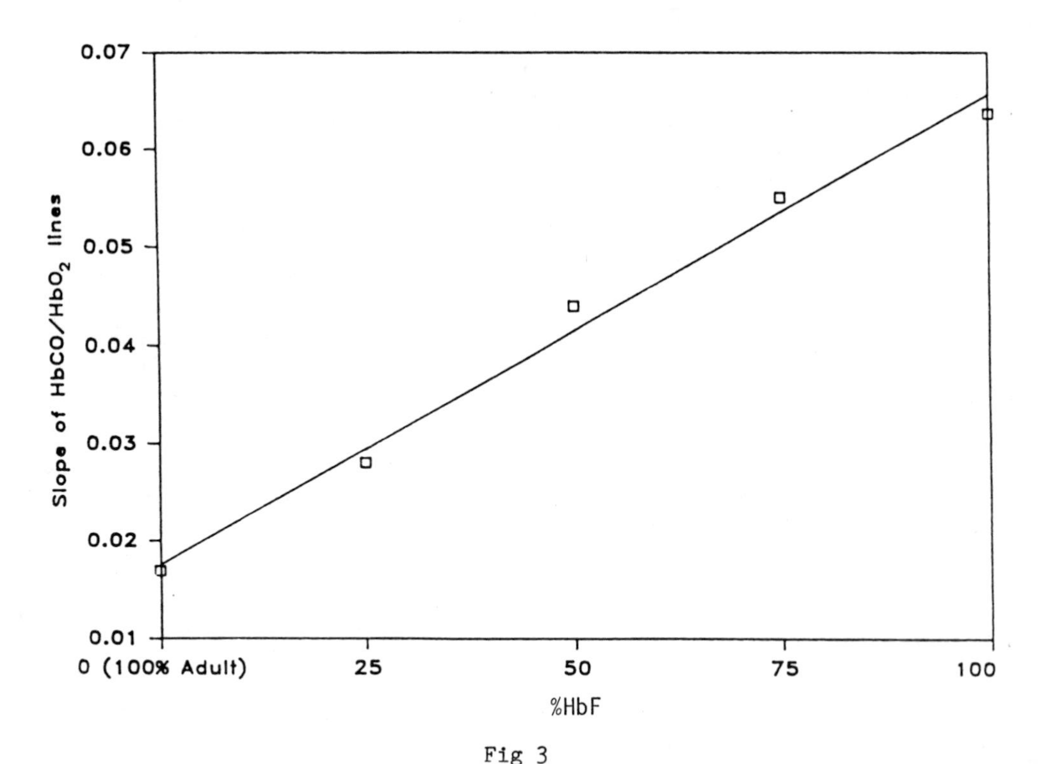


Fig 3
Relationship between slopes calculated from the relation between carboxyhemoglobin (HbCO) and oxyhemoglobin ($HbO_2$)- before and after deoxygenation- and percentages of fetal hemoglobin (%HbF).

## SUMMARY


Spectrophotometric methods with CO-oxymeters for measurements of carboxyhemoglobin and/or oxygen saturation in human blood include a systematic error depending on the percentage of fetal hemoglobin. It is of clinical importance to estimate the fetal hemoglobin to correct HbCO and $SO_2$ values respectively. The described method is simple and less time consuming than conventional methods like HPLC or electrophoresis. The two measurements of oxy- and carboxyhemoglobin in the same blood sample with different oxygen saturation are needed for the estimation of the HbF and can be performed, including the deoxygenation procedure, in about 40 minutes.


## REFERENCES


1. A. Zwart, A. Buursma, B. Oeseburg and W. G. Zijlstra, Determination of hemoglobin derivatives with the IL 282 CO-oxymeter as compared with a manual spectrophotometric five-wavelength method, Clin Chem 27:1903 (1981)
2. P. J. H. Cornelissen, C .L. M. van Woensel, W. C. van Oel and P. A. de Jong, Correction factors for hemoglobin derivatives in fetal blood as measured with the IL 282 CO-oxymeter, Clin Chem 29:1555 (1983)

3. R. Huch, A. Huch, P. Tuchschmid, W. G. Zijlstra and A. Zwart, Carboxyhemoglobin concentration in fetal cord blood, Pediatrics 71:461 (1983)
4. M. S. Jennis and J. L. Peabody, Pulse oximetry-an alternative method for the assessement of oxygenation in newborn infants, Pediatrics in press
5. P. Tuchschmid, Errors in the measurement of CO-hemoglobin in fetal blood by automated multicomponent analysis: cause and correction, This volume

# SKIN BLOOD FLOW

# CAPILLARY BLOOD PRESSURE

J.E. Tooke and S.A. Williams

Department of Physiology
Charing Cross and Westminster Medical School
Fulham Palace Road, London. U.K.

## SUMMARY

Human capillary blood pressure may be measured directly in nailfold capillaries of the fingers and toes. By applying servonulling pressure measuring techniques rapid fluctuations in capillary pressure may be recorded, opening the way to a greater understanding of capillary pressure control in health and disease. The estimation of mean pressure which may be accomplished manometrically is of value in determining the mechanism of oedema, identifying the site of raised peripheral resistance in disease states, evaluating the effects of vasoactive drugs on peripheral resistance, and investigating haemodynamic abnormalities associated with microangiopathies. Capillary pulse waveform analysis, made possible by servonulling techniques and computer analysis has already revealed important changes in hypertension.

## INTRODUCTION

Capillary blood pressure is a key determinant of the rate and direction of fluid flux across the capillary wall. An understanding of tissue fluid balance and its disturbance in oedema states relies therefore upon a knowledge of the quantity. In addition the capillary pressure gradient from arteriolar to venular end is the driving force that determines capillary flow rate. Using a microinjection technique that involved cannulation of single nailfold capillaries Landis[1] laid the foundation of our understanding of human capillary pressure control. This paper summarizes the modern application of this technique and the information that has been derived from its use.

## METHODS

Direct techniques rely on the introduction of glass microcannulae, tip diameter $\leq 5$ microns, into the lumen of nailfold capillary loops of the fingers or toes. The cannula holder is held within a Leitz micromanipulator and introduced at an angle of $40^{o}$ to the plane of the digit either under

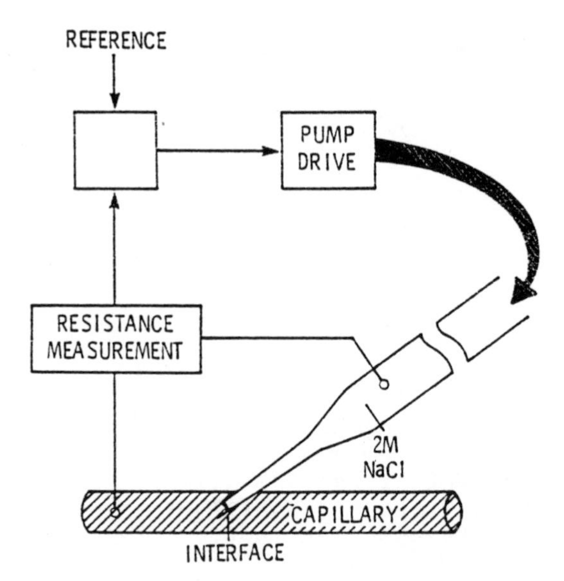


Figure 1. The principle of the servonulling system: Penetration of the pipette tip by capillary contents results in a rise in electrical resistance across the pipette tip. This leads to generation of a counter pressure equal to capillary pressure that restores the interface to the starting position. The counter pressure is measured with a conventional pressure transducer.

direct stereomicroscopic vision or using television microscopy to visualise the capillary loops. Pressure measurement may be either i) accomplished manometrically which permits the estimation of mean pressure or ii) performed using a dynamic servonulling technique that allows faithful registration of rapid changes in capillary pressure. The principle of the latter technique is illustrated in Figure 1.

## RESULTS

### Physiological studies

Mean capillary pressure is higher in the arterial limb of the capillary than the venular limb, such that these values straddle plasma oncotic pressure (particularly when allowance for venous pressure is made)(see Table 1).

Table 1. Capillary pressure obtained using a servonulling pressure measuring system (with the hand at the level of the sternal angle).

| | MEAN±SEM | RANGE | |
|---|---|---|---|
| Arterial limb | 37.7±3.7(n=12) | 17-60 | mmHg |
| Apex of capillary loop | 19.4±1.0(n=24) | 9-33 | mmHg |
| Venous limb | 14.6±0.5(n=38) | 5-22 | mmHg |

(Plasma oncotic pressure is approximately 24 mmHg).

Manometric studies of capillary pressure have demonstrated that the major determinants of mean capillary pressure are (i) skin temperature and (ii) the vertical distance between the heart and the capillary. Locally heating the skin or indirect heating to release sympathetic tone results in a fall in pre-capillary resistance and an anticipated rise in capillary pressure.

Capillary pressure in the foot of the standing subject reflects the hydrostatic pressure of the column of blood between the heart and the extremity, but nevertheless it does not rise as much as would be anticipated from a summation of the pressure recorded at heart level and vertical height . This suggests that an increase in the ratio of pre- to post-capillary resistance occurs in the dependent extremity, a response which operates as an important oedema-preventing mechanism in health.

The development of the servonulling method has not only confirmed the data obtained manometrically but the capacity to follow dynamic fluctuations in capillary pressure has revealed the marked variability of the quantity.

By simultaneously measuring heart rate, respiratory rate and capillary pressure it has been possible to demonstrate a respiratory fluctuation in capillary pressure (Figure 2). It is likely that this represents respiratory fluctuation in venous pressure as it is abolished if venous pressure is reduced to zero or artificially elevated by a venous occlusion cuff.

Faster fluctuations in time with the cardiac cycle are also visible, pulsatility being more marked on the arteriolar side of the capillary loop. The pulse waveform shows remarkable similarity to the arterial waveform and and occurs with a similar 200 ms delay after the 'R' wave of the ECG (Figure 3).

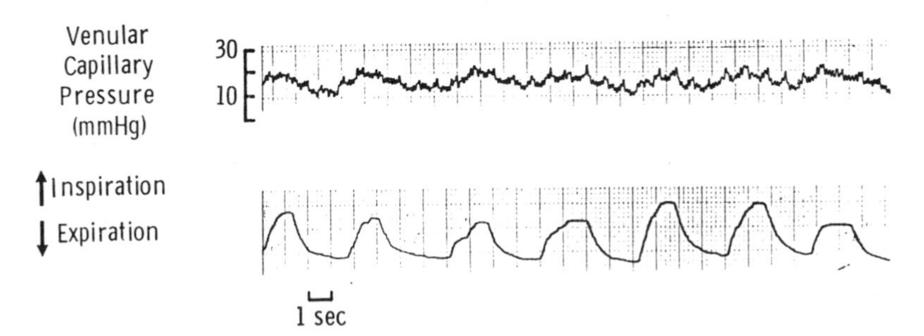


Figure 2. Synchronous recording of venular limb capillary pressure and respiratory rate. Respiratory fluctuations in capillary pressure are evident as well as cardiac pulsatility.

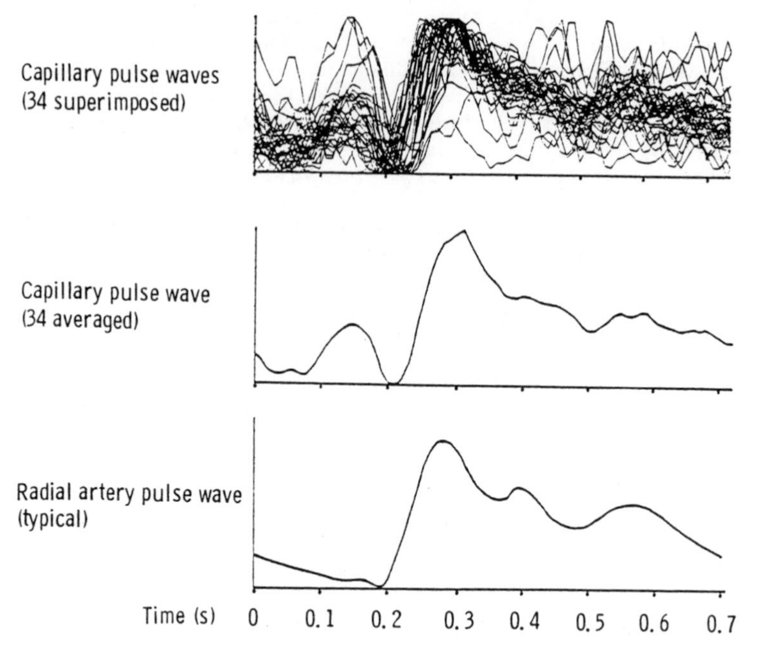


Figure 3. Computerized superimposition of numerous capillary pulse waveforms and resulting averaged normalized waveform compared with radial artery pulse.

Physiological fluctuations may be induced upon these background rhythms(Figure 4).

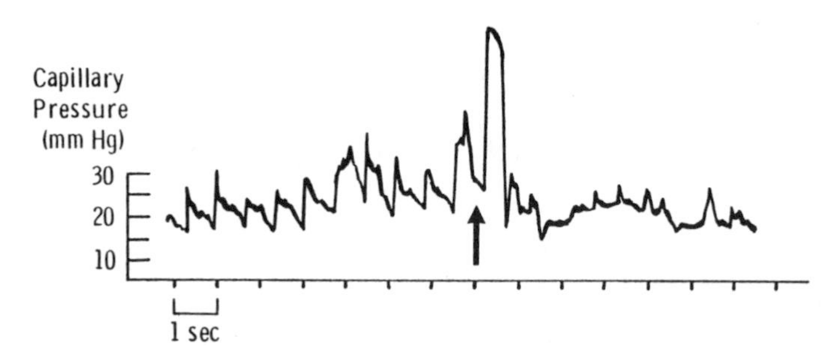


Figure 4. The response of capillary pressure to a sudden increase in sympathetic tone: The arrow corresponds to a sudden loud noise that is followed by a movement artefact, then a reduction in absolute pressure and pulse pressure.

Pathophysiological studies

The information derived from capillary pressure studies in disease states has been summarised in a recent review[5]. It has been demonstrated that capillary pressure is elevated in glomerulonephritis, cardiac failure as well as in late pregnancy, affording on explanation for the oedema common to these conditions.

Capillary pressure is reported as normal in arterial hypertension, providing direct evidence that the increased peripheral resistance observed in this disorder occurs at arteriolar level. However the implications of this finding are so great that further studies are required. In Raynaud's phenomenon arterial limb capillary pressure is low in comparison with normal women, who in turn exhibit lower pressures than normal men, an observation that possibly explains the female preponderance of the condition.

In diabetes it has been shown that toe capillary pressure is elevated compared with normal subjects, whereas there is a failure of maximum vasodilatation under some circumstances.

Little use has thus far been made of the tremendous potential of the servonulling method in clinical studies. We are currently engaged in the study of dynamic capillary pressure in hypertensive and control subjects with the aims of determining i) clarifying the issue as to whether capillary pressure is raised ii) determining whether pulse waveform propagation is normal and iii) whether maximum capillary pressure is limited by a 'fixed' pre-capillary resistance. Findings to date suggest that the systolic upstroke occurs earlier in hypertension although the dichrotic notch exhibits normal timing ie. the pulse wave is 'broadened'. Clearly such changes may have important consequences for moment to moment fluid flux and tissue nutrition.

## DISCUSSION

The recent application of the servonulling technique to the measurement of human capillary pressure has emphasized the variability of the quantity, a variability that could not be appreciated with the static manometric system of measurement. Though primarily a research tool, capillary pressure measurement has added to our understanding of oedema formation and oedema-preventing mechanisms, the site and reversibility of increased peripheral resistance as well as enhancing knowledge of the control of skin blood flow in general. Application of the servonulling technique to disease states promises to provide a much greater insight into the disturbance of microvascular control in such conditions.

## REFERENCES

1. Landis E M. Microinjection studies of capillary blood pressure in human skin. Heart. 15: 209-228 (1930).
2. Mahler F, Muheim M H, Intaglietta M, Bollinger A, Anliker M. Blood pressure fluctuations in human nailfold capillaries. Am J Physiol. 236: 888-893 (1979).
3. Levick J R, Michel C C. The effects of position and skin temperature on the capillary pressures in the fingers and toes. J Physiol. 274: 97-109 (1978).

4. Williams S A, Wasserman S, Tooke J E, Smaje L H. Dynamic human capillary pressure measurement: evidence for respiratory fluctuations. Int J Microcirc: Clin & Exp. 5: 254, M-154. (1986).
5. Tooke J E. The study of human capillary pressure. In: Tooke J E, Smaje L H, eds. Clinical Investigation of the microcirculation. Martinus Nijhoff, Massachusetts, USA.(1986).

MICROVASCULATORY EVALUATION OF VASOSPASTIC SYNDROMES

H. Saner, H. Würbel, F. Mahler, J. Flammer and P. Gasser

University Clinic of Medicine
Department of Angiology
Inselspital
CH-3010 Bern, Switzerland

INTRODUCTION

Application of a local cold exposure test to the finger nailfold capillaries reveals the functional difference between normal subjects and patients with Raynaud's syndrome at the level of the microcirculation. We developed a new method for local cold exposure which uses rapidly decompressed carbondioxide and requires considerably less effort experimentally than previously described systems (1). The effect of cold exposure on nailfold capillary blood flow was studied extensively in normal subjects and in patients with primary and secondary Raynaud's syndrome by means of television (TV-)microscopy. The cold exposure of the capillaries consistently resulted in a greater reduction of the erythrocyte flow velocity in patients with Raynaud's syndrome than in patients with normal conditions or other diseases. The flow stop was found to be a reaction specific to vasospastic disease (2-5).
The cause and pathogenesis of vasospasm remain enigmatic. We therefore were interested in the problem whether cold exposure in patients with vasospastic syndromes such as variant angina or reversible visual disorders of unknown origin would cause a vasospastic reaction of the nailfold capillary blood flow. This finding would indicate that a general tendency to vasospastic reaction may be present in such patients.
We performed the local cold exposure test of the nailfold capillaries in 12 patients with variant angina and in 25 patients with reversible visual disorders of unknown (probably vasospastic) origin and compared the results to the findings in normal controls.

METHODS

1. Cold exposure tests: The hands are warmed in 40°C water for three minutes prior to the test. The nailfold capillaries are then examined under TV-microscopy as previously described (3). Cold gas is directed at the finger for one minute. Erythrocyte flow-velocity is recorded before, during and after the cooling phase by using the frame-to-frame technique or by using a flying spot device in suitable cases. The flow stop duration is measured by a video-timer.
2. Patients with variant angina: We examined 12 patients (mean age 55 years) with variant angina who had coronary angiogramms within the last few years. The criteria for the definition of variant angina were angina pain at rest with transient ST-segment elevation and no evidence of myo-

cardial infarction. The results of this group were compared with the results of 12 age-and sex-matched normal controls. All tests were performed after discontinuation of previous medical treatment and repeated 30 minutes after 20 mg nifedipine orally.

3. Patients with visual disorders: We examined 25 patients (14 women and 11 men, mean age 49 years) with visual field defects of unknown origin. Patients with neither ophthalmologically nor neurologically explained visual field defects were included whose history of physical examination suggested vasospastic syndrome. The following tests were carried out: A. Visual field test on the automated perimeter Octopus, B. a detailed history considering evidence of vasospastic syndromes in different vascular beds and C. a cold exposure test of the nailfold capillaries before and 30 minutes after 10 - 20 mg nifedipine orally.

RESULTS

1. Patients with variant angina: In 9 of 12 patients with variant angina cold exposure caused a stop of flow of the nailfold capillary circulation (mean duration 23,4 s). 6 of 12 patients with stable angina had a flow stop (mean duration 11 s) as well as two normal controls (mean duration 1,25 s). The incidence of a flow stop was significantly higher in patients with variant angina ($p < 0.005$) and in patients with stable angina ($p < 0.025$) than in normal controls. Patients with variant angina tended to have an increased prevalence and longer duration of the flow stop than those with stable angina. This difference did not however reach statistical significance. When the test was repeated after administration of nifedipine patients with variant angina tended to have a lower incidence of flow stops of shorter duration.

Table I: Flow stop with local cold exposure in patients with coronary disease and in normal controls

| | No. of patients | Flow stop with local cold exposure | mean flow stop duration (s) |
|---|---|---|---|
| 1. variant angina | 12 | 9/12 pts* | 23,49* |
| 2. stable angina | 12 | 6/12 pts* | 11,08* |
| 3. normal controls | 12 | 2/12 pts | 1,25 |

* significant for control

2. Patients with visual disorders: 17 of the 25 patients had a history of frequent episodes of cold hands (ten women, seven men). 15 of the 25 patients had a flow stop in the nailfold capillaries with cooling. 13 of the 15 patients with a flow stop with cold exposure showed improvement with nifedipine.

Table II: Flow stop with local cold exposure in patients with reversible visual disorders of unknown origin

| | | microscopy of the capillaries | | |
|---|---|---|---|---|
| | | P | n | |
| patient history | + | 14 (82 %) | 3 (18 %) | $p < 0.01$ |
| | - | 2 (25 %) | 6 (75 %) | ($X^2$ - test) |

+ with the history of cold hands
- no history of cold hands

p blood flow stop in microscopy

n no flow stop in microscopy

DISCUSSION

The etiology and pathogenesis of vasospasm remain unclear. In variant angina, myocardial ischemia is caused by transient coronary-arterial spasm. Although altered adrenergic activity has been proposed as the cause of coronary spasm, and high circulating thromboxane levels are found in patients with variant angina, the real cause of the syndrome is unknown. Disturbances of the visual function may also be caused by vasospasm in some patients. In many of these patients the visual function deteriorated after immersion of one hand in cold water; after oral nifedipine the visual field improved. When asked specifically, the patients reported a high incidence of cold hands, and some of them suffered from migraine attacks. The biomicroscopic findings in these patients were mostly normal. Presently it is unknown where the vasospasm occurs in the eye.
The local cold exposure test of the nailfold capillaries which leads to a typical flow stop reaction in patients with Raynaud's syndrome reveals a high prevalence of vasospastic reaction in patients with variant angina and in patients with unexplained reversible visual field defects. This finding raises the possibility that a common underlying defect or mechanism may partially account for all these conditions. However, the clinical observation that vasospastic episodes in different arterial beds usually occur at different times and usually have different triggering mechanisms suggest that local factors may also play a part in the pathogenesis of these disorders.

REFERENCES

1. Mahler F., Saner H., Boss Ch. and Annaheim M. Local cold exposure test for capillaroscopic examination of patients with Raynaud's syndrome (Microvascular Research, in press).
2. Saner H., Würbel H., Foglia E., Linder M., Mahler F. Lokaler Kälteexpositionstest der Nagelfalzkapillaren bei Patienten mit vasospastischer Angina pectoris. Schweiz.med.Wschr. 1986; 116 (Suppl.20): 72
3. Mahler F., Saner H., Marth D., Ursenbacher B. Local cold exposure test for examination of the nailfold capillary circulation in Raynaud's syndrome. Inter Angiology 1983; 2 (4): 137-142.
4. Mahler F., Saner H., Annaheim M., Linder H.R. Capillaroscopic examination of erythrocyte flow velocity in patients with Raynaud's syndrome by means of a local cold exposure test. Progr.appl. Microcirc., vol. 11, pp 47-59, Karger, Basel 1986.
5. Gasser P., Flammer J., Guthauser U., Niesel P., Mahler F., Linder H.R. Bedeutung des vasospastischen Syndroms in der Augenheilkunde. Klin. Mbl. Augenheilk. (in press).

# INFRARED FLUORESCENCE VIDEOMICROSCOPY WITH INDOCYANINE GREEN (CARDIOGREEN®)

M. Brülisauer, G. Moneta, K. Jäger and A. Bollinger

Department of Internal Medicine, Policlinic
Angiology Division, University Hospital
Rämistr. 100, CH-8091 Zürich

## SUMMARY

After intravenous injection indocyanine green binds almost completely to the plasma proteins and may be detected in the skin capillaries by an infrared sensitive fluorescence videomicroscopy system. The technique opens a way to measure full capillary diameter and dimension of the plasma layer in an almost atraumatic way.

## INTRODUCTION

An intravital videomicroscopy technique is described using indocyanine green and an infrared sensitive videocamera (1). Comparison is made with fluorescence videomicroscopy using Na-fluorescein (2,3). With this dye (ca. 40 % protein bound) only the erythrocyte column and the surrounding pericapillary "halo" can be seen. It diffuses too rapidly to permit accurate localization of the capillary wall and the plasma layer can not be visualized. Indocyanine green is nearly 100% protein bound. This permits visualization and measurement of full capillary diameters.

## METHODS

The system for intracapillary detection of indocyanine green is similar to that previously described for fluorescence videomicroscopy using Na-fluorescein as a tracer (2,3). In contrast to Na-fluorescein indocyanine green fluoresces in the near infrared region with a peak absorption of 805 nm and a peak emission of 835 nm. Therefore, modification of light source, filters and videocamera were needed to visualize indocyanine green in the cutaneous capillaries. The apparatus consists of an incident light-fluorescence microscope (Wild-Leitz). Light energy for indocyanine green absorption is provided by a halogen light source. The halogen lamp is used because of the more favorable spectrum for inducing indocyanine green fluorescence. An excitation filter (780 nm) interposed between the lamp and the skin allows delivery of a narrow spectrum of light to the capillaries permitting indocyanine green absorption while minimizing reflecting light from the skin back towards the microscope ojectives. A barrier filter (peak 870 nm, Ditric Optics) between the skin and the camera lens permits transmission of indocyanine green fluorescent light while blocking light of lower wave-lengths, originally transmitted by the excitation filter. A microchip camera of appropriate spectral sensitivity is used (peak sensitivity 700 - 900 nm, Model SM 72, Kranz Electronics). The infrared camera is linked in series to a cathod ray oscilloscope, a television monitor with a 60 cm screen and a videorecorder with provision for single frame projections (BK 204, Grundig). A videotimer (VTG 22, For-A-Company) and a video scale marker (IV 600, For-A-Company) project running time and a scale on the monitor throughout the examination. With the use of a plane objective (10x0,3, Wild-Leitz) a final magnification of 740 x is achieved on the monitor screen. Accuracy of magnification was confirmed using a calibrated microscope slide (1 interval = 0.01 mm, Wild-Leitz).

With the above described system it is also possible to perform videodensitometric analyses of fluorescent light intensities, as introduced previously with Na-fluorescein as a tracer (2,3). The background light transmitted to the videocamera makes videodensitometric analysis more difficult. Densitometry using infrared fluorescence videomicroscopy with indocyanine green can therefore be facilitated by the addition of a video substaction device (Video Image Processor, Pie Data Medical). This helps to eliminate spontaneous infrared fluorescence of the skin as well as extraneously transmitted light. With the combination of a videodensitometer (Video Integrator 310, Colorado Video), a frequence filter (Model 3750, Krohn Hite), an x-y plotter (2000 Recorder, Houston Instruments) and the videosubstraction device, satisfactory densitometric recordings are made from single frames of the videotape at various times around isolated capillaries.

11 healthy people (6 men, 5 women, mean age 31 years, 22-54 years) were studied. The finger chosen for examination was positioned on the microscope stage at heart level and fixed with a plastic mass. Room temperature varied between 22°C and 24°C. Finger temperature was adjusted to 28° ± 0.5°C. Indocyanine green was given as a bolus via an antecubital vein at a dosage of 50 mg/l of estimated blood volume and a videorecording made continuously for 5 min after initial appearance of the dye in the capillaries under examination. Data analysis was done later during playback of the videotape. In the 11 volunteers we studied 29 nailfold capillaries, an average of 2.6 per person (range 1-4). Red cell column diameters were measured using incident white light and presumed full capillary diameters were measured using infrared light. Measurements were made from single frames of the videotape projected on the monitor.

RESULTS

The mean diameter of the red cell column on the arterial side was 12.1 ± 3.0 µm, the mean full diameter 17.7 ± 3.9 µm. The corresponding values on the venous side averaged 13.7 ± 4.1 µm and 20.1 ± 4.4 µm respectively. The difference between the red cell column size and the diameters measured with indocyanine green fluorescence were significant ($p<0.01$). The diameter of the red cell column amounted to 68 % of the full capillary width on both the arterial and venous side of the loop.

Densitometry was performed on an axis crossing the capillary loop 40 µm below the vertex. Two peaks appear in each individual curve. They correspond to the capillary's arterial and venous limbs while the dividing valley represents the area between the arterial and venous loop. In contrast the densitometer curves obtained after injection of Na-fluorescein show two peaks only during the first seconds after dye appearance. At later times the pattern is characterized by 3 pericapillary peaks (accumulation of dye in the pericapillary halo).

## DISCUSSION

A new technique is introduced to measure full capillary diameter including the plasma layer in a nearly non-invasive way (1). In addition, the videomicroscopy technique used previously with Na-fluorescein as a dye allows one only to visualize the pericapillary halo (2,3). By the combination of the two fluorescent tracers accurate diameter measurements of the red cell column, of the plasma layer and of the pericapillary halo are possible. Since indocyanine green binds almost completely to plasma proteins, it might be used in the future to detect large molecular permeability of skin microvessels.

## ACKNOWLEDGEMENT

This research supported by the Swiss National Science Foundation, grant Nr. 3.808-0.84.

## REFERENCES

1. G. Moneta, M. Brülisauer, K. Jäger, A. Bollinger, Infrared fluorescence videomicroscopy of skin capillaries with indocyanine green. Int J Microcirc Clin Exp (in print).
2. A. Bollinger, K. Jäger, W. Siegenthaler, Microangiopathy of progressive systemic sclerosis evaluated by dynamic fluorescence videomicroscopy. Arch Int Med 146:1541-1545 (1986).
3. A. Bollinger, J. Frey, K. Jäger, J. Furrer, J. Seglias, W. Siegenthaler, Patterns of diffusion through skin capillaries in patients with long-term diabetes. New Engl J Med 307:1305-1310 (1982)

MEASUREMENT OF TISSUE BLOOD FLOW BY

HIGH FREQUENCY DOPPLER ULTRASOUND

S. Basler, A. Vieli and M. Anliker

Institute of Biomedical Engineering
UNI/ETH Zurich
Zurich, Switzerland

## INTRODUCTION

Non-invasive blood flow measurements by means of Doppler ultrasound have been possible for a number of years in a variety of vessels. In contrast, the effective tissue blood flow has not been assessed by this method extensively so far[1].
Laser-Doppler systems are marketed for this purpose but these devices can only be used in superficial layers of about 1 millimeter thickness due to the opacity of the tissue.
To provide an alternative, a high frequency pulsed ultrasonic Doppler instrument capable of assessing blood flow in a depth range up to 15 millimeters has been developed at the Institute of Biomedical Engineering in Zurich. It operates in two different modes. In its single channel mode which the present study is based on it can be utilized to determine tissue blood flow in a depth range exceeding 1 cm. Alternatively, as a multichannel device it allows for flow profile evaluation in small vessels either intraoperatively or transcutaneously.

## MATERIAL AND METHODS

Depending on the requirements, the Doppler device can be operated either at 15 MHz or at 20 MHz [2]. Other operating parameters such as the transmit level, the burst length as well as the position and the duration of the receiver gates, can be adapted to the actual needs.

In order to obtain an adequate signal to noise ratio we used very low noise amplifiers combined with highly sensitive ultrasonic probes optimized for tissue blood flow applications.
The multichannel signal processing is based on a FFT (Fast Fourier Transform) processor yielding instantaneous velocity profiles in real time.
To assess tissue blood flow, however, the dynamic range had to be extended beyond the specifications necessary for macro vascular blood flow measurements. Thus a specific high performance channel was implemented, which allows for different signal processing options such as zero crossing, FFT analysis and first moment estimation. In addition a special analog first moment processor can be applied, which provides a direct measure of the tissue blood flow in the form of the product of the number of erythrocytes within the region of interest and their average velocity component in beam direction.

## RESULTS

In order to verify the performance of the system under controlled conditions a miniaturized flow phantom was developed which allows to physically model vessels with diameters in the range between 20 µm and a few mm. Human blood was pumped through these tubes with flow velocities ranging from 0.2 mm/s up to 100 mm/s.
The measurements proved to be reproducible for flow velocities higher than 0.5 mm/s with a maximal error of +/- 10%.
Initial in vivo studies confirmed that tissue flow applications require a dynamic range beyond the one provided by the multichannel subsystem. Thus, they were carried out exclusively with the aid of the high performance gate available in the instrument.
Moreover, serious problems were encountered in the form of artefacts, which are attributed to partially uncontrollable muscle activity. These artefacts have been reduced to an acceptable level by means of exponential decay averaging. Yet, their origin has not been fully clarified [3].
Fig.1 documents the frequency spectrum obtained with the transducer positioned above the flexor carpi ulnaris. Tracing #1 corresponds to normal blood flow, whereas #2 was recorded while the blood flow was inhibited by means of a cuff inflated to a pressure of 250 mm Hg.
At frequencies above 50 Hz the difference between the two curves amounts to 10 - 20 dB, whereas below 50 Hz i.e. at velocities below 3 mm/s it is relatively small.
It is assumed that motion associated with these low velocities is primarily due to permanent muscle activity.
This assumption is backed by Fig. 2 repeating curve #2 from Fig. 1 in tracing #1. Tracing #2 which shows a marked reduction of the signals below 50 Hz was obtained by lifting the probe from the forearm.
Fig. 3 shows a result obtained with the analog first moment processor in a reactive hyperaemia manoeuvre. The artefacts occurring at the cuff inflation and deflation time as well as the ones during the hyperaemic phase must be associated with involuntary motion of the arm.

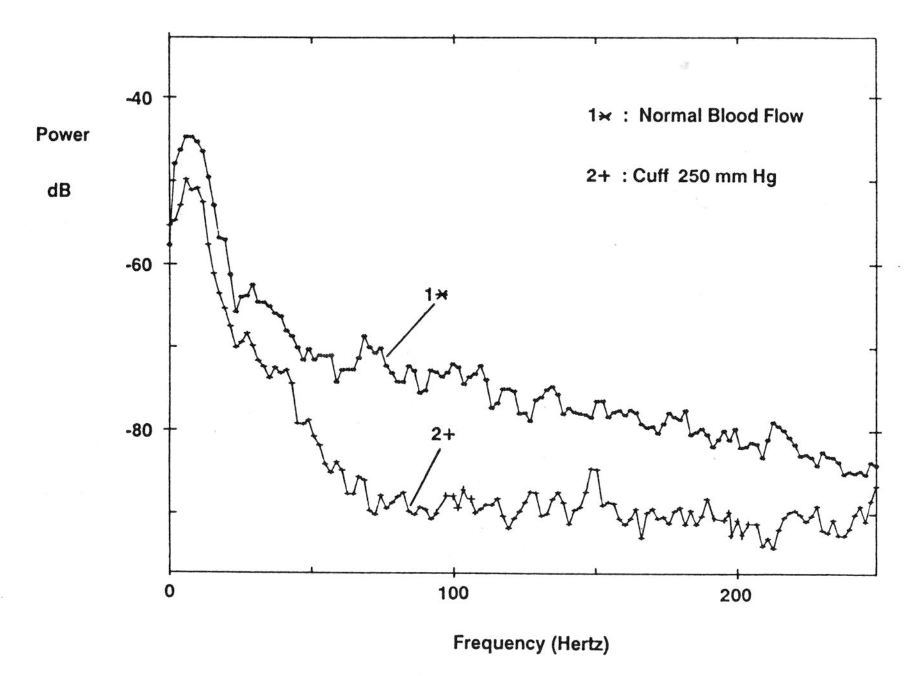


Figure 1

Frequency spectra of Doppler signals
Tracing #1 shows the spectrum corresponding to normal blood flow , tracing #2 was recorded while blood flow was inhibited.

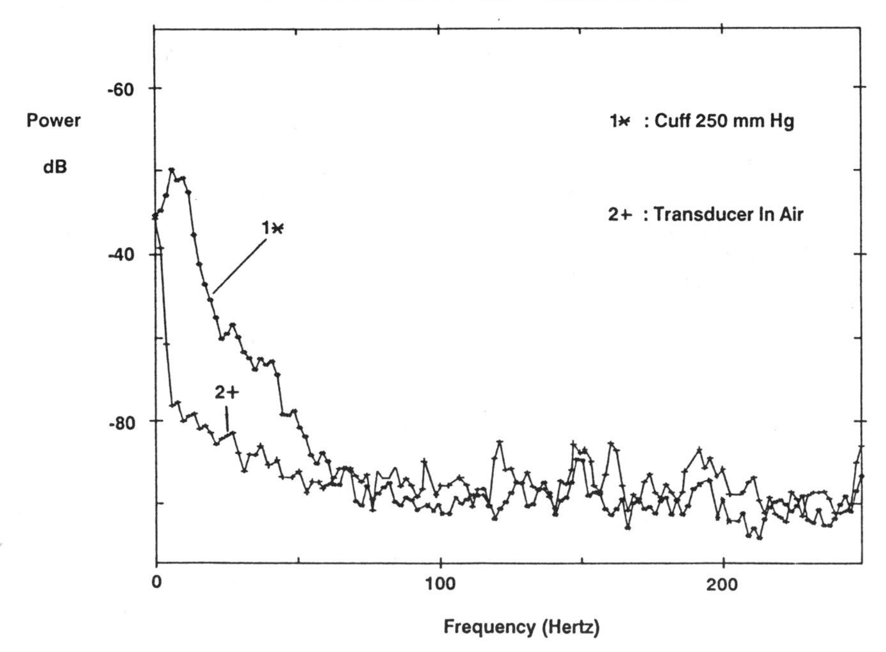


Figure 2

Frequency spectra of Doppler signals
Tracing #1 was recorded while blood flow was inhibited, tracing #2 shows the spectrum with the probe lifted from the forearm.

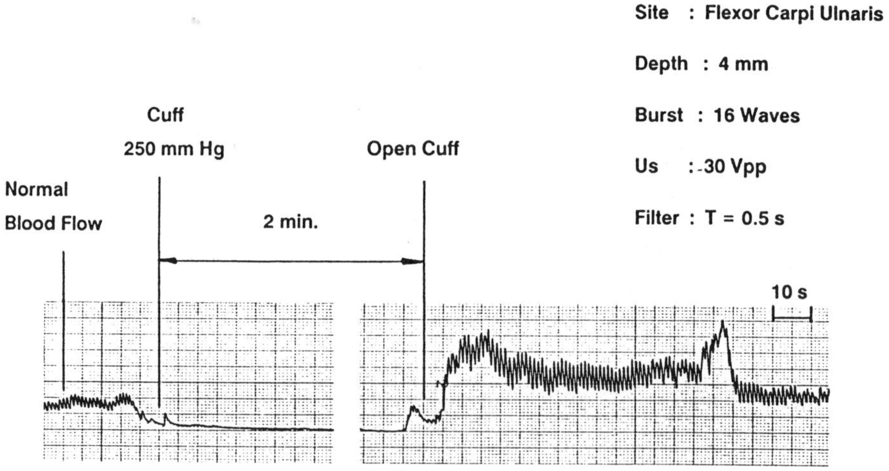

Figure 3

Measurement of a reactive hyperaemia

## CONCLUSION

The device utilized for this study is capable of detecting tissue blood flow in vivo and of measuring blood flow under controlled conditions with a maximal error of +/- 10% for flow velocities higher than 0.5 mm/s.
Artifacts due to microvibrations and other muscular activities should be reduced further by filter methods to be developed.

## REFERENCES

1. S. Dymling, measurement of blood perfusion in tissue using Doppler ultrasound
   Lund Inst. of Technology, Report, 1984
2. K.K. Shung, scattering of ultrasound by blood
   IEEE Trans. Biomed. Eng. Vol. BME-23, No. 6
3. G. Oster, muscle sounds
   Scientific America, March 1984

# LASER-DOPPLER PROBES FOR THE EVALUATION OF ARTERIAL ISCHEMIA*

H. Seifert, K. Jäger and A. Bollinger

Department of Internal Medicine, Medical Policlinic
Devision of Angiology, University Hospital Zürich
Switzerland

## SUMMARY

Rhythmic variations of microvascular flux have been studied at the forefoot of patients with arterial occlusive disease by laser-Doppler flowmetry. Two patterns of flow motion with characteristic amplitudes and frequencies could be observed. The prevalence of small waves with a mean amplitude of 0.21 $\pm$ 0.1 AU and a frequency of 21.7 $\pm$ 4.2 c/min increased with the degree of ischemia. Big flow motion waves with a mean amplitude of 0.77 $\pm$ 0.4 AU and a frequency of 3.0 $\pm$ 1.0 c/min were found in both controls and patients with different degrees of ischemia. The small waves may represent a compensatory mechanism in skin ischemia.

## INTRODUCTION

Laser-Doppler flowmetry permits transcutaneous noninvasive continuous measurement of microvascular dynamics. It records blood cell flux in small skin areas. We used the method to study rhythmic variations of microvascular flux related to the phenomenon of vasomotion in healthy controls and in patients with arterial occlusive disease with different degrees of ischemia.

## METHODS

Skin flux was measured with a commercially available laser-Doppler flowmeter (PeriFlux PF 2, Perimed, Sweden). This is a 2 mW helium-neon laser, producing light at a wave length of 632 nm. The light is brought to the skin with an optical fibre, where it is scattered and absorbed within a hemisphere with a radius of approximately 1 mm. The light scattered back from moving red cells undergoes a frequency shift (Doppler effect). The shifted as well as the unshifted backscattered light is led to two photodetectors where it is processed and amplified. The output signal corresponds to the red cell flux, expressed in arbitrary units (AU) (4).

The probe emitting the laser light was fixed in a thermostat probeholder heated to 32° C. It was placed on the forefoot of the patient 2 cm proximal to the base of the second and third toe (30 x gain, 0.2 s time constant). The flux values were continuously registered with a pen recorder using a paper speed of 6 cm/min.

Flow motions were analysed in 12 healthy controls (A), 12 patients with intermittent claudication (walking distance above 200 m, B), 12 patients with walking distance of 200 m or less (C) and 12 patients with rest pain or gangrene (D). Patients with diabetes mellitus, chronic venous incompetence or edema were excluded. Skin flux was measured at rest in supine position, during reactive hyperemia (after 3 min of arterial occlusion at the ankle by a pneumatic cuff) and during orthostasis.

## RESULTS

At rest two patterns of flow motions were observed: Big waves with a mean amplitude of 0.77 ± 0.4 AU and a frequency of 3.03 ± 1.0 cycles per min (c/min) and small waves with a mean amplitude of 0.21 ± 0.1 AU and a frequency of 21.7 ± 4.2 c/min. When flow motions were present there was no significant difference in amplitude and frequency of both patterns of flow motion waves between the 4 groups studied (Table I). The prevalence of big waves at rest in supine position was similar in the 4 groups, whereas the prevalence of small waves increased significantly ($p<0.01$) with more severe ischemia (Table II).

TABLE I
AMPLITUDE AND FREQUENCY OF FLOW MOTION WAVES

| group | A | B | C | D |
|---|---|---|---|---|
| big waves: | | | | |
| amplitude (AU) | 0.91±0.5 | 0.76±0.4 | 0.61±0.1 | 0.80±0.4 |
| frequency (c/min) | 3.71±1.3 | 3.18±0.4 | 2.67±1.0 | 2.53±0.6 |
| small waves: | | | | |
| amplitude (AU) | 0.24 | 0.27±0.1 | 0.22±0.1 | 0.19±0.1 |
| frequency (c/min) | 26.0 | 23.0±5.3 | 19.9±3.8 | 22.5±4.0 |

TABLE II
PREVALENCE OF FLOW MOTION WAVES

| group | A | | B | | C | | D | |
|---|---|---|---|---|---|---|---|---|
| | | % | | % | | % | | % |
| big waves | 10/12 | 83 | 10/12 | 83 | 10/12 | 83 | 9/12 | 75 |
| small waves | 1/12 | 8 | 4/12 | 33 | 9/12 | 75 | 11/12 | 92 |

During reactive hyperemia big waves were induced or enhanced in 69 % of the cases. The mean value of the maximal amplitude in this phase was 1.63 ± 1.2 AU, whereas the mean amplitude at rest was only 0.77 ± 0.4 AU. During orthostasis big waves often disappeared. We found them only in 31 % of the cases during orthostasis compared to 81 % in supine position. The small waves were less influenced by position or reactive hyperemia. They could be observed in 52 % of all cases in supine position and in 51 % during orthostasis.

## DISCUSSION

The phenomenon of spontaneous rhythmic vasomotion has been described both in men and animals (1-3), but no systematic studies have been performed to analyse flow motion in patients with arterial occlusive disease.

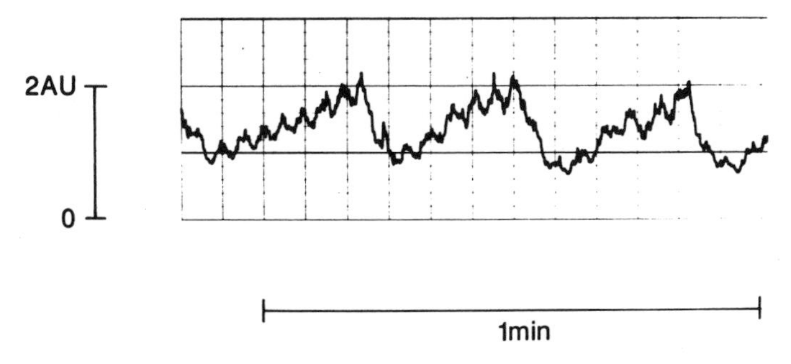


Superposition of big and small waves. Patient with intermittent claudication, supine position.

The main finding of this study is the presence of big and small flow motion waves with characteristic amplitudes and frequencies. The prevalence of small waves which have previously not been described increases with more severe ischemia. According to Colantuoni et al. (1) the amplitude of vasomotion waves is proportional to the vessel size, and the frequency augments with decreasing diameters (3). Therefore, the small flow motion waves probably originate in smaller arterioles. They might represent a compensatory mechanism in skin ischemia.

REFERENCES

1. S. Colantuoni, S. Bertuglia, M. Intaglietta, Quantitation of rhythmic diameter changes in arterial microcirculation. Am J Physiol, 246:H508-H517 (1984).
2. B. Fagrell, M. Intaglietta, J. Oestergren, Relative hematocrit in human skin capillaries and its relationship to capillary blood flow velocity. Microvasc Res, 20:327-335 (1980).
3. W. Funk, B. Endrich, K. Messmer, M. Intaglietta, Spontaneous arteriolar vasomotion as a determinant of peripheral vascular resistance. Int J Microcirc Clin Exp, 2:11-25 (1983).
4. T. Tenland, "On Laser Doppler Flowmetry". Dissertation. Linköping University (1982).

* Supported by Deutsche Forschungsgemeinschaft and Swiss National Science Foundation, grant Nr. 3.808-0.84.

# SKIN REACTIVE HYPEREMIA RECORDED BY A COMBINED $TcPO_2$ AND LASER DOPPLER SENSOR

U. Ewald[1], A. Huch[2], R. Huch[2] and G. Rooth[2]

Department of [1]Paediatrics, University Hospital
S-751 85 Uppsala, Sweden
Department of [2]Obstetrics, University Hospital
CH-8091 Zürich, Switzerland.

## SUMMARY

The $tcPO_2$ electrode used at 37° C monitors changes in cutaneous $PO_2$ which at this temperature is mainly determined by the changes in blood flow. Laser Doppler velocimetry (LDV) measures the product of red cell number times their velocity. A comparison of these methods for the detection of skin reactive hyperemia has been made with a combined $tcPO_2$ and laser Doppler sensor. The two parameters increased and decreased in parallel, the $tcPO_2$ signal being delayed 10-20 s compared to the LDV signal. A significant correlation between the two signals was obtained (r=0.73-0.93). In contrast, no significant correlation existed between the postocclusive peaks of the two signals recorded in 12 duplicate experiments. Comparing the peak amplitudes from the repeated recordings the $tcPO_2$ signal showed a higher correlation coefficient (r=0.91) and smaller intercept than the LDV signal (r=0.74). Both methods can be used for non-invasive recordings of skin reactive hyperemia. The $tcPO_2$ signal at 37° C reflects changes in blood flow of the most superficial capillaries and reactive hyperemia can be monitored with high reproducibility. Only this signal can be used for calculation of skin blood flow. The LDV signal conveys information of blood flow changes mainly of deeper dermal vascular beds and reacts virtually instantaneously, but is less reproducible.

## INTRODUCTION

The aim of the $tcPO_2$ technique, as traditionally used, is to increase skin blood flow by heating to such a level that $tcPO_2$ becomes independent of the local blood flow and approximates arterial $PO_2$. By measuring instead at 37°C $tcPO_2$ usually is less than 10 mm Hg, but increases rapidly with increase in skin blood flow, whereas it is not affected by changes in arterial $PO_2$. As $tcPO_2$ at 37°C approximates tissue tension of the epidermis it is termed $cPO_2$ where c stands for cutaneous (1). The laser Doppler technique measures velocities and shows good correlation to the flux (particle number times velocity) of red cells within a measuring volume, a hemisphere of about 1 mm (2). Postocclusive reactive hyperemia is an often used test of vascular capacity in clinical and experimental medicine and constitutes the physiological response to tissue hypoxia. Since the reaction occurs after denervation it is considered to be mediated locally in the tissue (3)

The aim of this study was to compare these two independent methods for the non-invasive recordings of reactive hyperemia in the skin.

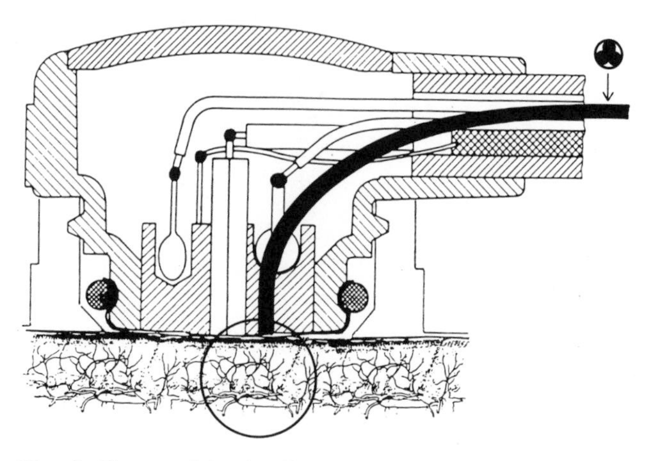

Fig. 1. The combined $tcPO_2$ and laser Doppler sensor.

METHODS

Twelve healthy volunteers were studied. A newly constructed combined $tcPO_2$ and laser Doppler sensor was made by two of us (H and H) and affixed to the inside of the forearm. The $tcPO_2$ equipment used was an Oxymonitor[R] (Hellige; Freiburg, FRG) and the laser Doppler equipment was manufactured by Perimed (Stockholm, Sweden). The sensor construction allowed identical temperatures on both measuring sites and was kept at 37° C and once the signals were stable a four minutes arterial occlusion of the forearm circulation was applied (Fig. 1). Simultaneous recordings of the subsequent reactive hyperemia was obtained on a Servogor 120 pen recorder. The experiment was repeated after 15 min under identical experimental conditions and unchanged skin site.

Permission for the study was given by the Ethics Committee of the Medical Faculty of Uppsala, Sweden.

RESULTS

The parameters increased and decreased in parallel, the $cPO_2$ signal being delayed 10-20 s compared to the LDV signal. Both parameters exhibited in their baseline recordings small rhythmical variations of the same frequency of 3-7 per min, a typical physiological vasomotor phenomenon. At the release of the cuff occlusion the LDV signal reacted virtually instantaneously whereas the $cPO_2$ began to increase with a delay of 10-20 s. The maximal postocclusive amplitude of the LDV signal was usually divided into a first peak within 5 s after cuff release and a second peak after 50-80 s. The $cPO_2$ signal exhibited only one single peak 60-90 s after the cuff release (Fig. 2).

A direct comparison between the two curves measuring in arbitrary units the amplitudes of the recorded reactive hyperemia with 30 s interval was performed. The coefficient of correlation varied between 0.73 and 0.93 between the experiments, which was significant at 0.01 and 0.001 levels. If a phase correction of 30 s between the two signals was made the mean correlation coefficient improved from 0.74 to 0.79.

When comparing only the postocclusive maximal peak amplitude of all experiments no significant correlation was found.

Comparing the peak amplitudes from the 12 duplicate experiments the $cPO_2$ signal showed a higher correlation coefficient ($r=0.91$) than the LDV signal ($r=0.74$). The equation of the linear regression line for $cPO_2$ was $Y= 4.13 + 0.75\ X$ whereas the LDV signal showed a larger intercept ($Y = 24.93 + 0.58\ X$).

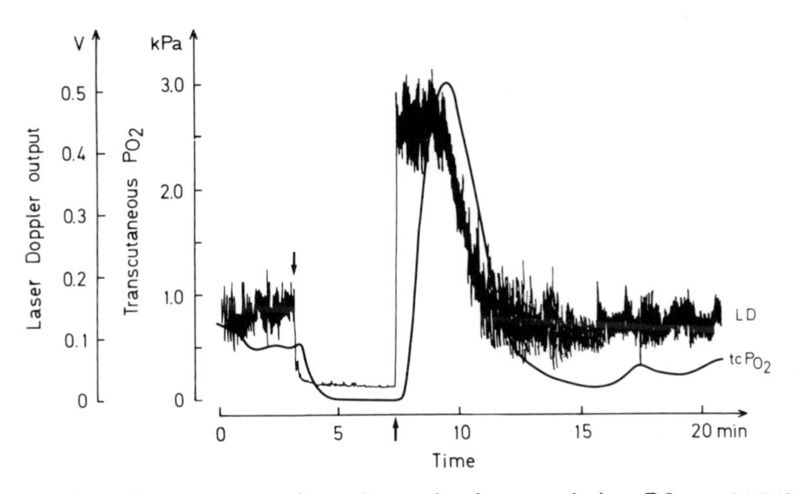


Fig. 2. Simultaneous recording of reactive hyperemia by $cPO_2$ and LDV.

## DISCUSSION

Previous comparisons between laser Doppler velocimetry and transcutaneous $PO_2$ and/or heating power as indicators of skin perfusion have all been performed using electrode temperature of 43-45° C. At this temperature the maximal heat induced vasodilation precludes normal vascular reactivity of skin vessels (4) and discrepancies between the two signals merely reflect changes in central circulation e.g. shock. No previous comparisons are available between the laser Doppler velocimeter which senses the velocity of the red cells of the blood and the transcutaneous $PO_2$ method at an electrode temperature of 37° C reflecting $PO_2$ in the outer epidermal layer, which is mainly affected by changes in the skin capillary blood flow. The two techniques convey information on blood flow changes in different vascular beds, LDV the product of erythrocyte number times mean velocities in a hemisphere down to a depth of 1000 um and $cPO_2$ changes in blood flow of the capillary loops at a depth of 50-100 um (Fig. 3). It might be expected that they behave differently in different physiological situations.

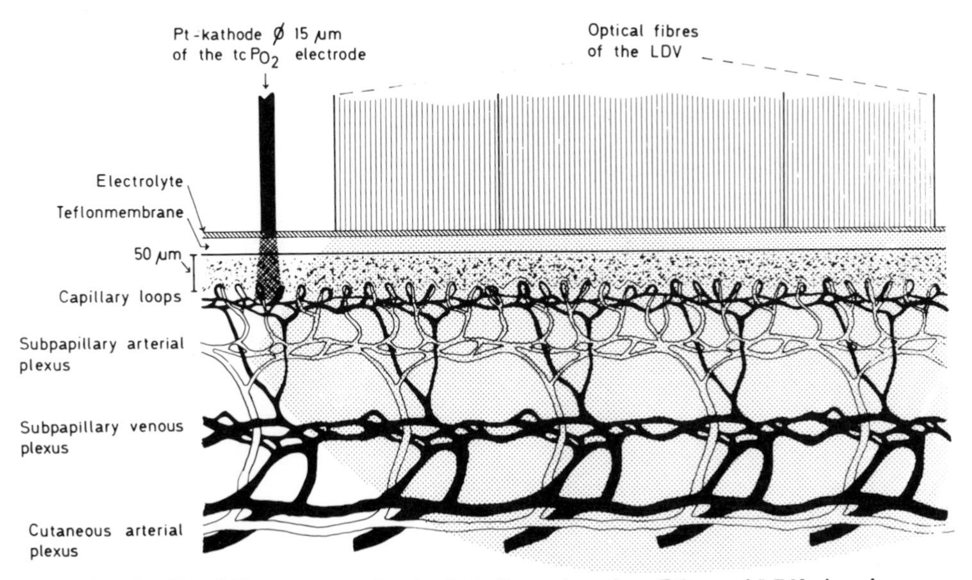


Fig. 3. The different vascular beds influencing the $cPO_2$ and LDV signal.

The laser Doppler meter reacts virtually instantaneously, whereas the $PO_2$ signal is delayed by 10-20 seconds. This difference is explained partly by differences in blood flow in the different vascular beds and partly by the time lag for a changing oxygen gradient to reach the electrode.

The difference in the postocclusive peak of the two signals can be explained by the sensitivity of the LDV signal to changing vascular diameter, flow direction, disaggregation of erythrocytes, acceleration of non-moving erythrocytes and of $cPO_2$ by the time lag caused by diffusion and metabolism. The second LDV peak represents the maximal flow increase due to vasodilation and corresponds to the $cPO_2$ peak.

To validate the comparisons of the blood flow measurements both should preferably be expressed in the same quantitative unit. The $cPO_2$ signal can be expressed in quantitative units by the use of a mathematical model, the finite element method and a fast computer or with a simplified method. With the recent calibration curves giving skin blood flow for different $cPO_2$ values, this becomes a more practical reality (1).

As Lübbers (4) stresses, $cPO_2$ changes are hyperbolically related to changes in skin capillary blood flow. When $cPO_2$ is about 20 mm Hg the changes in both variables are about equal. The laser Doppler signal increases linearly to changes in blood flow provided the vascular dimensions, flow directions and corpuscular number are constant. These facts together explain the lack of correlation between the methods comparing the post occlusive peak.

Steady state skin blood flow is extremely sensitive to changes in environmental and psychological conditions and is therefore not suitable for experimental studies. Instead provoked vascular reactions are used and postocclusive reactive hyperemia shows high reproducibility in short term experiments with different methods including $cPO_2$ (3, 5). The low reproducibility found with the LDV technique, also noted by (2), might at least partly be explained by the fact that even a minimal directional change of the laser beam results in large changes in amplitude of the signal due to its dependence on vessel geometry, flow direction and red cell number.

At 37° C both techniques can be used for non-invasive and continuous recordings of changes in skin blood flow. A combined sensor has a broader clinical and experimental applicability since different vascular beds can be compared from a simultaneous recording under identical conditions.

## REFERENCES

1. G. Rooth, U. Ewald and F. Caligara. Transcutaneous $PO_2$ and $PCO_2$ monitoring at 37° C, Cutaneous $PO_2$ and $PCO_2$. (In this volume)
2. G.E. Nilsson, T. Tenland and P.Å. Öberg. Evaluation of a laser Doppler flowmeter for measurement of tissue blood flow. IEEE Trans Biomed Eng 27: 597-604 (1980).
3. J.T. Sheperd. Reactive hyperemia in human extremities. Circ Res 14 Suppl. 1: 76-79 (1964).
4. R. Huch, A. Huch and D.W. Lübbers. Transcutaneous $PO_2$. Thieme Stratton Inc., New York (1981).
5. U. Ewald. Evaluation of the transcutaneous oxygen method used at 37° C for measurement of reactive hyperemia in the skin. Clin Physiol 4: 413-423 (1984).

# COMPARISON OF LASER-DOPPLER-FLUX AND $tcPO_2$ IN HEALTHY PROBANDS AND PATIENTS WITH ARTERIAL ISCHEMIA

Ludwig Caspary, Andreas Creutzig and Klaus Alexander

Dept. of Angiology, Medizinische Hochschule Hannover
Konstanty-Gutschow-Straße 8, D-3000 Hannover 61

## SUMMARY

Transcutaneous $PO_2$ ($tcPO_2$) and Laser-Doppler-Flux (LDF) were compared on the dorsum of the foot in 20 healthy probands and 35 patients with peripheral arterial occlusive disease at clinical stage II b or IV. The probes were kept at a temperature of 37°C. Using different procedures, we brought about dynamic changes and compared the reaction of the two signals. Venous occlusion resulted in a decrease of both signals to a similar extent in both groups. During leg dependency both signals decreased in the probands suggesting a normal vasoconstrictor response. In most of the patients an increase was observed, but some showed a decrease even at clinical stage IV. LDF had a stronger tendency towards a decrease. On leg elevation, LDF slightly increased in probands and decreased in patients. Here. the tendency towards a decrease was higher for $tcPO_2$. After arterial occlusion reactive hyperemia was more pronounced in probands. Differences between $tcPO_2$ and LDF seem to be mainly due to the different capillary systems contributing to the signal.

## INTRODUCTION

$TcPO_2$ measurements are usually performed at a hyperemization temperature of 44°C. However, under these circumstances vasoconstrictor activities are abolished. Microcirculatory responses to pressure variations, metabolic changes and pharmacological influences can better be studied at a modest hyperaemization obtained with an electrode core temperature of 37°C (1). The price for this advantage is a considerable variability in resting values which are usually between 1 and 8 mmHg and are mainly influenced by the local skin capillary flow.

Laser-Doppler flowmetry as a non-invasive optical method has its undoubted value in the rapid monitoring of flux changes in the skin capillary bed. The signal seems in part to derive from subpapillary venous plexus vessels; the proportion of the different contributing vessel systems is difficult to judge (2). Thus, similar to the $tcPO_2$ measurements, resting values given in mV or arbitrary units provide only limited information. We studied the relative changes of $tcPO_2$ and LDF signals on different manoeuvres in probands without and patients with severe impairment of the arterial inflow.

## PATIENTS AND METHODS

We examined 35 patients with peripheral arterial occlusive disease (PAOD). 17 of them had severe claudication (maximal walking distance shorter than 100 m, stage II b according to Fontaine) and 18 patients had foot gangrene (stage IV). All patients had angiographically proven occlusions of the superficial femoral artery and additional obstructions or occlusions of the calf arteries. Mean ankle pressure was 76 ± 16 mmHg in stage II b and 56 ± 17 mmHg in stage IV. Mean age was 67 years in both groups (range: 42 - 80 years in stage II b and 34 - 81 years in stage IV). 20 healthy volunteers with no apparent signs of arterial disease or diabetes, with a mean age of 26 years (range: 21 - 34 years) served as controls.

$TcPO_2$ was determined with the TCM-2-system (Radiometer, Copenhague) using a polarographically measuring platinum electrode with a tip diameter of 25 µm. The integral thermo-element was set to 37°C and a one-point-calibration was performed according to the manufacturer´s instructions. In addition, zero calibration was checked by a sodium thiosulfate solution. The electrode was fixed on the dorsum of the foot with the original fixation kit. Directly beneath it, at a maximal distance of 4 cm, the LDF probe was fixed with the original probe holder (PF-1-system, Perimed, Linköping) containing a heating element that was adjusted to 37°C, too. The signals were registered on a pen recorder and analyzed off-line.

The procedure included 20 min of measuring under resting conditions, the proband being in a comfortable supine position at a room temperature between 21°and 23 °C. Then venous circulation was occluded by a distal calf cuff with a pressure of 40 mmHg. After another 10 min of lying the proband sat up and brought the leg into a dependent position for 4 min, afterwards lying again for 10 min. The leg was then elevated to about 40 cm above heart level during 4 min. After 10 min of rest, a 3 min (healthy probands 6 min) arterial occlusion was performed by raising the cuff pressure to 40 mmHg above systolic pressure. After releasing of the pressure, the reaction was observed for another 20 min.

## RESULTS

### 1. Venous occlusion

The effect of a venous occlusion was very similar in both groups of patients and in the probands as well as in the two methods. We saw a decrease down to 60 - 70% of the initial value. After cuff release, the initial value was exceeded in most of the measurements (Table 1).

Tab.1: Effect of venous occlusion (cuff with 40 mmHg pressure at calf level) on the $tcPo_2$ and LDF signal, measured at 37°C on the forefoot

| | LDF | | | $tcPo_2$ | | |
|---|---|---|---|---|---|---|
| | before | during | after | before | during | after |
| | | - venous | | occlusion - | | |
| probands | 100 | 66.1 | 124.5 | 5.8 | 3.6 (66) | 3.6 (109) |
| (n = 20) | | ± 15.2 | ± 32.5 | ± 4.2 | ± 2.5 | ± 4.1 |
| patients | | | | | | |
| stage II b | 100 | 64.0 | 104.0 | 5.6 | 4.1 (73) | 5.9 (105) |
| (n = 17) | | ± 16.2 | ± 25.8 | ± 3.8 | ± 3.4 | ± 9.4 |
| stage IV | 100 | 65.5 | 111.2 | 4.9 | 3.0 (61) | 5.3 (108) |
| (n = 18) | | ± 19.2 | ± 17.6 | ±4.5 | ± 3.9 | ±4.5 |

LDF: % of initial value; $tcPo_2$: mmHg (% of initial value); $\overline{x}$ ± S.D.

2. Leg dependency

On leg dependency, the LDF as well as the $tcPO_2$ signal decreased in all healthy probands. In patients, in most of the cases a $tcPO_2$ increase was recorded. The increase was more marked in patients in clinical stage IV. However, some patients even in stage IV showed a signal decrease (Fig.1).

Similarly, the LDF signal decreased in all probands and as well in many of the patients at both stages. However, in a part of the patients (more often in stage IV) an increase of the signal was observed.

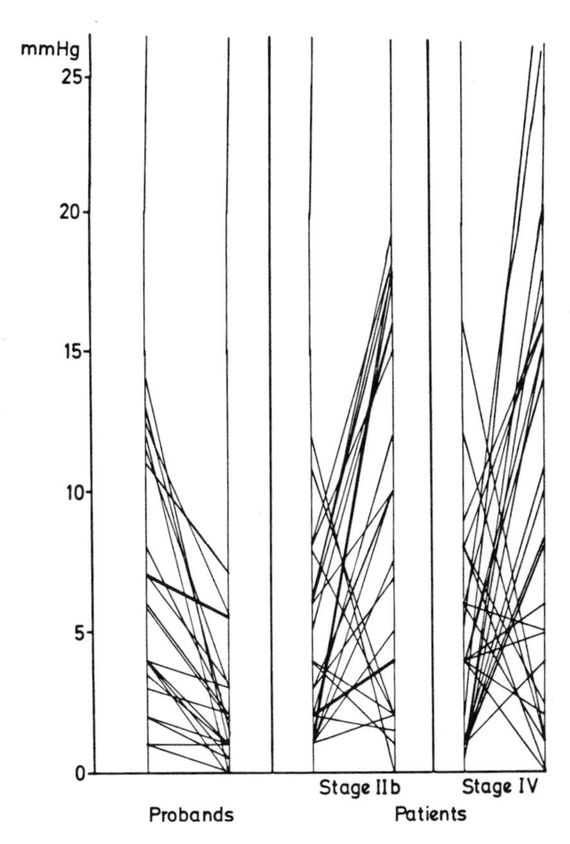


Fig.1: Reaction of the $tcPO_2$ signal on leg dependency (37°C, forefoot)

When the reaction of the LDF signal was compared to that of $tcPO_2$, we saw an increase of $tcPO_2$ in nearly all patients in whom the LDF signal increased, while a decrease of $tcPO_2$ was normally accompanied by a decrease of the LDF signal. Finally, for both stages a group of patients remains in whom LDF decreased and $tcPO_2$ increased. The tendency towards a decrease was higher for the LDF measurements (Table 2).

Tab.2: Direct comparison between LDF and $tcPO_2$ reaction on leg dependency

| | measurements | LDF | $tcPo_2$ |
|---|---|---|---|
| probands | n = 20 | ↓ 20 | ↓ 20 |
| patients | | ↓ 27 | ↓ 11<br>↑ 16 |
| stage II b | n = 40 | ↑ 13 | ↑ 12<br>↓ 1 |
| stage IV | n = 26 | ↓ 15 | ↓ 7<br>↑ 8 |
| | | ↑ 11 | ↑ 9<br>↓ 2 |
| ↑ signal increased | | ↓ signal decreased | |

## 3. Leg elevation

On leg elevation to a level of 40 cm above heart level, the reaction was much less uniform. The LDF signal was seen both to decrease and to increase. At the $tcPO_2$ electrode, a signal decrease or no change was observed. The tendency towards a signal decrease was higher in patients, especially for the more severe cases. Comparing the two methods, in general the decrease was more pronounced in $tcPO_2$.

## 4. Arterial occlusion

Arterial occlusion yielded a signal reduction to values near zero in all groups for $tcPO_2$.The LDF signal came to a rest level superior to zero which could be between 5 and 50 % of the initial value. There was no significant difference between patients and probands. When the cuff was released after 6 min of occlusion, in all volunteers a steep signal increase was seen that exceeded the initial value and can be interpreted as consequence of reactive hyperaemia. The mean amount of this postischemic overshoot was quite comparable for the two methods. However, the linear correlation coefficient was only r = 0.39.

In patients only a 3 min occlusion could be performed, because some of them reacted with an ischemic pain. After release of the cuff, the postischemic overshoot was smaller than for healthy probands and did not occur in all patients. At clinical stage II b there were some patients with a reactive overshoot in $tcPO_2$ who did not show this reaction with the LDF signal (Table 3).

Tab.3: Effect of arterial occlusion of 3 min duration (probands: 6 min) on $tcPO_2$ and LDF signal from the forefoot, measured at 37°C

| | L D F | | | $tcPo_2$ | | | | |
|---|---|---|---|---|---|---|---|---|
| | before | during | after | before | during | | after | |
| | | - arterial | | occlusion - | | | | |
| probands | 100 | 15.4 | 372.9 | 6.4 | 0.5 | (8) | 20.1 | (309 |
| n = 20 | | ± 12.6 | ± 239.8 | ± 4.1 | ± 0.6 | | ± 7.5 | |
| patients | | | | | | | | |
| stage II b | 100 | 21.9 | 118.7 | 4.7 | 0.3 | (6) | 9.4 | (200 |
| | | ± 16.3 | ± 12.0 | ± 4.8 | ± 0.4 | | ± 8.3 | |
| stage IV | 100 | 31.2 | 135.4 | 5.6 | 0.4 | (7) | 10.6 | (187 |
| | | ± 17.1 | ± 62.2 | ± 4.8 | ± 0.7 | | ± 7.2 | |

LDF: % of initial value, $tcPo_2$: mmHg (% of initial value); $\bar{x}$ ± S.D.

DISCUSSION

A comparison of the probe size itself demonstrates that the $tcPO_2$ signal is recorded from only a small number of skin capillary units while the LDF signal scans over an entire field of capillaries. The signal partly derives from the venous subpapillary plexus and perhaps to a certain amount even from underlying arterial plexus. As intensity of reflected light is decreasing with depth, the importance of the vessel system is decreasing exponentially. There is no exact knowledge about the contribution of the different vessel layers, but capillary loops and the venous plexus will take the major part. Some of the differences between the results obtained with the two methods may be explained by this assumption.

Leg dependency was shown to be an interesting test for the discrimination of patients, especially when $tcPO_2$ is concerned. The signal decrease in probands is interpreted as a vasoconstrictor reaction to the increased hydrostatic pressure. This reaction is absent in patients with microcirculatory disorders. However, some patients even in stage IV show a normal pattern indicating that PAOD does not uniformly disturb the microcirculation. The greater occurrence of signal decrease reactions in patients with the LDF measurements might reflect on vasoconstriction or pressure induced flux reduction on the venular site.

Contribution of the venous plexus layer may also explain the higher variation of the LDF response to leg elevation, when compared to $tcPO_2$. The $tcPO_2$ signal remained stable or decreased, only rarely showing an increase. This will be an effect of the decreased hydrostatic pressure. Leg elevation also produces a facilitation for the venous outflow so that the flux in th venous plexus may be increased. This increase will not take place when the arteriolar inflow is dominantly decreased.

Preliminary findings by comparison of LDF measurements at 37°C and at normal skin temperature revealed the importance of temperature influences. In healthy probands, the signal of the unheated probe showed a decrease in some cases of leg elevation when measurements at 37°C yielded an increase.

On arterial occlusion, the signal decrease of $tcPO_2$ to zero indicates a cessation of the oxygen supply. However, there is still a flux signal to be monitored. This is in accordance with microscopical findings that demonstrate a blood shifting within the capillaries when the arterial inflow is stopped.

The lower postischemic overshoot in patients can be expected in the presence of a vascular disease, reflecting the impaired reactive hyperemia on the part of the macrocirculation. However, it is interesting that there is a high variation in the amount of the overshoot and that not all capillaries take a part in it. The difference between the two methods was most obvious in the patients in clinical stage II b where the overshoot of the LDF signal was remarkably low, while the $tcPO_2$ signal showed a considerable overshoot. Possibly the challenge by 3 min of ischemia is not large enough to produce a high reactive hyperaemia in all vessels scanned by the LDF-probe while the nutritional skin capillaries (from which the $tcPO_2$ signal derives) have already to cover a supply deficit.

In conclusion, Laser Doppler flowmetry is an interesting complementary method to $tcPO_2$ measurements in studying skin microcirculation. Different results between the two methods seem to be mainly due to different vessel layers contributing to the signals.

REFERENCES

1. Creutzig, A., D. Dau, L. Caspary and K. Alexander, 1986: Transcutaneous

$PO_2$ measured at different electrode core temperatures in healthy volunteers and patients with arterial occlusive disease.
Int. J. Microcirc. Clin. Exp. , 5: in press
2. Tenland, T., E. Salerud, G.A. Nilsson and P.A. Ödberg, 1983: Spatial and temporal variations in human skin blood flow.
Int. J. Microcirc. Clin. Exp., 2: 81-90

THE USE OF THE HELLIGE OXYMONITOR TO STUDY SKIN BLOOD FLOW CHANGES

P.M. Gaylarde and I. Sarkany

Department of Dermatology
Royal Free Hospital
London, NW3 2QG U.K.

SUMMARY

Transcutaneous oxygen tension is a flow related parameter. Detailed analysis of the physiology and physical chemistry of oxygen consumption and diffusion indicates that $tcPO_2$ is not proportional to skin blood flow. Measurement of $tcPO_2$ at 37 °C allows changes in skin blood flow to be clearly demonstrated and is of use in many areas of clinical medicine.

Transcutaneous oxygen tension recorded using a polarographic oxygen electrode depends on skin blood flow, skin respiration, arterial oxygen concentration, temperature, skin and electrode permeability and the oxygen consumption of the electrode.

Applying clearance principles, the relationship between blood flow, respiration and the difference between arterial and venous oxygen concentrations is well established. This has long been used to measure cardiac output, since the other three parameters are readily determined. Arterial saturation is normally greater than 95% of maximum in subjects without lung disease and it may thus be considered to be constant. Tissue respiration is independent of oxygen concentration when $PO_2$ exceeds 2 mm Hg[1]. If skin respiration is invariant at constant temperature, when the tissue oxygen tension exceeds 2 mm Hg, then blood flow is inversely proportional to the difference in concentration between arterial and venous blood. It has recently been directly shown that the inference that tissue respiration is independent of tissue blood flow is accurate[2]. Correcting for the oxyhaemoglobin dissociation curve and for the deviation from zero order respiration kinetics when tissue $PO_2 < 2$ mm Hg, the relationship between venous oxygen tension and blood flow at 37 °C is shown when arterial oxygen concentration is constant (Figure 1).

The relationship shown in Figure 1 would also describe the $tcPO_2$ values observable if no countercurrent diffusion occurred between arterial and venous blood, and if no concentration gradient existed in the skin. The role of these two factors must be considered separately, since they will act in different ways on the ideal relationship illustrated. Under conditions of unvarying blood flow a constant concentration gradient will be established. If the oxygen consumption of the electrode is negligibly small, then the concentration gradient across the membrane will be small

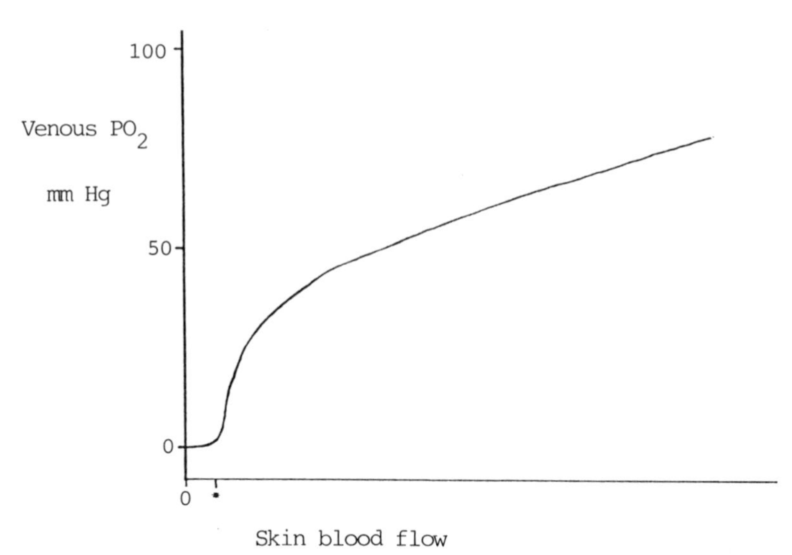


Fig. 1. Curve showing relationship between venous $PO_2$ and blood flow. When respiration is constant, flow is inversely proportional to the difference between arterial and venous oxygen concentrations. Respiration is independent of $PO_2$ when $PO_2 > 2$ mm Hg. Arterial oxygen concentration is assumed to be constant and curve corrected for oxyhaemoglobin dissociation.

and the oxygen tension on each side of the membrane will be equal. Diffusion across a membrane or through the skin is described by the equation

$$J = \frac{C_1 Kp_1 - C_2 Kp_2}{Ri_1 + Ri_2 + dRb}$$

where J = flux, C = concentration, Kp = partition coefficient of the permeant molecule between the membrane and the external phase, Ri = diffusional resistance of the membrane interface, d = membrane thickness, and Rb = bulk diffusional resistance (the reciprocal of Fick's permeability constant). This is an extension of the relationship first described by Barrer[3] in 1937 and subsequently rediscovered by numerous physiologists ignorant of his work. This equation shows that Fick's Law cannot be blindly used to describe diffusion across a series of phase changes at interfaces associated with membranes. When $d \rightarrow 0$, the sum of the interfacial resistances is large compared to the product dRb and the total membrane permeability is independent of d, the membrane thickness. The validity of this conclusion has been directly confirmed using a model system. For human skin, tissue permeability is only important in determining $tcPO_2$ at the skin interface when the vessels are 1 to 10 mm from the surface. Published data on the oxygen concentration gradient around capillaries obtained using microelectrodes indicates that the gradients observed must be induced by respiration rather than diffusional resistance.

Studies on response times in skin slices frozen and thawed x3 to minimise respiration and on *in vivo* response times[4] allow the depth of the $tcPO_2$ signal to be estimated. Figure 2 shows the relationship between skin thickness and response time. This relationship appears intuitively unexpected, but consideration of the equation shown above, in which permeability is a function of the solubility of oxygen in each of the tissue

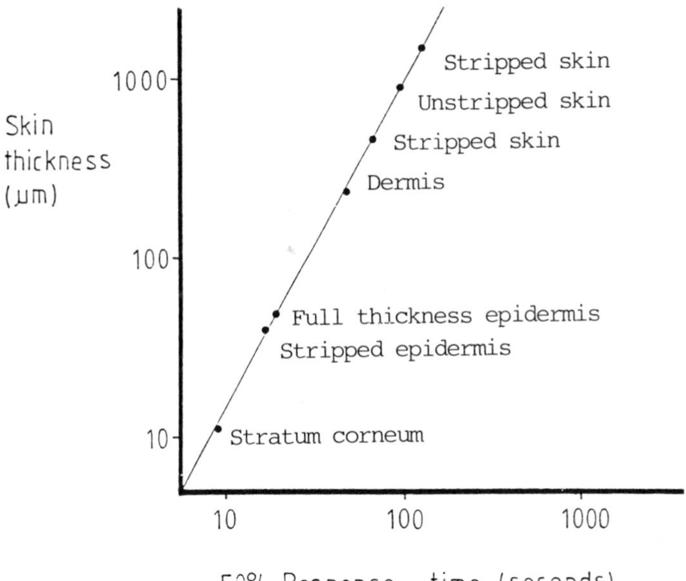


Fig. 2. The relationship between the rate of rise or fall in $PO_2$ following a stepwise change in oxygen tension and skin thickness. The result shows that for skin the time constant for any specific thickness is independent of composition. This is not true of other model systems investigated, even though this result is anticipated from the equation given in the text. The origin of this independence arises from the fact that oxygen solubility increases as permeability increases. The linear relationship will be observed when the interfacial resistances are constant and greater than the product dRb.

layers,explains why the time constant is independent of tissue composition, even though the permeability is dependent on this. The time-lag method has been used almost exclusively to determine Fick's diffusion constant of gases in membranes for the past 50 years; the rearrangement of Barrer's equation presented here shows that this method cannot be used to obtain this constant and explains also why the true relationship between membrane permeability and thickness has not been recognised. The *in vivo* time constants[4], together with data that shows fluctuations in $tcPO_2$ at a frequency of up to 0.25 Hz[5], indicate that the blood vessels within 100 μm of the epidermal surface largely determine $tcPO_2$. Analysing published data, the $PO_2$ difference between the surface of the dermal capillaries and the stratum corneum is probably less than 10 mm Hg, but direct confirmation of this deduction remains to be produced. Since respiration is independent of oxygen tension when it exceeds 2 mm Hg, this difference will remain constant and the curve illustrated in Fig. 1 will be shifted to the right. Skin blood flow at the point of inflection when $tcPO_2$ = 2 mm Hg may be estimated by comparison with data obtained using xenon clearance; on standing, $tcPO_2$ falls from a median of 6.2 mm Hg to 0.6 mm Hg (N = 52), whereas skin blood flow measured by xenon clearance falls by only 25 to 50%. The point of inflection is therefore approximately 2 $ml.100\ ml^{-1}.min^{-1}$. This data therefore provides a calibration point. Using other parallel methods, it should be possible to provide a complete calibration of the relationship between flow and oxygen tension. Studies on gut oxygen tension[6] and blood flow support the derived relationship shown in Fig. 1.

Countercurrent diffusion between arterial and venous blood lowers the concentration of oxygen in the blood feeding the superficial dermal blood vessels. Since this is a dynamic process, the loss of oxygen from small arteries and arterioles will be dependent on the rate of flow as well as the concentration gradient and the bulk permeability of the intervening tissue. The effect of this process will be to alter the shape of the curve shown in Fig. 1. Precise modelling of the relationship between $tcPO_2$ and skin blood flow will depend on detailed knowledge of respiration and its variation throughout the epidermis and dermis, the role of counter-current exchange and the constants determining permeability. The problem of providing numerical solutions from an exact model are that the relevant parameters will vary at different parts of the body and will be temperature dependent. At present, only an empirical solution using a direct method of flow determination to calibrate the relationship appears feasible.

Although the transcutaneous oxygen electrode cannot yet be used to record absolute blood flow, it is an excellent means of recording changes in skin blood flow, since it is tolerant of subject movement and responds rapidly. As shown above, when the $tcPO_2$ lies between 2 and 20 mm Hg, values typically observed in normal skin at an electrode temperature of 37 °C, the transcutaneous oxygen electrode is very sensitive to small changes in blood flow. When blood flow is increased by factors such as local heating and the oxygen tension exceeds 40 mm Hg, the variation of oxygen tension with blood flow is small and the system should be regarded as being relatively insensitive to changes in skin blood flow. In addition, elevated skin temperatures interfere with important physiological responses of clinical interest. We have used the oxygen electrode at 37 °C and at elevated temperatures to investigate the effects of vasodilators on skin blood flow in patients with Raynaud's disease[7], the treatment of hypertrophic scars by compression therapy[8], the mechanism of pharmacological suppression of alcohol-induced flushing[9], the cause of leg ulcers[10], the mode of action of U.V. light on the regulation of skin blood flow[11], the role of pharmacological agents on vascular reflexes[12], the design of an improved drug delivery system (in press), neurological abnormalities in the control of blood flow in diabetics, and the etiology

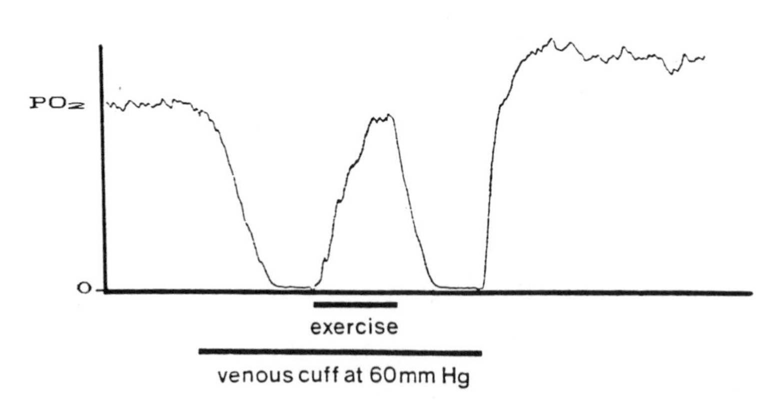


Fig. 3. Effects of venous pressure elevation and exercise on $tcPO_2$ in a recumbent normal subject. Note rapid fall in $PO_2$ to values near zero and periodic fluctuations in $PO_2$ which remain apparent even during the initial fall following venous pressure elevation. Electrode (37 °C) applied directly to stratum corneum without electrode membrane.

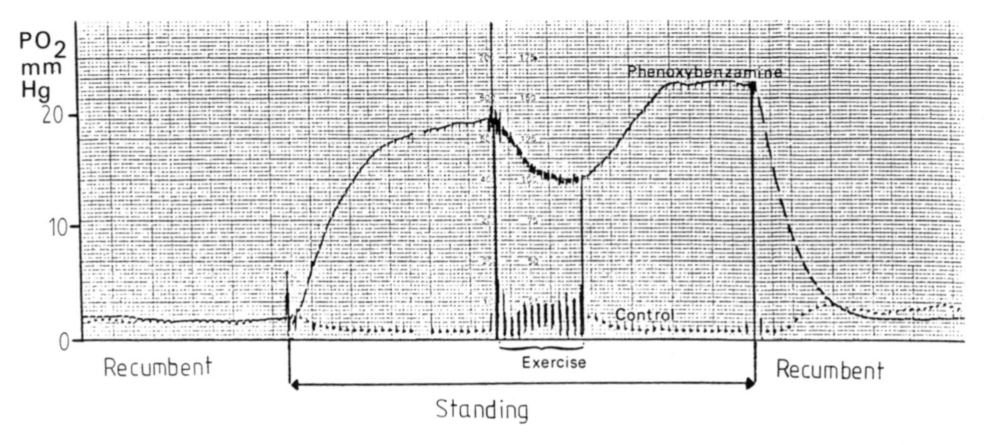


Fig. 4. Effect of phenoxybenzamine (20 µg) injected intradermally 24 h previously in a normal recumbent subject compared to normal adjacent control site 5 cm from the treated site. The untreated control skin shows a normal fall in $PO_2$ induced by venous pressure elevation on standing and a rise on exercise. On the phenoxybenzamine treated site, there is a very marked rise in $tcPO_2$ and a fall during exercise. The responses seen after phenoxybenzamine at this time (together with others not illustrated) show a striking resemblance to those seen in patients with severe atherosclerosis. This pattern of response to phenoxybenzamine is both time and concentration dependent, and is completely different both earlier and later at the dose illustrated. When the dose is varied, the effects also vary in a manner that is consistent with a simple model in which the concentration of the bound drug declines with time. The increase in $tcPO_2$ on standing does not occur after a dose of 200 ng, the responses at 1 h after injection resembling those seen 5 days after 20 µg. Electrode temperature, 37 °C. Electrode membrane present.

of port wine stains (in press). Fig. 3 shows the normal postural and exercise response, and Fig. 4 shows the effect of 20 µg given intradermally 24 h previously, together with a recording from an adjacent control site. The abnormal vasodilator response seen over the phenoxybenzamine treated site, together with other abnormal responses observed, are similar to the altered pattern of responses found in patients with atherosclerosis[12,13,14]. These traces illustrate some of the potential areas of use of the oxygen electrode in clinical medicine and show especially the value of the technique in the study of changes in skin blood flow.

The principle requirements for the accurate use of the transcutaneous oxygen electrode are adequate gain to detect small changes in oxygen tension at low partial pressures and good zero stability. This latter characteristic is of great importance when oxygen levels are close to zero. Other valuable attributes are rapidity of response, accurate temperature control of the oxygen electrode, and low electrode oxygen consumption. The Hellige machine has proved to be satisfactory in all these aspects.

REFERENCES

1. Jöbsis, F.F. Basic processes in cellular respiration, in "Handbook

of Physiology, Section 3: Respiration". W.O. Fenn, H. Rahn, eds., American Physiological Soceity, Washington (1964).

2. Kvietys, P.R., and Granger, D.N. Relation between intestinal blood flow and oxygen uptake. Am J. Physiol. 242:G202 (1982)
3. Barrer, R.M. XII. Gas flow in solids. Phil. Mag. 28:148 (1937)
4. Tan, O.T., Gaylarde, P.M., and Sarkany, I. Skin oxygen tension and blood flow changes in response to respiratory manoeuvres. Clin Exper. Dermatol. 7:33 (1982)
5. Gaylarde, P.M., and Sarkany, I. Periodic skin blood flow. N. Engl. J. Med. 312:1194 (1985)
6. Piasecki, C. First experimental results with the oxygen electrode as a local blood flow sensor in the canine colon. Br. J. Surg. 72:452 (1985)
7. Sarkany, I., Tan, O.T., and Gaylarde, P.M. Vasodilator refractoriness in Raynaud's syndrome. Clin. Exper. Dermatol. 7:679 (1982)
8. Berry, R.B., Tan, O.T., Cooke, E.D. et al. Transcutaneous oxygen tension as an index of maturity in hypertrophic scars treated by compression. Br. J. Plastic Surg. 38:163 (1985)
9. Tan, O.T., Stafford, T.J., Sarkany, I., et al. Suppression of alcohol-induced flushing by a combination of $H_1$ and $H_2$ histamine antagonists. Br. J. Dermatol. 107:647 (1982)
10. Dodd, H.J., Gaylarde, P.M., and Sarkany, I. Skin oxygen tension in venous insufficiency of the lower leg. J. Roy. Soc. Med. 78:373 (1985)
11. Dodd, H.J., Tatnall, F.M., Gaylarde, P.M., and Sarkany, I. The effect of ultraviolet irradiation on skin oxygen tension and its potential role in the management of venous leg ulcers, in: "Phlebology 85", D. Negus and G. Jantet, eds. John Libbey, London 601-4 (1986)
12. Eickhoff, J.H. Forefoot vasoconstrictor response to increased venous pressure in normal subjects and in arteriosclerotic patients. Acta Chir. Scand. 502:7 (1980)
13. Caspary, L., Creutzig, A., Höppner, L., Katsis, C., and Alexander, K. Comparison of laser-Doppler and $tcPO_2$ measurements in healthy probands and patients with arterial ischemia. 3rd International Symposium on Continuous Transcutaneous Monitoring, Zurich (1986)
14. Gaylarde PM, Dodd HJ, Sarkany I. Pharmacological aspects of the control of blood flow in the lower leg, in: "Phlebology 85", D. Negus and G. Jantet, eds. John Libbey, London (1986)

## CALCULATION OF SKIN BLOOD FLOW

# ESTIMATED PERIPHERAL BLOOD FLOW IN PREMATURE NEWBORN INFANTS

W.A. van Asselt, J.J. Geerdink, G. Simbruner and A. Okken

Dept. Pediatrics, Div. Neonatology, University Hospital
Groningen, The Netherlands

## INTRODUCTION

Premature infants are very susceptible to changes in their physical environment. It is well known i.e. that a relatively small fall in incubator temperature may result in a substantial fall in body temperature, which apparently is not compensated for a sufficient rise in metabolic rate[1]. Harpin and Rutter[2] have demonstrated that, if humidity in the incubator is high, very premature infants have a higher and more stable body temperature. This is probably related to a reduced evaporative heat loss at the high humidity.

As is known from studies reported by Brück[3], premature infants do have the ability to change their peripheral blood flow in response to a change in environmental temperature. Less is known about the effect of a change of environmental humidity on peripheral blood flow. In premature infants a change in peripheral blood flow in response to a change in environmental humidity would indicate that apart from temperature, humidity is an important part of their physical environment. We have studied the effect of a change in humidity on peripheral blood flow in premature newborn infants.

## METHODS

The estimation of peripheral blood flow in this study is based on the determination of heat loss from the limbs, the flow of heat from head and trunk to the limbs and total heat production. We have measured core temperature, limb skin temperature and operative temperature, and simultaneously performed indirect calorimetry. Heat loss from the limbs is calculated from the difference between skin temperature and operative temperature in the incubator. Heat flow from head and trunk to the limbs is calculated from the difference between core temperature and limb skin temperature, combined with the specific heat of blood[4]. Heat production is calculated from oxygen consumption and carbondioxide production (indirect calorimetry)[5].

All infants were studied in forced convection incubators in the nursery. During the study the humidity in the incubator was changed at random, from low to high in half of the infants, and from high to low in the other half of the infants. Incubator temperature was kept at the same level throughout each study.

Table 1. Incubator temperature (Tinc,°C) and water vapour pressure ($P_{H_2O}$,mmHg) at low (L) and high (H) humidity for different age groups.

| Mean | Tinc (°C) | | $P_{H_2O}$ (mmHg) | |
|---|---|---|---|---|
| age (d) | L | H | L | H |
| 3.3 | 35.0 | 35.2 | 17 | 31 |
| 10.7 | 34.5 | 34.4 | 16 | 30 |
| 17.0 | 33.2 | 33.2 | 19 | 25 |

Table 2. Estimated peripheral blood flow (PBF) at low (L) and high (H) humidity (ml/min/m²) for different age groups.

| Mean | Relative humidity | |
|---|---|---|
| age (d) | L | H |
| 3.3 | 205 ± 111 | 491 ± 264 |
| 10.7 | 337 ± 151 | 693 ± 454 |
| 17.0 | 623 ± 283 | 577 ± 81 |

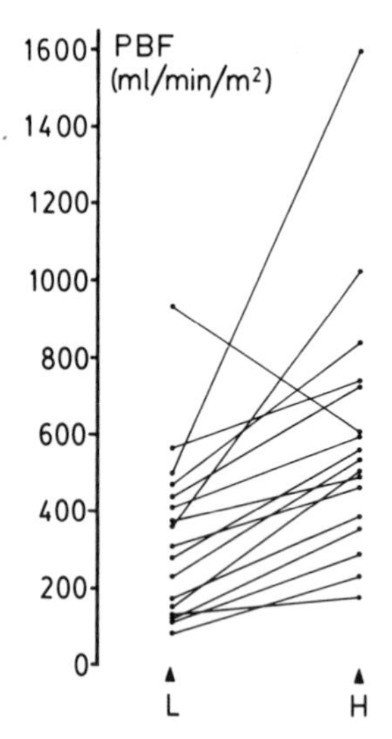


Fig. 1. Estimated peripheral blood flow (PBF) at low (L) and high (H) humidity (ml/min/m²).

PATIENTS

Ten infants were studied, some of them several times. Gestational age ranged from 27 to 32 weeks (mean 29.1 weeks), birth weight from 990 to 1675 grams (mean 1295 grams), and age at the time of the study from 1 to 18 days (mean 8.8 days). Arbitrarily we have divided the infants into two different age groups. Table 1 shows the environmental conditions during the study for each age group.

RESULTS

Results are presented in figure 1 and table 2. The estimated peripheral blood flow in all infants but one is higher at the high humidity.

CONCLUSIONS

In this study we have used a relatively simple method to estimate peripheral blood flow from temperature differences mainly, and it shows that peripheral blood flow of very premature newborn infants is affected by humidity in the incubator. In all infants, except one, peripheral blood flow was significantly increased at a higher humidity in the incubator, even though the incubator temperature was not changed. This, most probably, is the result of the reduced evaporative heat loss under these circumstances. This study thus confirms the observations of Harpin and Rutter that humidity is an important part of the physical environment of very premature newborn infants, and it further demonstrates that studies concerning peripheral blood flow measurements in these infants should not only include measurements of temperature, but also of humidity.

REFERENCES

1. P.J.J. Sauer, H.J. Dane, H.K.A. Visser, New standards for neonatal thermal environment of healthy very low birth weight infants in week one of life, Arch Dis Child 59:18 (1984).
2. V.A. Harpin, N. Rutter, Humidification of incubators, Arch Dis Child 60:219 (1985).
3. K. Brück, Temperature regulation in the newborn infant, Biol Neonate 3:65 (1961).
4. G. Simbruner, Thermodynamic models, Facultas Verlag, Wien (1983).
5. A. Okken, J.H.P. Jonxis, P. Rispens, W.G. Zijlstra, Insensible water loss and metabolic rate in low birth weight newborn infants, Pediat Res 13:1072 (1979).

# SKIN BLOOD FLOW CALCULATIONS FROM TRANSCUTANEOUS GAS PRESSURE MEASUREMENTS

F. Caligara[1], G. Rooth[2,3] and U. Ewald[3]

[1]Commission of the European Communities, Joint Research Centre, Karlsruhe Establishment, Institute for Transuranium Elements, Postfach 2340, D-5700 Karlsruhe, FRG
[2]Department of Obstetrics, University Hospital, CH-8091 Zürich, Switzerland
[3]Department of Paediatrics, University Hospital, S-751 85 Uppsala, Sweden

## INTRODUCTION

It is the aim of this study to show how relevant information concerning skin blood flow and skin metabolism may be obtained by means of transcutaneous gas pressure measurements.

## CALCULATION OF SKIN BLOOD FLOW

The blood flow through a tissue is calculated from:

$$\Phi = \dot{V} / \Delta O_{2b} \qquad (1)$$

where $\dot{V}$ = the oxygen consumption of the tissue
$\Delta O_{2b}$ = the amount of oxygen given out by the blood.

The equipment used for transcutaneous monitoring of $Po_2$ and $Pco_2$ indicates different skin parameters if used at 37°C, hence these parameters will be called cutaneous ($cPo_2$ and $cPco_2$ respectively).

$\Delta O_{2b}$ and $\dot{V}$ for the skin tissue may be obtained from cutaneous $Po_2$ and $Pco_2$ if some physiological and anatomical constants are known.

Fig. 1 shows a section of the skin tissue, which consists of:
- layer a, living tissue perfused by the capillaries
- layer b, living but not perfused tissue
- layer c, keratinized tissue, in which no metabolism occurs.

The electrode is applied to the outer side of c. The oxygen and carbon dioxide tensions are also shown in Fig. 1.

Under steady state conditions the slope of the oxygen profile is such that each tissue section is crossed by the amount of oxygen consumed downvalley; the oxygen concentration profile decreases therefore from the left to the right.

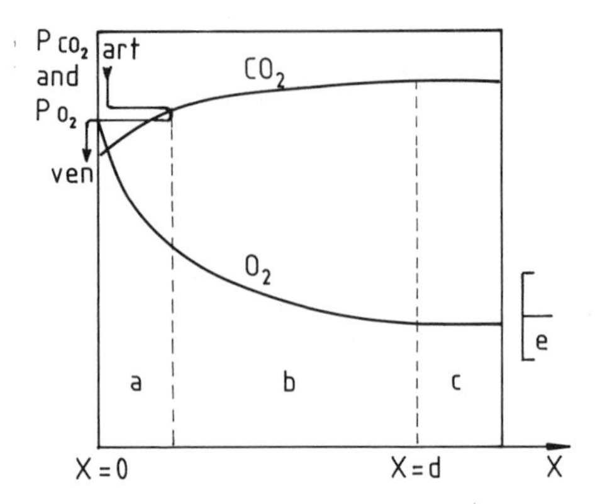


Fig. 1. Gas pressure profiles across the skin tissue. a = living perfused tissue, b = living tissue, c = keratinized tissue,d = thickness of the living skin tissue, e = electrode.

The countercurrent $O_2$ flow due to the close contact between the venous and arterial capillary branches and the high permeability of the capillary membrane to $O_2$ tend to equalize the $Po_2$ in the two branches for the same value of x, but capillary $Po_2$ falls along x and reaches a nadir at the capillary dome.

The shape of the $Po_2$ profile is described by the differential equation:

$$D \frac{\delta C}{\delta x} = - \dot{V} (d-x) \tag{2}$$

where D = $O_2$ diffusion coefficient
C = $O_2$ concentration
d = thickness of living tissue.

$C = Po_2 \cdot \alpha$ ($\alpha$ = $O_2$ solubility).

The integration of (2) gives:

$$P_vo_2 - cPo_2 = \Delta Po_{2v} = \tfrac{1}{2} \cdot V \cdot d^2 / (D \cdot \alpha) \tag{3}$$

where $\Delta Po_{2v}$ is the oxygen pressure drop of metabolic origin. Equation (3) provides the relationship between $Po_2$ drop in the tissue and the metabolic $O_2$ consumption of the tissue itself. In order to solve (3) for V, $cPo_2$ is measured and $P_vo_2$ must be independently obtained.

For the $Pco_2$ profile, the following considerations hold:

- $CO_2$ is produced in the tissue and carried away by the blood, its profile therefore mirrors that of $O_2$, $cPo_2$ being higher than $P_vco_2$.
- the diffusion coefficient of $CO_2$ is about 20 times larger than that of $O_2$, $\Delta Pco_2$ is therefore $\cong$ 20 times smaller than $\Delta Po_2$[1].
- $\Delta Pco_2$ is further reduced by the fact that the amount of $CO_2$ transported is lower than the amount of $O_2$, since the respiratory quotient RQ is less than 1.

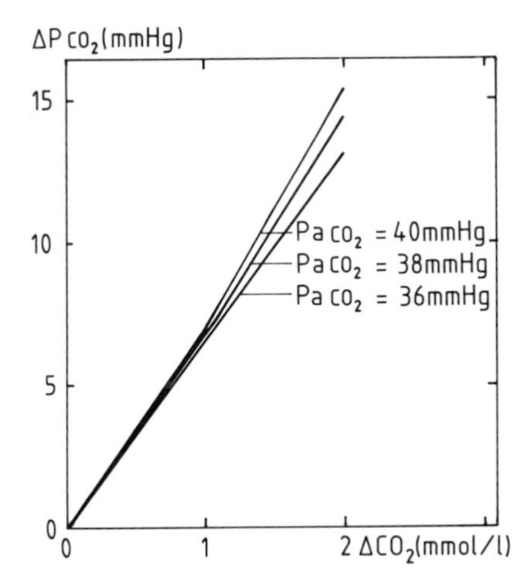


Fig. 2. $\Delta Pco_2$ against $\Delta CO_2$ for different arterial $Pco_2$ (calculated from the Siggaard-Andersen alignment nomogram).

From our experience a typical value for $\Delta Po_2$ is = 45 mm Hg, 2 mm Hg can therefore be taken as a standard value for $\Delta PO_{2v}$ and

$$cPco_2 - P_vco_2 = \Delta Pco_{2v} = 2 \text{ mm Hg} \tag{4}$$

may be used.

## CALCULATION OF $P_vo_2$ FROM $cPco_2$

The calculation is based on the assumption that the skin metabolism is aerobic and its rate is constant, then:

$$\Delta O_{2b} = \Delta CO_{2b} / RQ \tag{5}$$

where $\Delta CO_{2b}$ = the $CO_2$ carried away by the blood,
$\Delta O_{2b}$ = the $O_2$ given out by the blood to supply the tissue.

$\Delta CO_{2b}$ may be read out from Fig. 2, which has been reconstructed from the Siggaard-Andersen[2] alignment nomogram and shows $\Delta Pco_{2b} = P_vco_2 - P_aco_2$ against $\Delta CO_{2b}$. Venous $Pco_2$ is given by (4) and the arterial $Pco_2$ may be taken as 40 mm Hg in absence of hyperventilation or other abnormalities. Thus from Fig. 2 and equation (5) one has the volume of $O_2$ lost by the blood to the tissues.

Fig. 3 shows the blood oxygen dissociation curve plotted as venous $Po_2$ against vol % $O_2$ lost, assuming arterial 99 % saturation, pH = 7.4 and Hb = 160 g/l. From it $P_vo_2$ may be read out when $\Delta O_{2b}$ is known and then (3) is solved for $\dot{V}$, if D, $\alpha$, d are known. $\Phi$ is then calculated from (1).

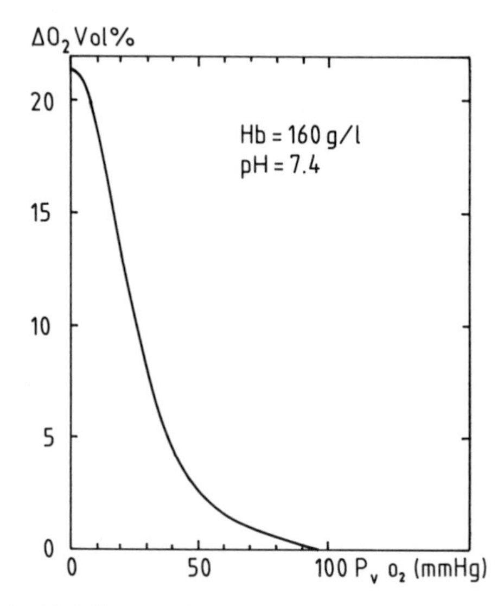


Fig. 3. Vol % $O_2$ given out by the arterial blood against the resulting venous $Po_2$.

## EXAMPLE OF CALCULATION

Fig. 4 gives $cPco_2$ and $cPo_2$ for a subject who at time 0 started to perform work on an ergometer. A gradual increase in $cPo_2$ will be seen, indicating the increase of the skin blood flow needed for the thermoregulation.

The constants of the calculations are:

$O_2$ diffusion coefficient $= D = 1.35 \cdot 10^{-5}$ cm$^2$ / sec
$O_2$ solubility $= \alpha = 1.48 \cdot 10^{-9}$ mol / cm$^3$ / mm Hg
skin tissue thickness $= d = 3 \cdot 10^{-5}$ m

We measured[3] D in vivo in human skin, and found $0.98 \cdot 10^{-5}$ at 28°C: assuming a 3.5 % per degree increase, we calculated the above value for 37°C. The value for d is given by Severinghaus[4], and the $O_2$ solubility is that of water. Arterial $Pco_2$ was assumed to be 36 mm Hg, because the subject hyperventilated breathing through a mouthpiece.

We obtained the following result:

Calculated $\dot{V} = 1.7 \cdot 10^{-9}$ mol / sec / g
The steady state skin blood flow was 5.3 cm$^3$ / m / 100 g tissue
The peak skin blood flow was 13.5 cm$^3$ / m / 100 g tissue

Calculating $\dot{V}$ from the slope of the $tcPo_2$ (t) curve Severinghaus[4] found $\dot{V} = 2 \cdot 10^{-9}$ mol /sec / g in the skin kept at 43°C. The results on skin blood flow are in agreement with those obtained by Sejrsen[5] with the Xenon clearance technique.

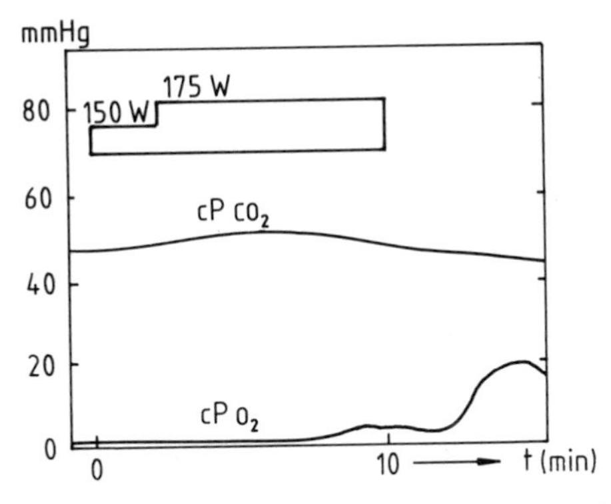


Fig. 4. Experimental $cPco_2$ and $cPo_2$ during and after work of a healthy male volunteer using a bicycle ergometer.

ACKNOWLEDGEMENTS

The study was sponsored by the Joint Research Centre of the Commission of the European Communities. Dräger AG placed the $tcPco_2$ equipment at our disposal.

REFERENCES

1. E.P. Hill, G.G. Power, L.D. Longo. A mathematical model of carbon dioxide transfer in the placenta and its interaction with oxygen. Am. J. Physiol. 224: 283 (1973).
2. O. Siggaard-Andersen. "The acid base status of the Blood". Munksgaarden, Kopenhagen 1974.
3. F. Caligara, G. Rooth. Measurement of the diffusion coefficient of the oxygen in the subcutis of man. Acta Physiol. Scand. 53: 114 (1961).
4. J.W. Severinghaus, M. Stafford, A.M. Thunstrom. Estimation of skin metabolism and blood flow with $tcPO_2$ and $tcPCO_2$ electrodes by cuff occlusion of the circulation. Acta Anaesth. Scand. suppl. 68: 9 (1978).
5. P. Sejrsen. Blood flow in cutaneous tissue in man studied by washout of radioactive Xenon. Circ. Res. 25: 215 (1969).

# THE PERCENTUAL INITIAL SLOPE INDEX OF $tcPo_2$ AS A MEASURE OF THE PERIPHERAL CIRCULATION AND ITS MEASUREMENTS BY DIFFERENT $tcPo_2$ ELECTRODES

R. Lemke[1] and D.W. Lübbers[2]

[1]Städtische Kliniken Dortmund, Kardiologische Abteilung
[2]Max-Planck-Institut für Systemphysiologie
Rheinlanddamm 201, 4600 Dortmund, FRG

Transcutaneous $Po_2$ measurements ($tcPo_2$) do not only reflect the arterial oxygenation of blood, but they may also demonstrate changes of peripheral circulation[1,2]. Two methods are available to measure the influence of peripheral blood flow on $tcPo_2$: The arterial $tcPo_2$ index ($tcPo_2/P_ao_2$)[2,3] and the percentual initial slope index[4].

Since it was observed that different types of $tcPo_2$ electrodes can give somewhat different results, we tested two commercially available $tcPo_2$ electrodes (Draeger and Radiometer-Company) in healthy volunteers in order to compare the behaviour of these two electrodes under different flow conditions.

Six healthy volunteers were studied. The $tcPo_2$ electrodes of the two companies were attached one near the other to the inner side of the forearm. The first measurement was taken with the arm elevated above the head. The arm was kept elevated till a constant $tcPo_2$ reading was achieved. Then the arm was lowered until a constant $tcPo_2$ reading was obtained on the hanging arm. Fig. 1 shows an example of a measurement of the initial slope index. The solid line is the trace of the Draeger electrode, the dotted line the trace of the Radiometer electrode. The Draeger electrode starts (lower part of Fig. 1) with a $tcPo_2$ of 85 mm Hg. Then a blood pressure cuff was wrapped around the upper arm and the cuff was inflated up to suprasystolic values (arrow) till the $tcPo_2$ electrode reading dropped to zero. Then the inflation was quickly released in approximately 4 seconds. The maximal velocity of the $tcPo_2$ increase was determined by laying a tangent on the initial steepest part of the recovery curve. Then the $\Delta tcPo_2$ /min (initial slope) was determined graphically. To obtain a slope index independent of the starting $tcPo_2$ value, the index was calculated in percent of the starting $tcPo_2$ ($tcPo_2(st)$) using the equation:
$[(\Delta tcPo_2(in)/min):tcPo_2(st)] \times 100$.
This index was called "percentual initial slope index" (PIS index). To calculate the $tcPo_2$ index, a capillary $Po_2$ from the hyperemic ear lobe was drawn. Finally, the Doppler pressure of the radial artery was measured while the arm was elevated or hanging down.

The $tcPo_2$ electrode of the Draeger Company has 3 platinum wires of 15 $\mu$m diameter and a membrane of 25 $\mu$m Teflon. The $tcPo_2$ electrode of the Radiometer Company has a single 25 $\mu$m platinum wire and a 15 $\mu$m polypropylene membrane. 44°C was used as heating temperature for both electrodes.

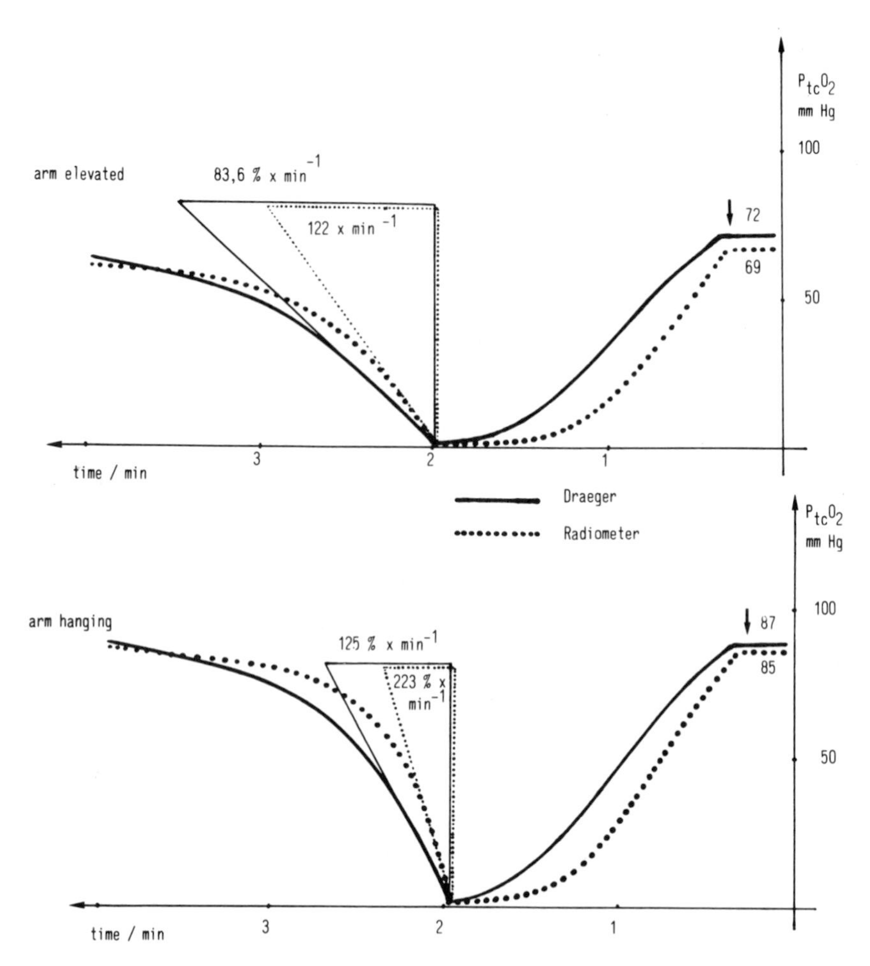


Fig. 1. Example of the measurement of the percentual initial slope index of the $tcPo_2$ with the arm elevated and hanging down. Solid line: Draeger electrode. Dotted line: Radiometer electrode.

RESULTS

The values measured are listed in Table 1. The Doppler measurements showed a significant difference between the elevated and the hanging arm. This demonstrated a definite decrease of the peripheral flow with the arm elevated. Consequently, the arterial $tcPo_2$ index and the PIS index of both electrodes were significantly higher with the arm hanging down. While the flow conditions were good, the Radiometer electrode measured only slightly lower $tcPo_2$ values compared to the Draeger electrode; also the arterial $tcPo_2$ indices between the two electrodes differed only slightly. With the arm elevated, the Radiometer electrode showed distinctly lower values than the Draeger electrode, whereas the mean percentual slope index of the Radiometer electrode was considerably higher for the arm hanging down as well as elevated, compared to the Draeger electrode. The difference in the electrode behaviour is clearly shown in Fig. 1, which demonstrates that the reaction of the Radiometer electrode is always quicker than that of the Draeger electrode.

These results demonstrate that a different design of the electrode (diameter of Pt wire, membrane properties, heating area) had distinct ef-

Table 1

$P_aO_2$ = 93.7 ± 3.14 mm Hg (n = 6)

| arm | D-flow (mm Hg) | $tcPo_2$ (mm Hg) | | $tcPo_2$ index | | PIS index | |
|---|---|---|---|---|---|---|---|
| | | D | R | D | R | D | R |
| hang. | 121.7 | 86.0 | 82.5 | 0.92 | 0.88 | 103.2 | 167.3 |
| | (14.4) | (8.2) | (4.6) | (0.06) | (0.04) | (29.1) | (35.8) |
| elev. | 89.2 | 65.0 | 53.0 | 0.70 | 0.52 | 71.7 | 95.9 |
| | (8.0) | (10.8) | (16.3) | (0.09) | (0.18) | (20.4) | (21.8) |
| p | | | | 0.001 | 0.005 | 0.005 | 0.005 |

hang.: arm hanging down; elev.: arm elevated; D.flow: Doppler "flow" of the radial artery (mm Hg); $tcPo_2$ index: $tcPo_2/P_ao_2$; PIS index: $[(\Delta tcPo_2(in)/min):tcPo_2(st)] \times 100$; n: number of volunteers; numbers in brackets: standard deviation; p: significance.

fects on the measured $tcPo_2$ value. For theoretical reasons such an influence had to be expected, but it was surprisingly large. The influence of the design was small at good flow conditions (i.e. in the upper branch of the circulatory hyperbola), but became large at reduced flow (middle and lower part of the circulatory hyperbola). Considering the percentual initial slope index both electrodes are very sensitive towards a change in the peripheral flow by demonstrating a larger PIS index of the arm hanging down compared to the arm elevated.

## CONCLUSION

Electrodes from different producers may show clinically important differences of the $tcPo_2$ values depending on the flow condition. We found that compared to the Draeger electrode the Radiometer electrode was more sensitive towards changes in peripheral flow, especially if the flow conditions worsened. Both electrodes demonstrated significant changes of the percentual initial slope index depending on the flow changes. The Radiometer electrode showed a significantly faster response compared to the Draeger electrode. Our results indicate that measurements with electrodes of different producers cannot be compared without knowing the influence of the different design. To obtain comparable results a better standardization of the electrodes would be helpful.

## REFERENCES


1. R. Huch, A. Huch and D. W. Lübbers, "Transcutaneous $Po_2$," Thieme-Stratton Inc., New York (1981).
2. D. W. Lübbers and U. Großmann, Gas exchange through the human epidermis as a basis of $tcPo_2$ and $tcPCO_2$ measurements, in: "Continuous Transcutaneous Blood Gas Monitoring." R. Huch and A. Huch, eds., M. Dekker, New York-Basel, pp. 1-34 (1983).
3. K. K. Tremper and W. C. Shoemaker, Transcutaneous oxygen monitoring of critically ill adults, with and without low flow shock. Crit. Care Med. 9:706-709 (1981).
4. R. Lemke, D. Klaus and D. W. Lübbers, Experiences with the commercially available $tcPo_2$ electrodes in adults, in: "Continuous Transcutaneous Blood Gas Monitoring." R. Huch and A. Huch, eds., M. Dekker, New York-Basel, pp. 143-151 (1983).

EXAMINATIONS ON THE BLOOD FLOW DEPENDENCE OF $tcpO_2$

USING THE MODEL OF THE "CIRCULATORY HYPERBOLA"

Jürgen M. Steinacker, Wolfgang Spittelmeister and
Reinhard Wodick

Department of Applied Physiology
University of Ulm
D-7900 Ulm, Fed. Rep. of Germany

SUMMARY

The relation of transcutaneous $pO_2$ ($tcpO_2$) and cutaneous blood flow (CBF) was measured on the forearm of 19 healthy volunteers by use of a $tcpO_2$ electrode heated to 45°C. CBF was estimated indirectly from the heating power of the electrode (HP) and with a 8 MHz bidirectional ultrasonic probe by Doppler shift in a fingertip warmed to 45°C (DF). Arterial blood flow was regulated by a cuff on the upper arm. The arterial flow was reduced in 10-15% stages of effective perfusion pressure $P_{eff}$.

There was a decrease in $pO_2$ when CBF was restricted in stages as suggested by the model of the "circulatory hyperbola" according to Lübbers. A linear dependence between $P_{eff}$, HP and DF was observed. These results indicate, that there is no autoregulation in the hyperemizied capillary bed. During respiration of air mean $tcpO_2$ was 86.0 Torr (±6.2) in normal blood flow conditions and reflects well $p_aO_2$. Transcutaneous $pO_2$ may also be used as a measure of CBF. To distinguish between these two modes, the determination of $p_aO_2$ in capillary blood probes is necessary for calculating the transcutaneous index $tcpO_2/p_aO_2$.

INTRODUCTION

The method of measuring $tcpO_2$ is based on the fact that, in the case of a heat-induced hyperemic area of the skin, the oxygen consumption of the skin and of the electrode is considerably less than the amount of oxygen supplied by the blood. At the end of the heating-up period, an equilibrium is reached between the oxygen transport by the blood and the oxygen extraction by skin and electrode. After that, $tcpO_2$ follows the fluctuations of the arterial $pO_2$ ($p_aO_2$). In this state, the measuring system electrode/skin/capillary is not dependent on small fluctuations of the local cutaneous blood flow (CBF), since the total flow far exceeds normal CBF (Lübbers and Grossmann 1983, Huch et al.1983).

As long as the circulation is not restricted, this measuring system can be used to continuously check changes in $p_aO_2$. However, if CBF is low, oxygen consumption of the skin and the electrode will exceed the oxygen supplied to the skin and $tcpO_2$ will be lower than $p_aO_2$.

Therefore, $tcpO_2$ can also be an indication of CBF. The dependence of $tcpO_2$ on blood flow was described theoretically in the "circulatory hyperbola" (Lübbers and Grossmann 1983).

In the present study, the dependence of $tcpO_2$ on CBF and $p_aO_2$ will be examined on the forearm, which is taken as a skin model wherein the variables are blood flow and $p_aO_2$.

METHODS

A modified electrode of the Clark-type, heated to 45 °C (15 µm platinum cathode, 25 µm Teflon membrane, heated area 72.5 $mm^2$) was used for the measurement of transcutaneous $pO_2$ on the skin of the forearm of 19 healthy, male volunteers. The skin was carefully cleaned with benzine before fixation of the electrode. The measurements were taken under quiet conditions at room temperature (21 °C).

The systolic blood pressure $P_{sys}$ was measured according to Riva-Rocchi by an upper arm cuff on one arm. On the other arm with the electrode the blood flow was regulated by a second cuff. The pressure of this cuff ($P_{cuff}$) was regulated in similar percentages of $P_{sys}$ (15, 30, 45, 60, 80, 90, 100%). Having reached a relative steady state condition, cuff pressure was released to return to normal conditions. When steady-state was attained again, the cuff was reinflated. The $P_{cuff}$ will be equal to venous pressure $P_{ven}$ in the forearm in the steady state condition. Therefore, the effective perfusion pressure $P_{eff}$ was calculated as $P_{eff} = P_{sys} - P_{ven}$.

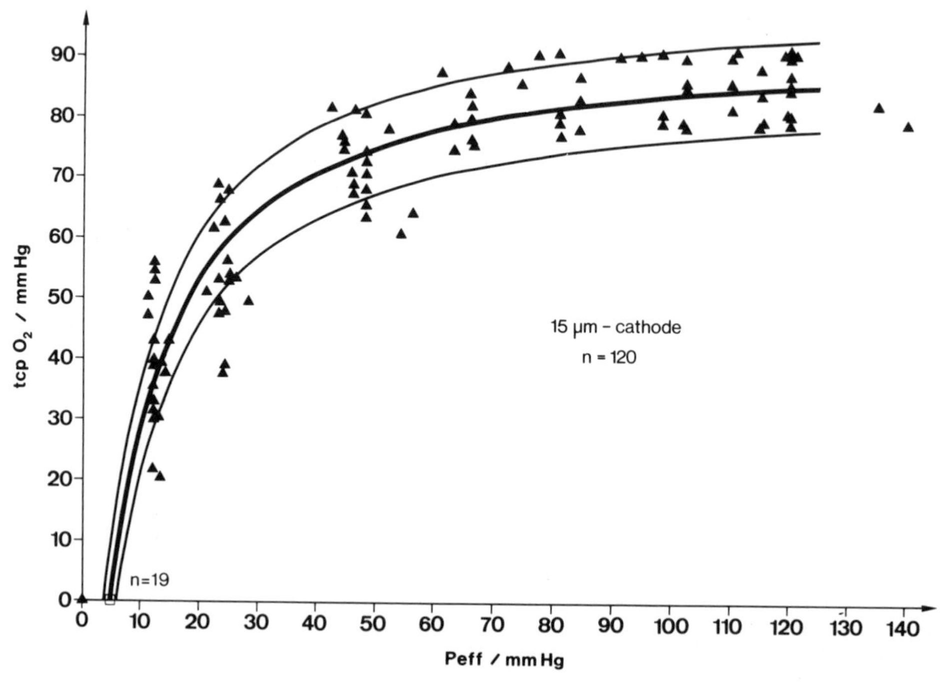


FIG.1: Transcutaneous $pO_2$ ($tcpO_2$) and effective perfusion pressure $P_{eff}$ during several stages of perfusion reduction and the corresponding hyperbolic regression curve (thick line) and the estimation error (thin lines)

The heating power (HP) of the electrode needed for keeping the point of measurement constant at $45^{\circ}C$ was monitored continuously. The relative HP was proposed as the difference between the basic HP in no-flow conditions and HP in the actual conditions.

On the arm on which the electrode was placed the middle fingertip was warmed to $45^{\circ}C$ in a water bath. During steady state, blood flow in the fingertip was determined with a bidirectional 8 MHz ultrasonic probe by Doppler shift. Forward and backward flow as well as the sum of both were registered. If the sum signal was 0, the area below five pulse curves of the forward flow signal was measured planimetrically and the average area of these pulse curves was taken as relatively dependent on CBF (Doppler flow, DF).

In each steady state capillary blood probes were taken from the hyperemized earlobe for evaluating $p_aO_2$. The steady state values of $P_{eff}$, $tcpO_2$, HP and DF were measured for each stage of perfusion reduction. Mean values and standard deviations were calculated for each stage. A hyperbolic regression curve was calculated by the method of least squares for $tcpO_2$ and $P_{eff}$. An offset was used for the calculation, which was 5 mmHg for $P_{eff}$ at zero level of $tcpO_2$ (Lübbers, personal communication).

In a second experiment 12 healthy persons inhaled oxygen via a nose probe during normal flow and restricted flow conditions ($P_{cuff}$= 60 and 80% of $P_{sys}$) and the same variables were measured.

## RESULTS

If there was a reduction in $P_{eff}$ of more than about 30 % $tcpO_2$ was affected, whereas HP and DF decreased as soon as $P_{eff}$ was decreased (fig.2). After 2-7 min of each stage, a relative steady state for $tcpO_2$ and HP was observed.

In normal conditions during respiration of air in these healthy volunteers mean $tcpO_2$ was 86.0 (±6.2) mmHg. The mean transcutaneous index $tcpO_2/p_aO_2$ in these conditions was 1.01 (±0.08). A hyperbolic regression curve was fitted to 120 transcutaneous $pO_2$ values and the corresponding $P_{eff}$ for all stages of perfusion reduction:

$$tcpO_2 = (-1076.74 / (P_{eff} + 6.58)) + 93.86$$

and an estimation error of ± 7.5 mmHg was calculated (fig.1).

An almost linear dependence of HP and DF on $P_{eff}$ was found. The following regression lines were calculated for the mean values of HP (in mW) and $P_{eff}$:

$HP = 0.55\ P_{eff} + 239$ (r=0.99) and for DF (in $mm^2$) and $P_{eff}$:

$DF = 3.98\ P_{eff} + 3.7$ (r=0.98).

The results of the experiment are plotted in fig.2 for the stages in $P_{cuff}$ and the corresponding values for $tcpO_2$, HP and DF.

The results of the second experiment are shown in fig.3, $tcpO_2$ and $tcpO_2$ index are plotted against the heating power in conditions of normoxia (room air) and slight hyperoxia. In hyperoxia (mean $p_aO_2$ = 156 (±29.3) mmHg) mean $tcpO_2$ was 140 (±14.9) mmHg and mean $tcpO_2$ index was 0.9 (±0.13).

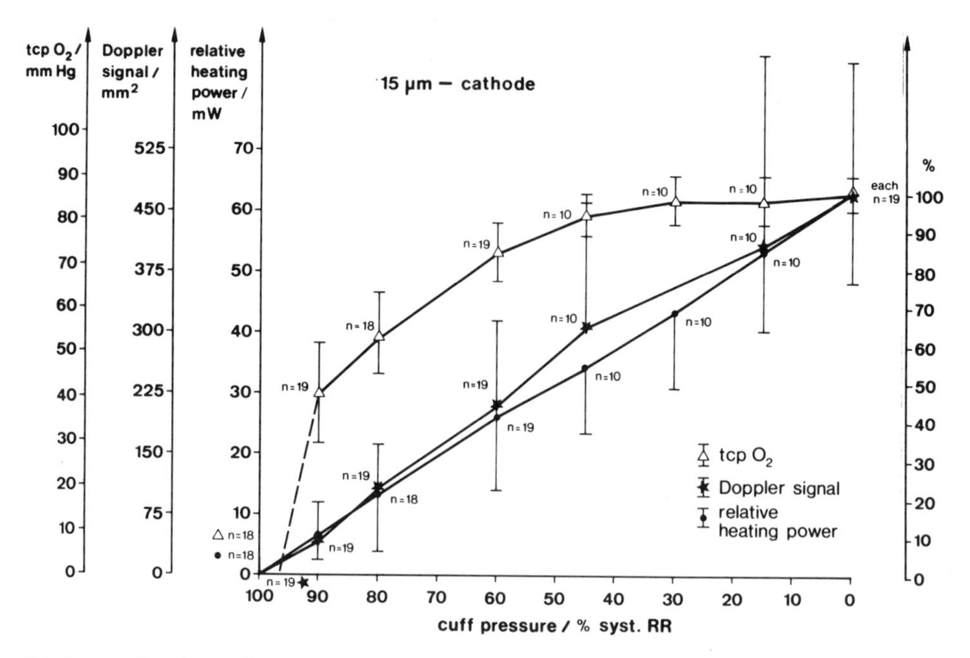


FIG.2: Relation of $tcpO_2$, the heating power of the transcutaneous electrode at $45^0C$ and the Doppler flow signal during several stages of perfusion reduction. Average values and standard deviations for each stage. Dotted line for $tcpO_2$ from 90% to the calculated $pO_2$ offset.

DISCUSSION

The determinants for transcutaneous $pO_2$ measurements are proposed as blood flow and $p_aO_2$. These factors were examined in this study in a model under laboratory conditions. In fig.1, the $tcpO_2$ values show a hyperbolic dependence on $P_{eff}$, as suggested by the hypothesis of the "circulatory hyperbola" for the dependency of $tcpO_2$ on blood flow (Lübbers und Grossmann 1983, Grossmann et al. 1984). The calculated curve agrees well with the observed values.

A direct, accurate and non invasive measurement of blood flow in men is not possible. Therefore, indirect methods, HP and DF, were chosen for the measurement of CBF. $P_{eff}$ was proposed to be the determining factor for the blood flow in hyperemia. In these experimental conditions, the linear correlation of HP and DF to $P_{eff}$ confirm the hypothesis, that the CBF under the electrode is not influenced by autoregulation in local blood vessels and that there are no changes in capillary diameter (Steinacker et al. 1985).

The deviations in the values are worth noting. The scattering of $tcpO_2$ is probably due to the inter-individual variations in the skin's structure and capillary density and of varying degrees of preparation of the measurement site. Another variable is different oxygen consumption in the skin (Lübbers und Grossmann 1983, Huch et al. 1983, Grossmann et al.1984, Steinacker et al. 1985). This scattering of $tcpO_2$ is not important in comparison to that of HP and DF.

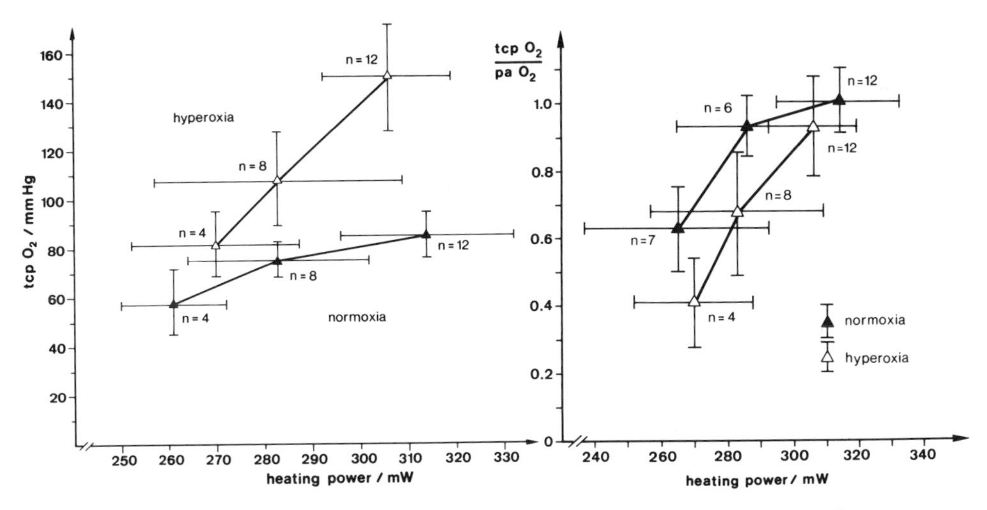


FIG.3: $tcpO_2$ and HP in normoxia and hyperoxia (left side) and transcutaneous index $tcpO_2/p_aO_2$ (right side) during 60% and 80% reduction of effective perfusion pressure $P_{eff}$.

HP is mainly influenced by the surrounding temperatures of skin, blood and the environment. These variables could not be kept constant for all subjects. HP gives additional information, because of its action in relation to blood flow, which is linear for one measurement site and a single experiment.

The variations in DF are mainly due to different angles of the ultrasonic beam, which were not constant for all persons and all sites of measurements (Flax et al. 1979). In this study, the ultrasonic Doppler signal of 8 MHz may not only be dependent on CBF, but also on the flow in subdermal plexus. This is possible, because the mean penetration of the ultrasonic device will be 1.5cm. This range is limited by absorption by the end-phalanx of the finger. The evaluation of forward and backward flow helps to register the flow in the capillary loops of the skin and to minimize signals from small arterioles, which give a longitudinal flow signal owing to the structure of the skin (Flax et al. 1979).

The expected hyperbolic dependence of $tcpO_2$ on $P_{eff}$ and on CBF was proven. If there is a high CBF, as in heat-induced hyperemia under the $tcpO_2$ electrode, $tcpO_2$ will reflect changes of $p_aO_2$. This is the case during changes of 100-70% of $P_{eff}$ observed in the arm model. Slight changes in CBF will not influence $tcpO_2$ as was found when estimating $p_aO_2$ by $tcpO_2$ (Lübbers 1979, Huch et al.1983). When $P_{eff}$ and therefore CBF become more and more restricted (fig.2), as in the model investigated, or during pathophysiological alterations, $tcpO_2$ will decrease due to greater arterio venous $O_2$ differences in the capillaries of the skin and will then become an indicator of local blood flow.

An offset value for $P_{eff}$ of 5 mmHg was used for the calculation of the hyperbola in fig.1. This was necessary, because it was not possible to measure exactly in the steady state values at very low $P_{eff}$ in this experimental setup. This offset is due to the oxygen consumption of the electrode and the skin and will be typical for an individual $tcpO_2/P_{eff}$

calibration curve. Wyss et al. (1981) measuring on the foot found an offset value of about 22 mmHg. In this study mean $tcpO_2$ on the arm was 54.8 mmHg at a $P_{eff}$ of 23.7 mmHg. These findings can be explained by methodological differences in the studies, but they can be also explained by the different vascular structure of the skin of feet or of arms. Therefore, following a suggestion of Lübbers, a $tcpO_2$ offset of 5 mmHg was used for $P_{eff}$, which was calculated from regression computations. This seems to be a practical approach for the problem. The exact value of the $tcpO_2$ offset could be measured in a modified experiment. The offset point implies, that a $tcpO_2$ of zero does not necessarily mean, that CBF and $P_{eff}$ equals zero (Wyss et al. 1981, Lübbers and Grossmann 1983, Grossmann et al. 1984).

In hyperoxia (fig.3) $tcpO_2$ seems to be virtually linearly dependent on HP and therefore on CBF. This effect is reduced by calculating the transcutaneous index $tcpO_2/p_aO_2$. This index is smaller in hyperoxia for the same CBF. These findings are due to $O_2$-shunts in the capillaries, but also to the influence of the hemoglobin binding curve and the apparent solubility of oxygen in the blood (Grossmann et al. 1984). In normoxia the apparent solubility of oxygen will be about 0.5, in hyperoxia it decreases and $tcpO_2$ is more linearly dependent on CBF. In restricted flow conditions the apparent solubility will increase and the $tcpO_2$/CBF relationship is steeper. The $tcpO_2$ index is helpful for the indirect evaluation of local oxygen availability and CBF if $p_aO_2$ is changing (Huch et al. 1983, Steinacker et al. 1985).

ACKNOWLEDGEMENT

This study was supported by the Deutsche Forschungsgemeinschaft. We wish to thank Prof.Dr.D.W.Lübbers and Prof.Dr.R.Huch for their helpful suggestions.

REFERENCES

Flax S.W., Webster J., Uplike S.J., 1970, Statistical evaluation of the Doppler ultrasonic blood flow meter. Biomed Sci Instr 7: 201

Huch R., Huch A., Lübbers D.W., 1983, Transcutaneous $pO_2$. Thieme, Stuttgart New York

Grossmann U., Winkler P., Lübbers D.W., 1984, The effect of different parameters (Temperature, $O_2$ consumption, blood flow, hemoglobin content) on the $tcpO_2$ calibration curves calculated by the capillary loop model. in: Bruley D., Bicher H.I., Reneau D. ed, Oxygen Transport to Tissue, Adv. Exp. Med. Biol. 180, Plenum, New York London

Lübbers D.W., Grossmann U. , 1983, Gas exchange through human epidermis as a basis of $tcpO_2$ and $tcpCO_2$ measurements. in: Continuous transcutaneous blood gas monitoring, Huch R., Huch A. ed., M.Deccer, Basel New York

Steinacker J.M., Spittelmeister W., Wodick R.E., Fallenstein F., Lübbers D.W., 1985, Examinations on the dependence of transcutaneous $pO_2$ on cutaneous blood flow. Pflügers Archiv 405: Suppl.2: R63

Wyss C.R., Matsen III F.A., King R.V., Simmons C.W., Burgess E.M., 1981, Dependence of transcutaneous oxygen tension on local pressure gradient in normal subjects. Clin Sci 60: 499.

## ESTIMATION OF THE DETERMINANTS OF TRANSCUTANEOUS OXYGEN TENSION USING A DYNAMIC COMPUTER MODEL

A. Talbot-Pedersen, M.R. Neuman, G.M. Saidel and E. Jacobsen

Case Western Reserve University, Cleveland Ohio, USA and Rigshospitalet, Copenhagen Denmark

SUMMARY

A dynamic model for oxygen transport within the outer layers of the skin and a $tcPO_2$ sensor has been developed. By comparing model simulations with clinical measurements from adults, it is possible to analyze some of the physiologic and physical determinants of the $tcPO_2$ measurement. The results indicate that the permeability of the epidermis is a significant parameter when using a microcathode sensor. Using this model, the values obtained for normalized stratum papillare blood flow and metabolic oxygen consumption of the dermis are higher than those seen for other models.

### INTRODUCTION

Transcutaneous oxygen tension ($tcPO_2$) measurements from adults have been difficult to apply in the clinical setting due to poor correlation with arterial oxygen tension ($PaO_2$). The objective of this work was to develop a theoretical basis for designing a system for real-time evaluation of $tcPO_2$. A dynamic, distributed model for transport of oxygen in the skin and the $tcPO_2$ electrode has been developed and used to analyze 1) how parameters of the skin and 2) how the cathode size and membrane of the sensor affect the measured $tcPO_2$.

### METHODS

#### Mathematical Model

The model system (Fig. 1) consists of: 1) the $tcPO_2$ sensor including the cathode, electrolyte, and membrane; 2) the skin including dead epidermis, viable epidermis, capillary loop and tissue of the stratum papillare; and 3) a connecting contact liquid layer. The 7 layers are divided into an arterial region centered around the arterial limb of the capillary loop and a venous region around the venous limb. Starting with dynamic, two-dimensional oxygen mass

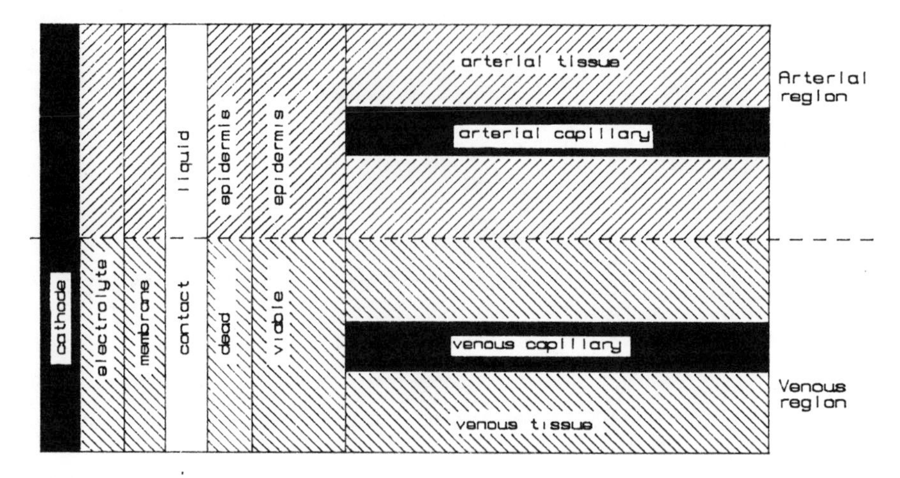


Fig. 1. Geometric structure of the dynamic oxygen model.

balances for each of these 14 sections, we obtain one-dimensional equations describing the dynamics for the oxygen concentration distributed in the axial direction perpendicular to the skin surface. In this model, we account for the blood flow in the capillary loop, the metabolic oxygen consumption of the epidermis and the dermis, and the permeabilities of the epidermis, the sensor membrane and the sensor electrolyte. The hemoglobin-oxygen equilibrium is represented by a double exponential model that includes the effect of temperature and pH. The temperature dependence of permeability and metabolic oxygen consumption is calculated by the Arhenius formula. The system behavior is simulated by numerical solution of the model equations.

## Clinical Studies

Five male and five female healthy Caucasian subjects ranging in age from 25 to 39 years were studied in the supine position. $TcPO_2$ was measured using a Radiometer prototype $tcPO_2$ electrode with a 25 m diameter cathode covered by a 15 m polypropylene membrane fixed on the forearm. Online measurements were taken for 8 minutes after stabilization of the electrode at 44 °C, and at inspired $O_2$ levels of 21, 40 and 60%. Following the 8-minute period for each $FiO_2$, $PaO_2$ was obtained from arterial blood samples. Also, $tcPO_2$ measurements were made during arterial occlusion proximal to the sensor by inflation of a sphygmomanometer cuff. Separate in vitro experiments were carried out with the sensor alone to evaluate its characteristics and determine the parameters of the sensor part of model.

## RESULTS

### Steady-state

Fig. 2 shows a simulated steady-state skin oxygen profile. The $PO_2$ of the blood entering the arterial limb of the capillary loop is assumed equal to $PaO_2$ corrected to 42 °C. Values used in the model are: $O_2$ conductivity = 4.5

(dead epidermis) and 9.0 (viable epidermis) $\mu m^2$- mol/ml-s-kPa; thickness = 15 (dead epidermis) and 50 (viable epidermis) m; and metabolic oxygen consumption = 3.7 nmol/ml-2.

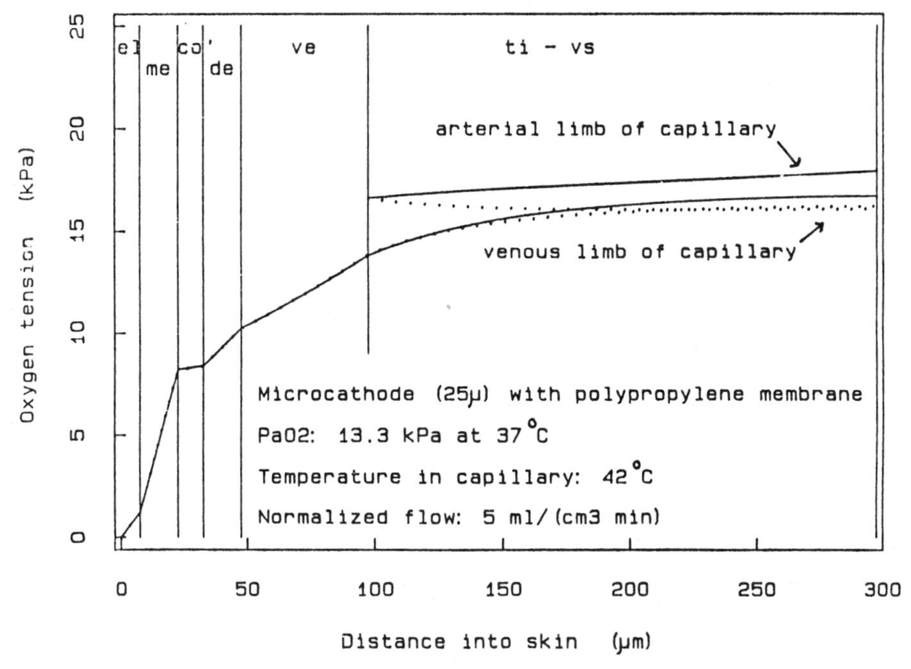


Fig. 2. Skin oxygen profile with layers as shown in figure 1.

Blood perfusion was found to be a significant determinant of $tcPO_2$ as shown in Fig.3, where simulated $tcPO_2$ approximates clinical results. Simulations show that the $tcPO_2$ value saturates when normalized flow reaches approximately 10 $ml/cm^3$-min for $PaO_2$ 13.3 kPa. Normalized flow is defined as the volumetric flow in the capillary loop divided by the volume of the stratum papillare.

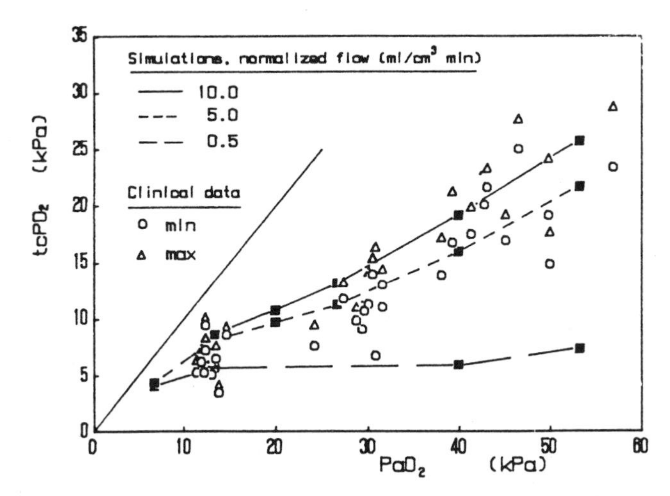


Fig. 3. Effect of blood flow on $tcPO_2$-$PaO_2$ difference. Microcathode at 44°C.

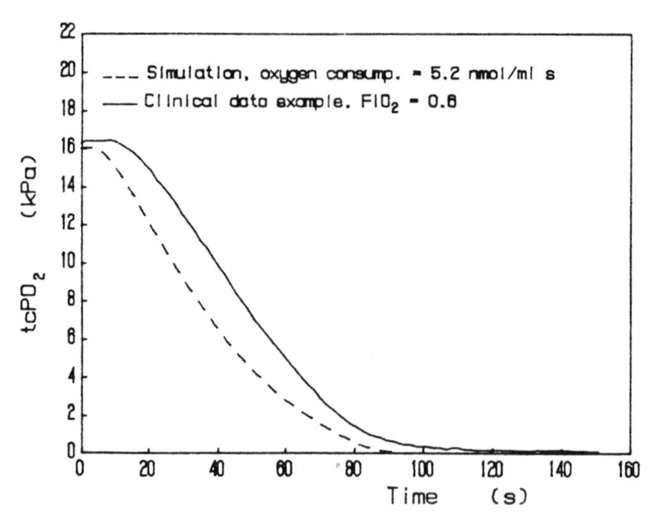


Fig. 4. Arterial occlusion, actual data compared to simulation.

## Dynamics

Simulated dynamic results with a metabolic oxygen consumption of 5.2 nmol/ml-s (or 0.7 ml/100g-min) and a solubility 8.9 nmol/ml-kPa were compared to actual data following arterial occlusion for one subject (Fig. 4). The simulated and experimental maximum slopes are 18.2 and 16.4 kPa/min, respectively.

## DISCUSSION

This one-dimensional model has advantages of a three-dimensional model[1], by allowing shunt diffusion between the arterial and venous parts of the capillary loop. In addition, it can simulate different types of $tcPO_2$ sensors. Assuming the same parameter values and non-oxygen consuming sensor at the skin surface instead of a microcathode, we obtain steady-state results with our model, that are similar to the results reported by Lubbers and Grossman.[1]

Comparison of the model simulations with clinical measurement from adults suggests that the ratio of the permeability of the skin to the permeability of the transducer membrane and electrolyte is lower than previously reported.[2]

This is supported by studies showing increased $tcPO_2$ values when the skin is stripped. We conclude that the characteristics of the microcathode $tcPO_2$ electrode have a significant influence on the $tcPO_2$ results. Observed $tcPO_2$ dependence on skin location could be partly due to local differences in skin permeability.

In making a comparison of parameter values among models, differences may result because of the nature of the models and the definition of model parameters. For example, we find normalized flow values one order of magnitude higher than usually reported. This normalized flow would be 25% lower if the epidermis was included in the volume for normalization. All methods for skin perfusion measurements estimate flow based on special models and assumptions. However, with measurement variables other than $O_2$ (e.g. Xe or heat flux), perfusion of the skin layers deeper than the stratum papillare are of importance.[3] The high flow values obtained here are produced by the combined effects of the skin permeability and properties of the sensor. Further studies with sensors having a lower $O_2$ consumption (e.g. lower permeability membrane or smaller cathode area) are necessary to confirm this prediction.

The maximum slope seen during occlusion is dependent on both the permeability of the epidermis and the metabolic oxygen consumption (Fig. 4). Since the model includes the effect of other permeabilities, the estimates for metabolic oxygen consumption are higher than for other methods.[2]

ACKNOWLEDGEMENTS

Supported by the Danish Technical Research Council, U.S. NIH grants RR 00210 and RR 2024, and Radiometer A/S.

REFERENCES

1. D.W. Lubbers, U.Grossman, Gas Exchange Through the Human Epidermis as a Basis of $tcPO_2$ and $tcPCO_2$ Measurements, In: Huch R., Huch A. ed., Continuous Transcutaneous Blood Gas Monitoring. Marcel Dekker, New York (1983).

2. J.W. Severinghaus, M. Stafford, A. M. Thunstrom, Estimation of Skin Metabolism and Blood Flow with $tcPO_2$ and $tcPO_2$ Electrodes by Cuff Occlusion of the Circulation, Acta Anaesthesiol Scand., Suppl 68: 9-15 (1978).

3. P. Jaszczak, P. Sejrsen P., Determination of Skin Blood Flow by 133Xe Washout and by Heat Flux from a Heated $tcPO_2$ Electrode, Acta Anaesthesiol Scand., 28: 482-489 (1984).

# TECHNICAL ASPECTS OF CONTINUOUS MONITORING

# QUALITY AND SAFETY ASPECTS IN THE DEVELOPMENT AND FABRICATION OF TRANSCUTANEOUS SENSORS

Ewald Konecny and Uvo Hölscher

Research and Development
Drägerwerk Aktiengesellschaft
Moislinger Allee 53 - 55, 2400 Lübeck

## SUMMARY

The safe and reliable application of transcutaneous sensors requires their construction and fabrication to be embedded in a system of external regulation and internal rules.

## INTRODUCTION

External regulations must be fulfilled, internal rules within the product must be surveyed with respect to both safety - i.e., what is the consequence of an error of a certain type - and reliability - i.e., what is the probability of such an error -, and within the factory a system must be installed to ensure that the necessary steps are carried out at an early enough state of the technical development.

As external rules, specifically in Germany, the fabrication and use of all energy-driven medicotechnical apparatus is governed by a regulation "Medizingeräteverordnung" issued by the Federal Ministry of Labour which classifies such apparatus in 4 different classes, class 1 being reserved for essential life support equipment explicitly named in an appendix, class 2 for implantable items, class 3 for "energy-driven" apparatus, and class 4 for all other medicotechnical apparatus. Since transcutaneous monitors are classified in class 3 only, the main requirements for the instruments are to be compliant with up-to-date technical standards and to ensure that a technical instruction by a qualified person from the manufacturer has been carried out before the instrument is used with a patient.

For the US, the FDA requires premarket approval. Surprisingly this applies to $tcpCO_2$ and combined sensors, not, however, to $tcpO_2$ sensors.

The relevant technical standards are issued by essentially two bodies, the International Organization for Standardization ISO, and the International Electrotechnical Commission IEC, both located at Geneva. Within them there cooperate representatives from all associated countries in order to set standards and to help to harmonize those rules with the respective national standards.

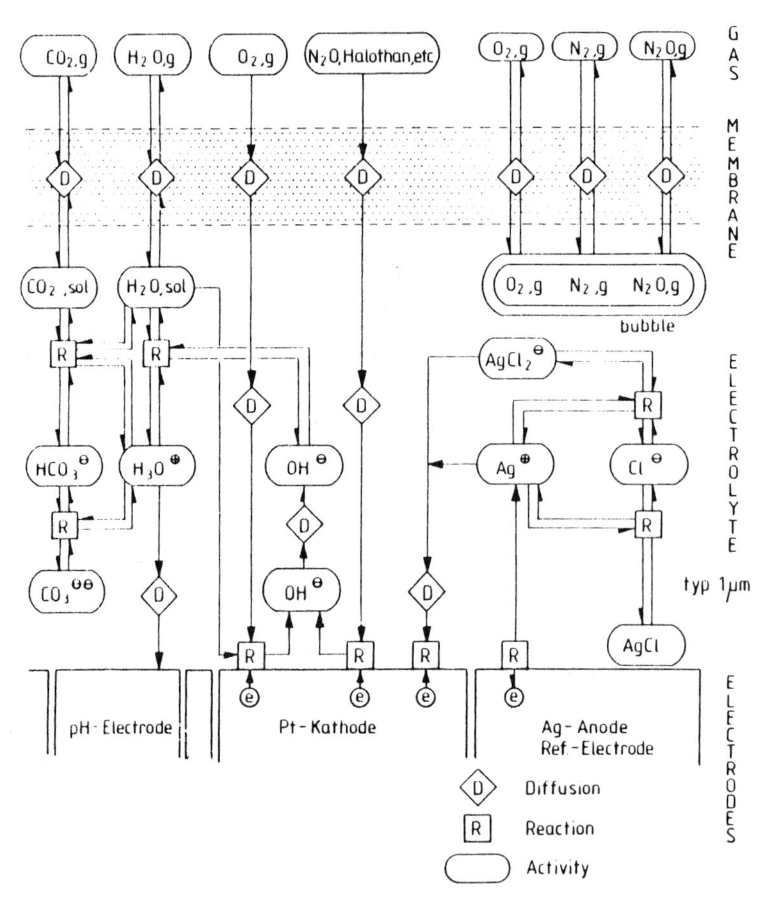


Fig. 1: Electrochemistry of the combined $tcpO_2/tcpCO_2$-Sensor

Among them, the standard IEC 601-1 is of central interest for all electrical apparatus used in medicine. The specific requirements for specific types of instruments are regulated by IEC 601-2 which is under preparation. Thus a committee is preparing to issue specific requirements for transcutaneous sensors and monitors, probably to be published in 1987. These rules contain only the minimum safety requirements to protect user, patient and environment in single fault condition. To enable the doctor to select and order the appropriate equipment available on the market for his application in a transparent way, a third standard IEC 601-3 is under preparation in which standardized rules are given for describing the specific instrumental properties like rise time, linearity, drift etc. The present draft contains many inputs from the corresponding draft of the American Society for Testing and Materials ASTM.

## Methods

The technical layout of a combined sensor for $tcpO_2$ and $tcpCO_2$ is a rather complicated system with many possible interactions of different functional parts like the membrane, the electrolyte, the electrodes (Fig. 1). The whole system is so complicated that it cannot be described satisfactorily in an analytical way. Therefore, a lot of empirical tests

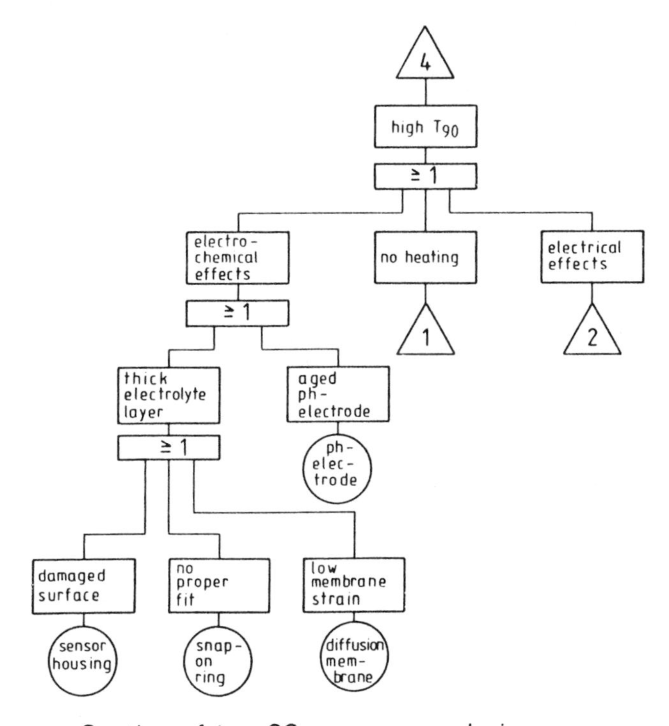


Fig. 2: Failure Tree Analysis: Analytical method for determination of failure reasons which lead to a system failure

are required to obtain a reliable set of data for making a complete safety analysis of the system. Fig. 2 shows, as an example, a small portion of such a scheme of individual faults resulting in an intolerable increase of the risetime of a $tcpCO_2$ sensor. In consequence, the design must be laid out in such a way that a single fault must not result in a hazard for the patient (nor for the user).

Another important criterion is reliability. In general, an increase in safety decreases reliability (e.g.: paralleling a decisive component by a redundant one may increase the safety considerably, but will decrease its reliability by a factor of 2).

The parameter $\lambda$ describing the number of faults per time unit for each component i follows mathematically the Weibull distribution

$$\lambda_i(t) = \frac{\beta_i}{\tau_i} \cdot \left(\frac{t}{\tau_i}\right)^{\beta_i - 1}$$

where $\beta < 1$ describes the portion of early failures from weak components (e.g. defective and therefore unreliable pH-glass electrodes), $\beta = 1$ describes the random failures (e.g. of resistors, capacitors, integrated circuits) for which

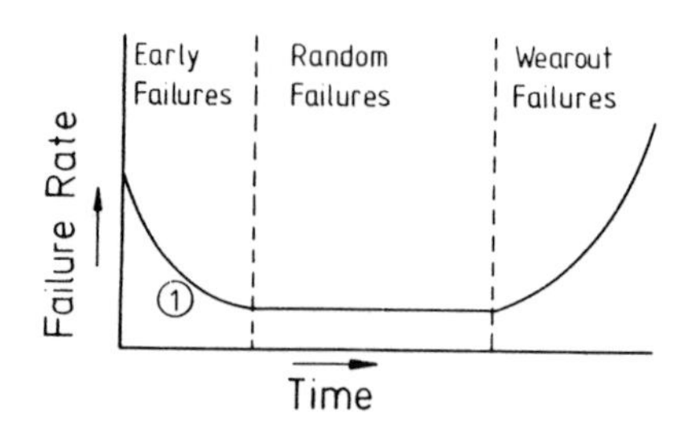


Examples:
- Switches
- Plugs
- Transistors
- Sensors

① to be eliminated by BURN-IN

Fig. 3: Failure Rate of Components

the failure rate is constant and $\beta > 1$ describes failures due to wearout (e.g. cable wear, epoxy). Each $\tau_i$ denotes an average life time. This distribution results in the typical bath tub curve (Fig. 3). The individual $\lambda_i$'s of all components can be estimated either from published tables or from own experience with similar parts and can be combined to an overall failure rate $\lambda$

The estimated failure rates are compared with those observed on the real instruments at the lab and in the field. This procedure requires a quantitative, detailed, quick report back to the factory on the sort of problem and on the time the instrument had been in use. Although this procedure is tedious, it enables to counteract quickly and to discriminate weak components or weak points of the design at a rather early stage.

These theoretical considerations on safety and reliability must result in practical consequences:

First, a system must be installed during all development phases to ensure communication between the different departments involved.

Fig. 4 describes different stages of the development procedure. After the product idea, the conceptual design and the construction of a functional model - which should verify clearly that the most important intellectual steps of the project have been done. Typically, within university research one would probably move to the next subject at this point. Nevertheless, from the industrial point of view, at that point on the average only about 10 to 15 % of the overall engineering work has been carried out on the way to a marketable product. For finally arriving at that goal, cooperative

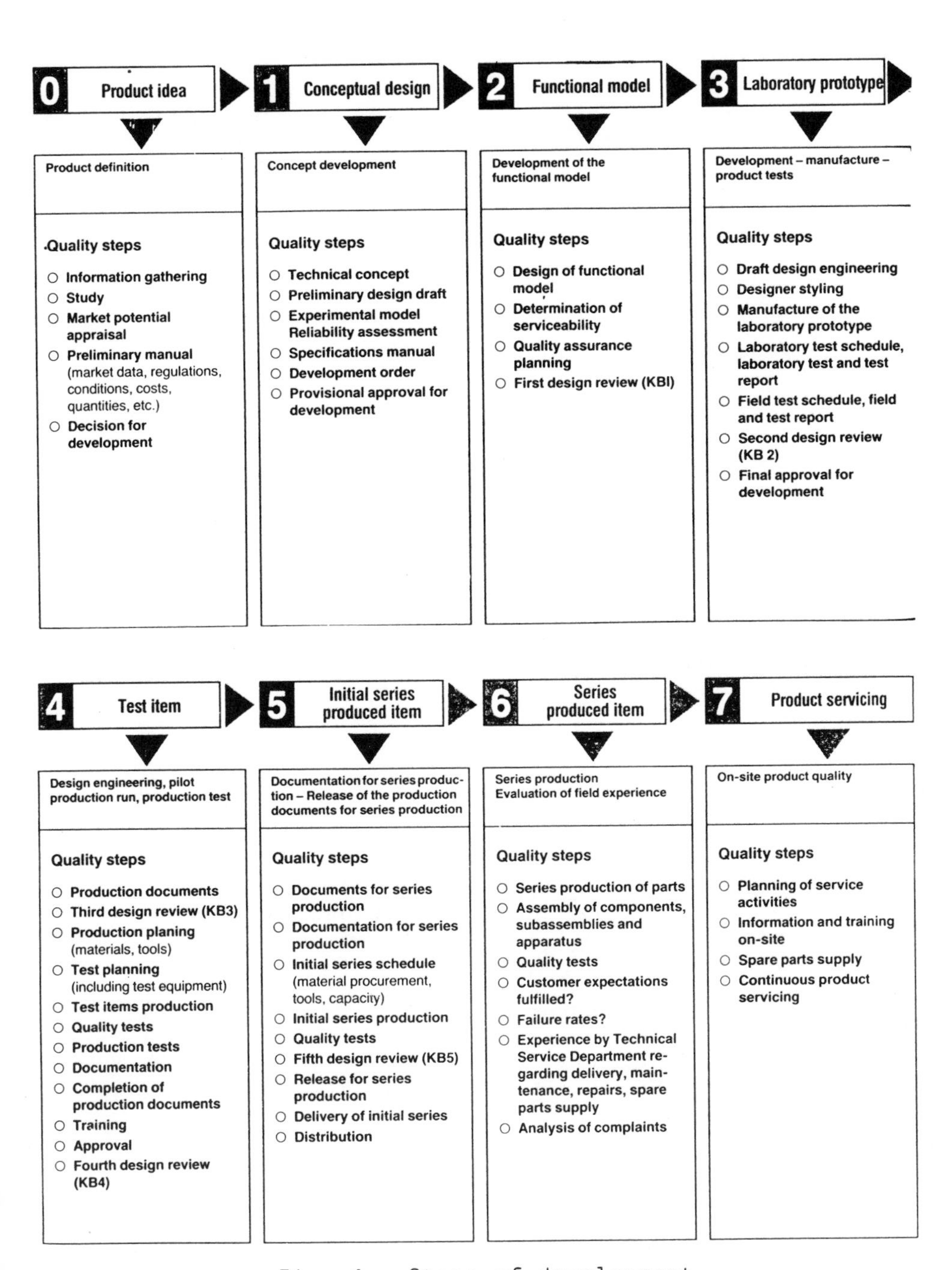


Fig. 4: Steps of development

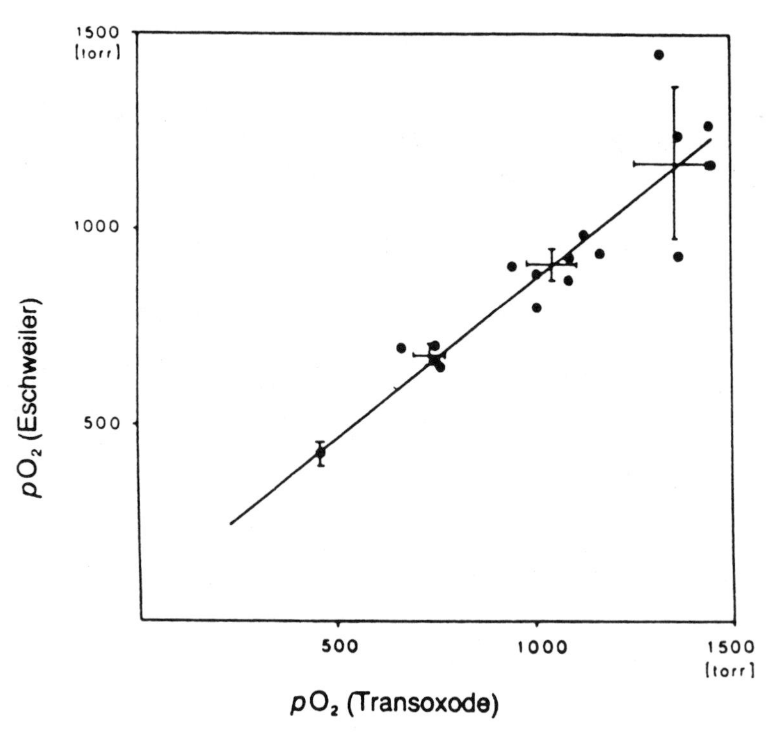


Fig. 5: Correlation between transcutaneous data and blood gas data under hyperbaric conditions

contributions from different departments are necessary. Therefore, at special points in the development, short but formal design review meetings are foreseen, which should ensure that the people involved in the project next are informed early enough about the state of the project.

Second, we found it necessary that people in research and development work in close local proximity to manufacturing and quality control.

Third, special measures must be taken to increase reliability such as burn-in and stress testing to pick out "ill" components or to reduce faults induced by dust particles using special installations for the sensor assembly and for thick film electronical circuits.

## Discussion

Careful differentiated analysis of environmental effects on the different components allows application of transcutaneous sensors also under unusual environmental conditions, e.g. under hyperbaric pressures. Hyperbaric conditions with oxygen-enriched atmospheres are used for therapy of CO-poisoning and of gaseous oedema in some special clinics which are equipped with the hyperbaric chambers necessary. At total pressures between 3 and 16 bar the $pO_2$ is kept around 2 bar (Ref. 1).

As is demonstrated in the relevant literature, tissue $pO_2$ as a function of time indicates the functional state of organs and provides an early warning for potential hazards of elevated oxygen levels. The sensor demonstrates good linearity even in the range of high $pO_2$'s which exceed the physiological range under normal conditions considerably (Fig. 4).

We anticipate that sensors for $tcpO_2$ and $tcpCO_2$ developed and fabricated according to rather strict professional standards and applicable even at extreme conditions will find access to even much wider fields of application, such as monitoring during anesthesia (Ref. 2).

## References

1. Tirpitz, D. Behaviour of $pO_2$ in tissue under hyperbaric condition - first results in transcutaneous measurement, in Proc. 8th Congr. European Undersea Biomedical Soc., Lübeck-Travemünde, Germany, 1982, 469 f.
2. Tremper KK, Shoemaker WC: Transcutaneous oxygen monitoring of critically ill adults, with and without low flow shock. Crit Care Med 1981; 9: 706-709.

# A NEW SYSTEM FOR $tcPo_2$ LONG-TERM MONITORING USING A TWO-ELECTRODE SENSOR WITH ALTERNATING HEATING

F. Fallenstein, P. Ringer, R. Huch and A. Huch

Perinatal Research Unit
University of Zürich (Switzerland)

## INTRODUCTION

It would be an advantage to reduce the thermal stress of skin tissue in $tcPo_2$ monitoring. With this in mind we have developed a new technique which uses two small separate $tcPo_2$ electrodes assembled in one common sensor housing. The basic idea is to heat only one electrode at a time and process its signal. After a given period of time, control is switched over to the other electrode. In this way, a quasi continuous recording of $tcPo_2$ can be obtained by reading the data from the active channel alternatingly.

## THE SENSOR AND THE ELECTRONICS

We designed and built a sensor incorporating two independent complete $tcPo_2$ electrode systems in one housing and with one cable (Fig. 1). The size of this device is the same as a common single electrode device.

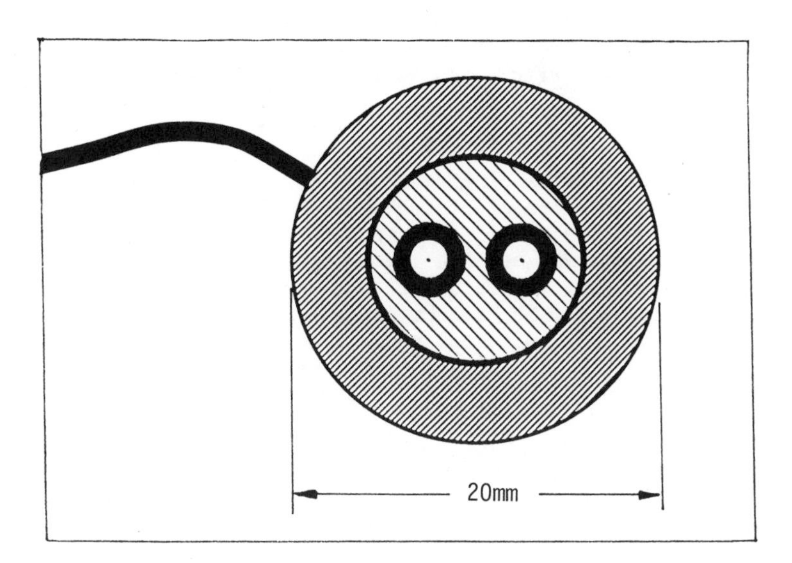


Fig. 1: Outline of the two-electrode $tcPo_2$ sensor used for alternating heating

Since the heating systems are repeatedly switched on while the probe is on the skin, the temperature control electronics must guarantee that there is absolutely no overshoot in temperature when one or the other heating is warming up. For this reason, we developed a microcomputer based temperature control system. In this way, it also became easy to implement the time control for the alternating heating sequence, to perform the calibration of the probe and to display the monitored $tcPo_2$ values on-line on the screen. Temperature and $tcPo_2$ data from both channels can be stored on magnetic disks for further evaluation. In our study, we used a Personal Computer (Texas Instruments Inc.) running under MS-DOS 2.11 (Microsoft Corp.) with a special self-developed add-on board to interface with the sensor. The source code of the software was written in Turbo Pascal Version 3.0 (Borland Int.).

## ASPECTS OF MAIN INTEREST IN THE NEW SYSTEM

Three questions had to be answered in order to prove whether the basic principle of the system would work:

1. How much does the temperature decrease in an electrode system during the inactive interval?

2. How long is the time required to re-establish sufficient hyperemia when an electrode system changes from the inactive to the active interval?

3. What is the thermal stress on the skin caused by the application of alternating heating compared to that from a conventional one-system electrode with continuous heating?

A further question applies to all three points of interest: Are the observations influenced by the length of the switch-over interval?

## THE FIRST STUDY WITH 4 h MONITORING TIME

Our first series of measurements with the new system was performed on 10 healthy adult volunteers. The total monitoring time was 4 hours for each one. The temperature of the active electrode was set at 45 °C. Each person was monitored twice: (1) with sequence intervals of 20 minutes and (2) with sequence intervals of 40 minutes.

Fig. 2a and 2b show the results from one of these measurements obtained with a digital plotter from the disk-stored data. In Fig. 2a are the temperature tracings of both electrode systems. It can be seen that the active system reached the specified temperature of 45 °C without any overshoot already a few seconds after being switched on and maintained this value within $\pm 0.1$ °C. The temperature of the unheated electrode settled to a level between 39 and 40 °C. In Fig. 2b are the $tcPo_2$ tracings. The $tcPo_2$ signal was stable in the physiological range from the second active interval on. As expected, the values during the unheated intervals fell to about 2/3 of the normal level due to the reduced temperature of the inactive electrode.

Table 1 gives the mean and the variability of the temperature in the passive electrode. Because a systematic difference could not be found between the 20-minute and the 40-minute intervals, the numbers in Tab. 1 were calculated from all 20 experiments of this study. We believe that the mean temperature drop of 5.5 °C is sufficient for the skin area to "recover" from the previous heating interval to a certain extent.

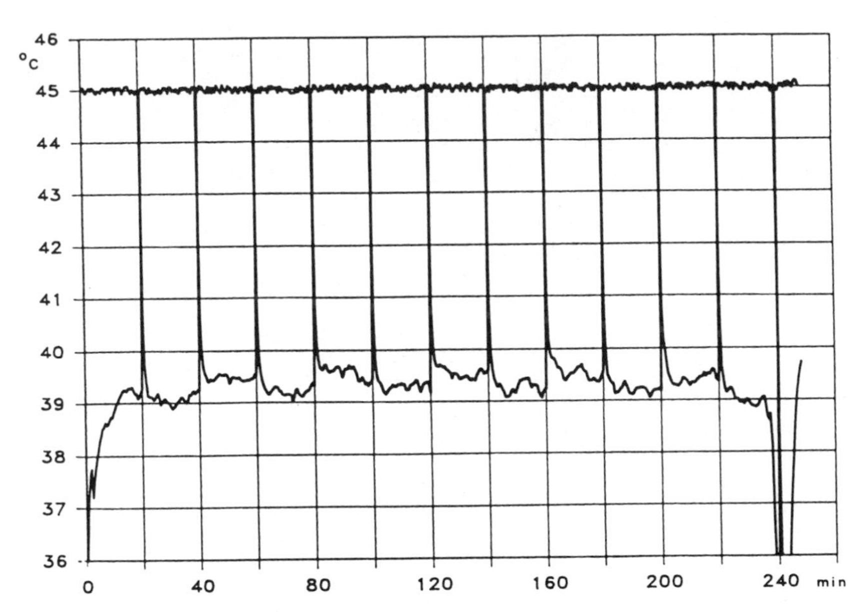


Fig. 2a: Temperature curves in the alternately heated two-electrode system

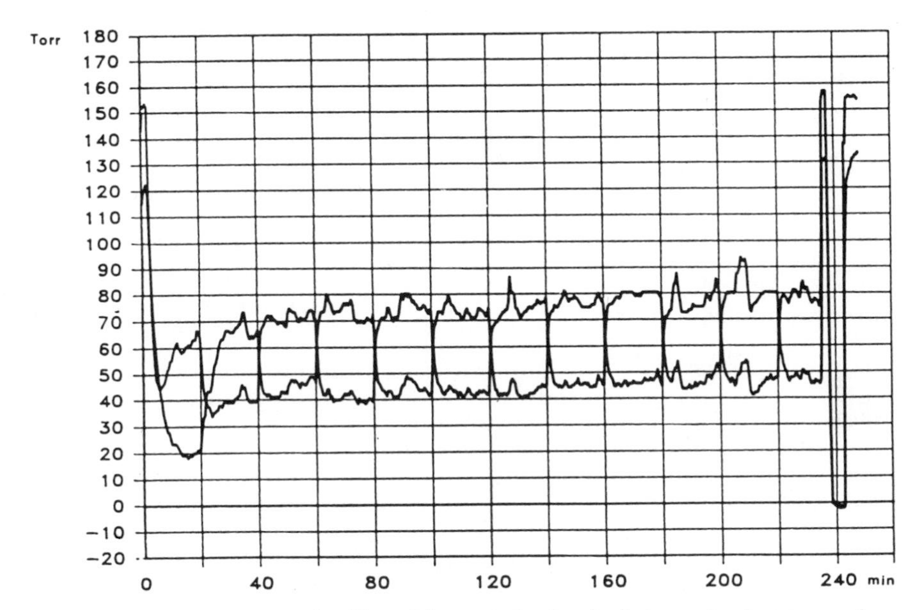


Fig. 2b: $tcPo_2$ curves in the alternately heated two-electrode system

Table 1: Temperature of unheated system

| | |
|---|---|
| mean | 39.48 °C |
| SD | 0.38 °C |
| range | 39.0 ... 40.1 °C |

Table 2: Time until $tcPo_2$ steady state after switching over

| |
|---|
| $t(50\%) < 30$ sec in all measurements |
| $t(95\%) \approx 4 \times t(50\%) < 2$ min |

The response time of the $tcPo_2$ signal when the electrode is switched on from the inactive to the active state was determined as the time delay between reactivation and reaching 50% of the active/inactive $tcPo_2$ difference. Table 2 shows that this was less than half a minute in all measurements. Assuming an exponential course of the tracings, the 95%-time can be estimated as 4 times the 50% time, giving less than 2 minutes for our data. Again, a difference between the 20-minute and 40-minute intervals could not be seen. In further studies we will try to overlap the heating intervals in order to get a 100% continuous monitoring.

After each measurement and also 24 hours later, we classified the state of the skin surface which had been exposed to the sensor. This was done by assigning the most appropriate score number from Table 3. The individual results of all measurements are given in Table 4. Also here, there was no difference in regard to the interval length. No blisters or burns were observed and in most cases - directly after the measurements as well as 24 hours later - only two red spots developed. With a conventional one-system electrode, a continuous 4-hour monitoring period at 45°C generally produces a skin irritation of grade 2 or 3 on adults. So we conclude that there is a significant improvement of thermal stress with the new system.

Table 3:
Classification of
skin irritation

0 - no reddening

1 - red spots only

2 - red spots in a lightly redded field

3 - intensively reddened field around the sensor site

4 - moderate blisters

5 - burns

Table 4:
Total monitoring time: 4 h
Sensor temperature: 45°C

| Subject | 20 min periods | | 40 min periods | |
|---|---|---|---|---|
| | end | +24 h | end | +24 h |
| 1 | 2 | 0 | 3 | 0 |
| 2 | 1 | 1 | 1 | 1 |
| 3 | 1 | 1 | 1 | 1 |
| 4 | 1 | 1 | 1 | 1 |
| 5 | 1 | 0 | 1 | 0 |
| 6 | 1 | 1 | 1 | 0 |
| 7 | 1 | 1 | 1 | 1 |
| 8 | 1 | 1 | 1 | 1 |
| 9 | 1 | 0 | 1 | 1 |
| 10 | 1 | 1 | 1 | 1 |

One particular result of this study to point out is that in the range of 20 to 40 minutes, the length of the alternating heating sequence seems to be uncritical.

## TWO FURTHER STUDIES WITH EXTENDED MONITORING TIME

From the results of the first experimental series, we were encouraged to extend the total monitoring time. We started a second pilot study with 4 volunteers. These measurements were performed overnight and lasted 8 to 10 hours. Table 5 shows the classification of the the skin surface. The results were unsatisfying at the first glance: in two cases blisters occurred, in one case even a real burn. In the latter case the $tcPo_2$ signal started to decrease gradually from the 6th hour on and approached zero with increasing harm to the skin (Fig. 3).

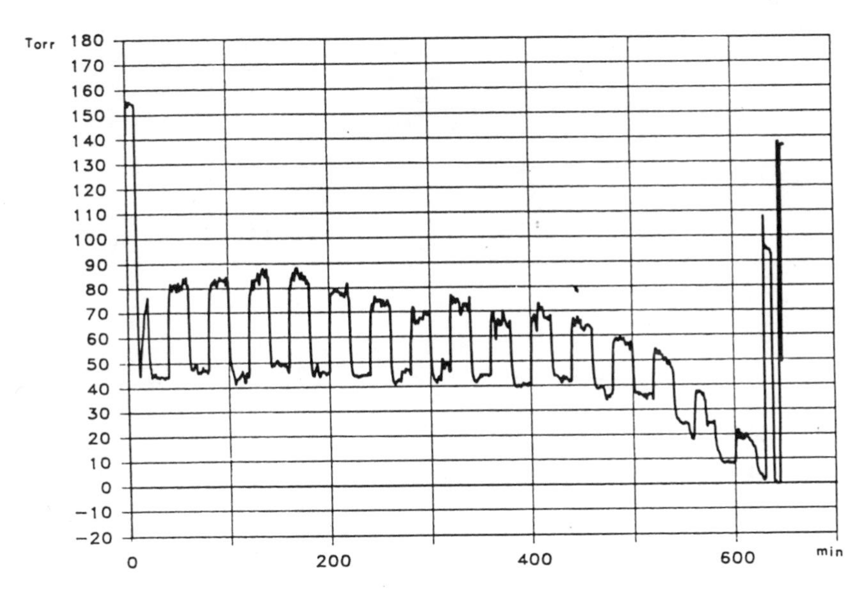


Fig. 3: Decreasing $tcPo_2$ signal caused by thermal overstress of the skin

Table 5:
Total monitoring time: 8-10 h
Sensor temperature: 45 °C

| Subject | 20 min periods end | +24 h |
|---|---|---|
| 1 | 4 | 4 |
| 2 | 1 | 4 |
| 3 | 2 | 1 |
| 4 | 5 | 5 |

(for score see Table 3)

Table 6:
Total monitoring time: 8 h
Sensor temperature: 44.5 °C

| Subject | 20 min periods end | +24 h |
|---|---|---|
| 1 | 1 | 1 |
| 2 | 1 | 1 |
| 3 | 1 | 1 |

(for score see Table 3)

To be on the safe side, we controlled the temperature of the two electrode systems and in fact we found a slight deviation of +0.1 and +0.3 °C respectively, probably caused by drift of the temperature sensors.

In a last step, we recalibrated the temperature carefully to 44.5 °C and performed three additional measurements, again overnight on adult volunteers and for exactly 8 hours. As one can see in Table 6, the results related to thermal stress are comparable to those from the 4-hour study.

## CONCLUSION

It was shown that - using a periodically interrupted heating system - the thermal stress of skin in long-term $tcPo_2$ monitoring can be markedly reduced. With a sensor incorporating two identical $tcPo_2$ electrode systems and operating in an alternating sequence, quasi continuous $tcPo_2$ measurements can be provided. Our pilot study with extended monitoring time showed that while continuous monitoring for 8 hours at 45 °C is not to be recommended, at 44.5 °C nearly no problems arise. This clarifies that the limits of tolerance and intolerance with respect to thermal stimulation of the skin are very close together in any one individual but can vary widely from person to person. Further studies will show whether 6-hour periods can be carried out at 45 °C without any harm even to subjects with sensitive skin.

# A NOVEL APPROACH FOR AN ECG ELECTRODE INTEGRATED INTO A TRANSCUTANEOUS SENSOR

Uvo Hölscher

Research and Development Electrochemical Sensors
Drägerwerk Aktiengesellschaft
Moislinger Allee 53 - 55, 2400 Lübeck

## SUMMARY

The integration of an ECG-electrode into a common transcutaneous sensor allows simple handling and leads to a reduction of the physiological stress of pre-term infants. Furthermore it may allow future replacement of an invasive method to measure the ECG under labour by a non-invasive one.

## INTRODUCTION

The non-invasive signals of heart rate (HR) and transcutaneous oxygen partial pressure ($tcpO_2$) give the basic information about the cardio-pulmonal status of pre-term infants. For this purpose, three ECG-electrodes and a transcutaneous sensor are applied via adhesives. In particular, the skin of prematurely born infants is stressed merely by contact with the adhesive. The object of our work was the combination of a transcutaneous sensor and an ECG-electrode to reduce the number of transducers applied to neonates and in consequence to reduce their iatrogenic stress.

## METHODS

The signal to noise ratio of ECG-electrodes is increased by enlarging the electrode surface and choosing an Ag/AgCl-system. Arranging the elements of the tc-sensor and the ECG-electrode simply side by side would enhance the over all size of the transducer. Keeping the transducer size constant would reduce the available space for the sensitive parts or for the fixation means. All three possibilities show disadvantages in handling. Therefore, we developed a solid-state device, arranging the ECG-electrode and the sensitive part of the transcutaneous sensor behind each other.

In Fig. 1 a sectional view through a transcutaneous sensor for oxygen and carbon dioxide is shown. The ECG-electrode covers the whole outer surface of the diffusion membrane. The electrode is deposited as a microporous, gas-permeable silver layer with a thickness of only some thousand atoms. The blood gases can easily penetrate the ECG-electrode and diffuse through the membrane into the electrochemical sensor behind. Using the specially composed contact cream guarantees a quality of the ECG-signal comparable or better to that of disposable electrodes. In this design, neither the space for application nor the size of the electrochemical system is reduced. This guarantees a similar performance for the transcu-

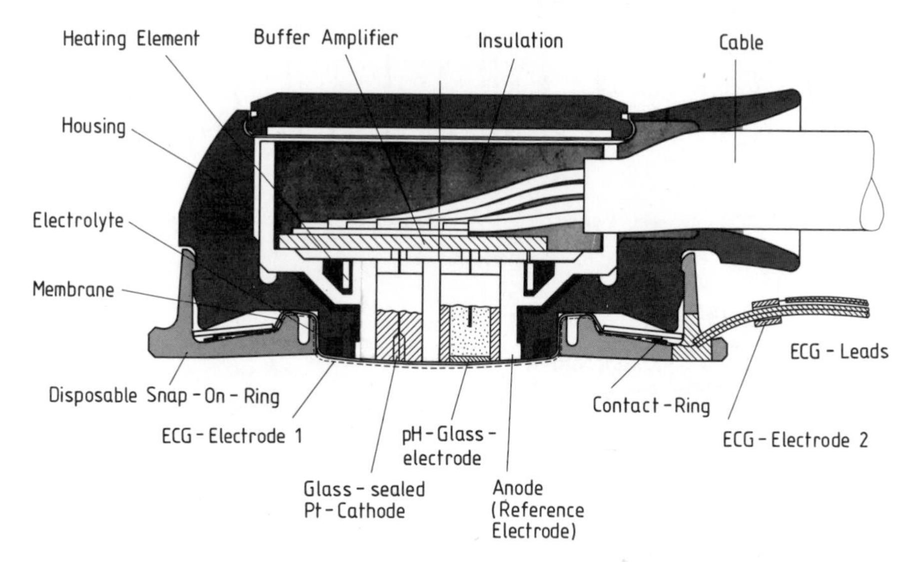


Fig. 1: Combined $tcpO_2/CO_2$-sensor with integrated ECG-electrode, pat. pend.

taneous sensor with or without ECG-electrode. The snap-on ring with the ECG-electrode may be used on $tcpO_2$, $tcpCO_2$ and combined $tcpO_2/CO_2$-sensors alternatively.

RESULTS

Fig. 2 shows the protocol of the heart rate signal. The signal was recorded from a healthy 30-year-old man via a combined ECG-$tcpO_2$ transducer. A complete ECG-signal (Fig. 3) was gained, using three identical thin-film ECG-electrodes only. Noise level and impedance of the thin-film ECG-electrodes are low and artefacts by movements are small. Simultaneous measurements of ECG and transthororacic impedance have been tested successfully under clinical conditions.

During labour usually the heart rate signal of the cardiotocogram is obtained directly by a screw- or clip ECG-electrode after the membranes have been ruptured. In case of suspicious CTG-patterns an invasive fetal blood gas analysis or a non-invasive transcutaneous gas analysis should be performed. This gives information as to whether or not an acute fetal hypoxia or acidosis is present and to determine which clinical actions have to be taken.

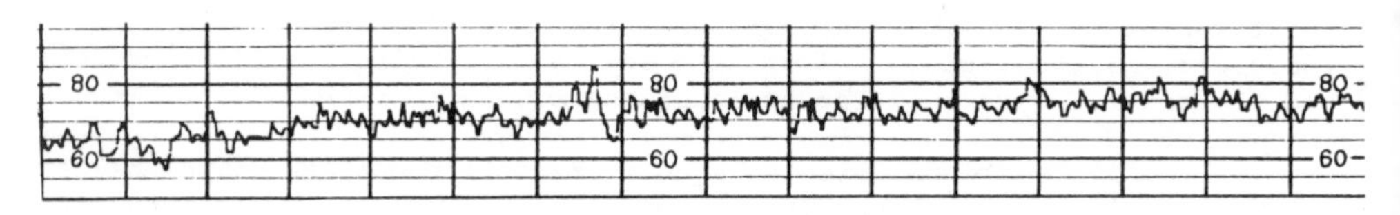


Fig. 2: HR-signal recorded by thin-film ECG-electrode integrated into a tc-sensor

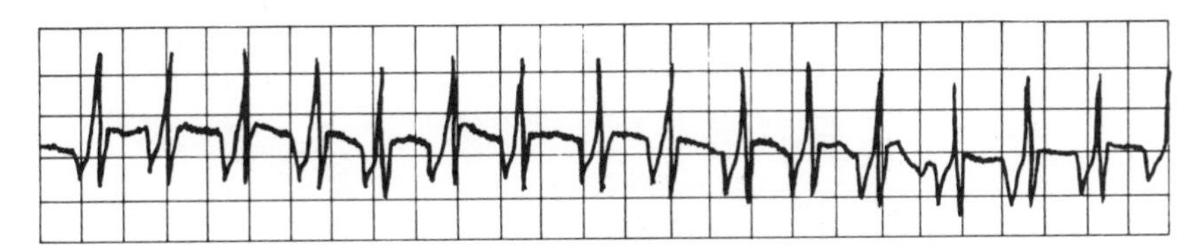

Fig. 3: ECG-signal recorded by thin-film electrodes integrated into tc-sensors

The non-invasive direct ECG-electrode combined with a tc-sensor allows a reduction of the statistically occurring complications of the invasive method. Furthermore, possible mental reservations of the mother against an invasive technique may become obsolete. For this purpose a special contact cream is required to convert e. g. the remainder of vernix caseosa, so that the electrical isolation by lipids is reduced. The composition is still under investigation.

## DISCUSSION

The novel ECG-electrode integrated onto the diffusion membrane of a tc-sensor yields excellent signals from neonates and adults. It also allows thorax impedance measurement simultaneously. The stress on prematures by application of numerous transducers can be reduced by using this combined sensor. The ability of the transducer to monitor the ECG of fetuses during delivery depends on an adequate electrode contact cream which is still under investigation.

MICROELECTRONIC SENSORS FOR SIMULTANEOUS MEASUREMENT OF $PO_2$ AND pH

C. C. Liu, M. R. Neuman, L. T. Romankiw*, and E. B. Makovos

Case Western Reserve University, Cleveland, Ohio
*IBM Watson Research Center, Yorktown Hts., New York

SUMMARY

Sensors capable of simultaneous measurement of pH and $PO_2$ in an aqueous solution have been fabricated in planar form using microelectronic technology. Separate potentiometric metal-metal oxide pH and three electrode amperemetric $PO_2$ sensors have been fabricated together on a silicon chip with active surface of 2 x 3 mm. Rhodium-rhodium oxide electrodes operating in a cyclic voltammetry mode have also been studied and show sensitivity to both pH and $PO_2$ in different regions of the voltage swept curve.

INTRODUCTION

Traditional electrochemical sensors can be miniaturized and applied to biomedical sensing problems by using microelectronic technology. This offers the potential of fabricating small, reproducible, planar structures of metals and insulators on flat surfaces such as glass or polished, oxidized silicon. Microelectronic technology offers the potential of mass production as well as the possibility of low per unit cost. The dimensional reproducibility may allow simpler calibration procedures for such sensors to be developed.

METHODS

Two types of sensors are reported. A microelectronic version of previously reported $PO_2$ and pH sensors[1,2] has been fabricated using thin film deposition techniques and photolithography on silicon-silicon dioxide substrates. The oxygen sensor is of the three electrode potentiostatic design in which gold films are used as both working and counter electrodes, and an electrolytically chlorided silver film serves as the reference electrode. One hundred forty individual 25 micrometer diameter microcathodes are defined by precise holes in a silicon dioxide film covering a gold layer. These cathodes are separated by 30 micrometers so that there is no interaction of their diffusion gradients, and they are connected electrically in parallel so that the current from each is summed. A single gold counter electrode having a larger surface area than the ensemble of cathodes returns current to the cell. The reference electrode is used to establish the working electrode bias of 700 mV with respect to the electrolytic solution it contacts.

The reference electrode also serves as a part of a two electrode palladium-palladium oxide potentiometric pH sensor as previously reported in a larger form.[2] A palladium strip is deposited and oxidized by reactive sputtering to make the indicating electrode.

The entire structure is formed upon a 5 x 20 x 0.5 mm silicon bar. The actual sensor is at one end of this bar and occupies a region 2 x 3 mm. Electrical conductors are brought along the bar to connect lead wires at the opposite end to the sensing elements.

A second electrochemical sensor for measurement of pH and $PO_2$ is based upon a Rhodium-Rhodium oxide potentiostat system similar to the oxygen sensor described above. Instead of applying a fixed bias voltage, however, the system is cycled linearly from -.8 to +1.0 volts at rates ranging from 50-100 mV/s. Voltammograms were obtained with the electrodes in a physiologic Ringer's solution buffered with Tris and containing 3.4 g/l bovine albumin. The solution pH was changed over the physiologic range of from 6.9-7.8.

## RESULTS

The microelectronic combination $PO_2$-pH sensor chips were found to give linear responses over the physiologic

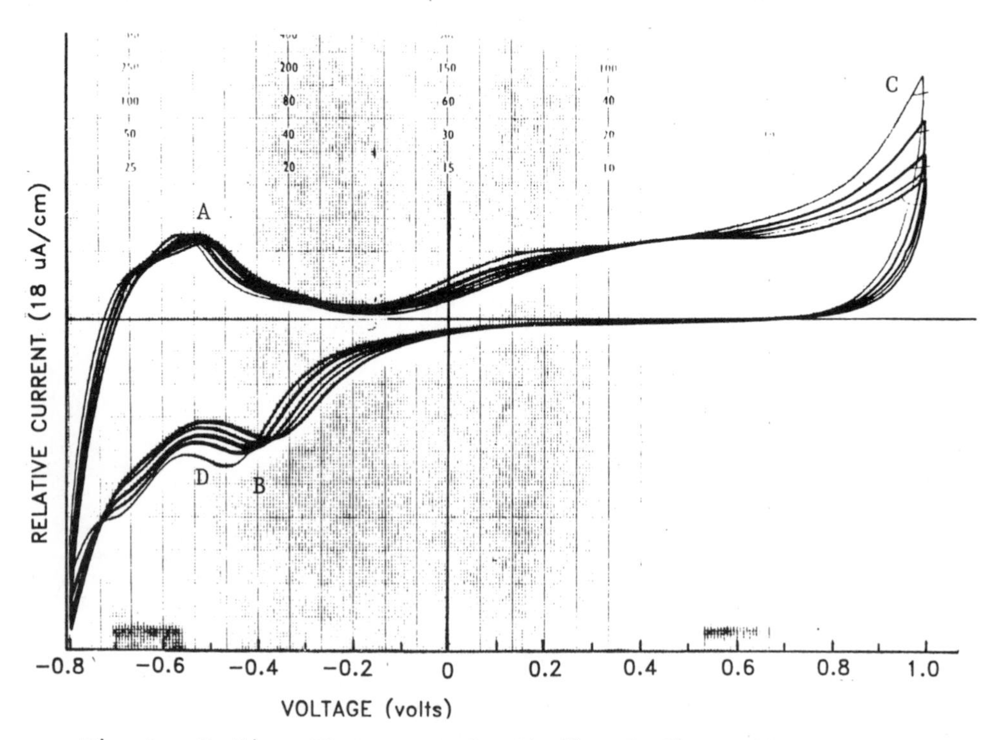


Fig. 1. Cyclic voltammograms for Rhodium-Rhodium oxide sensors.

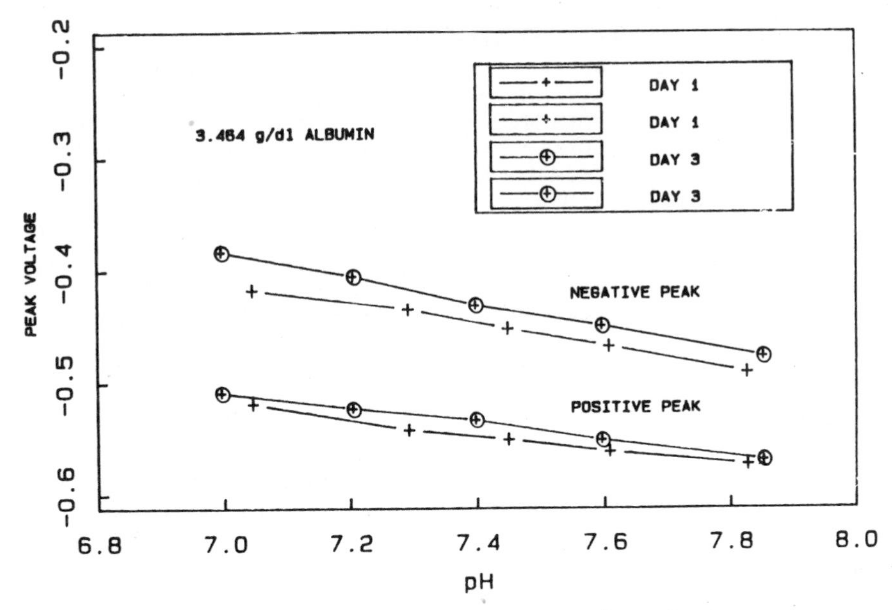


Fig. 2. Peak voltage as a function of pH.

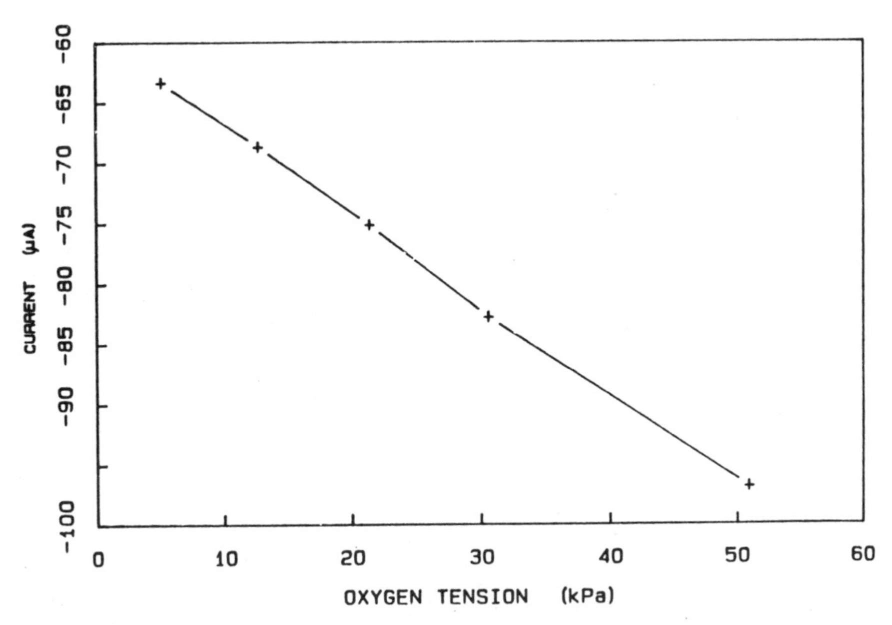


Fig. 3. Current as a function of $PO_2$

range of values. The $PO_2$ sensor had a sensitivity of 42 nA/kPa, and the pH sensitivity was 55 mV/pH at 25°C.

A set of cyclic voltammograms from the Rhodium-Rhodium oxide sensor is shown in Fig. 1. The different curves correspond to solutions of different pH. It is noted from the figure that the curves separate in cathodic regions A and B and anodic region C. One can plot the peak voltage of, for example, regions A and B as a function of pH as shown in Fig. 2 or the current at point C as a function of pH. The former is preferred for stability, and as seen from Fig. 2 has linear pH relationships with slopes of 69.1 and 95.7 mV/pH.

Oxygen sensitivity is seen in region D (Fig. 1) which represents the current seen in a conventional two electrode polarographic oxygen sensor. Fig. 3 shows the current as a function of $PO_2$ to be linear, although there is a higher offset current than seen with equilibrium sensors.

DISCUSSION

The above results show that the same electrode system can be used for measuring both pH and $PO_2$. Any cross sensitivity between these two variables can be removed by solving the simultaneous equations

$$V_p = a(pH) + b(PO_2)$$
$$I_p = c(pH) + d(PO_2)$$

where $V_p$ is the peak voltage and $I_p$ is the peak current and a-d are empirically determined constants.

By solving these equations in near real time, the independent pH and $PO_2$ can be determined.

The 50-100 mV/s scan rate is required for these systems so that there is sufficient time for the reactions to take place. The peaks can be lost or distorted at higher scan rates. A complete cycle, therefore, takes from 36 to 72 s. Thus, if one is interested in a particular peak, the system response time could be quite long. Faster responding electrode systems can be obtained by out of phase scanning of several pairs of working and counter electrodes formed using microelectronic structures such as described above. For example, by having two sets of working and counter electrodes scanned 180° out of phase, the system response time could be improved by a factor of two.

We can, therefore, conclude that these microelectronic techniques have potential application in fabricating the sensor elements of a combination $tcPO_2$ and $tcPCO_2$ sensor.

Supported by N.I.H. grants HD 19552, RR02024 and RR00210.

REFERENCES

1. C.C. Liu, M.R. Neuman, K.L. Montana, and M. Oberdoerster, Miniature Multiple Cathode Dissolved Oxygen Sensor for Marine Science Applications, Marine Technol, 16: 468 (1980).

2. C.C. Liu, B.C. Boccicchio, P.A. Overmyer, and M.R. Neuman,, A Palladium-palladium Oxide Miniature pH Electrode, Science, 207: 188 (1980).

# A MODIFIED ELECTRODE RING FOR USE IN TRANSCUTANEOUS MEASUREMENT OF $Po_2$

Harald Schachinger and
Thomas Lauhoff

Kinderklinik der Freien Universität
Heubnerweg 6
D-1000 Berlin 19 (West)/Germany

## INTRODUCTION

The transcutaneous $Po_2$ method ($tcPo_2$) can only reflect $P_ao_2$, if there is maximal or close to maximal perfusion in the skin below the electrode. Hyperaemia can only be induced by heating the electrode. It has been shown that at temperatures of 45° C and even 44° C values agree well with the $P_ao_2$ values (3). However, these temperatures may have an adverse effect on the skin in form of blemishes or even II° burns, in particular in patients with circulatory disorders (e.g. heart defect, sepsis) (1, 2).

In order to reduce skin irritation, the electrode may be attached to different parts of the skin during the course of measurement. If then a lower temperature (e.g. 43° C) is used, a poor correlation with the arterial values must be expected. Alternatively, measurement may be performed on the buccal mucous membrane or on the conjunctiva (4, 5, 6).

Non-pathological hyperthermia together with sufficient hyperaemia was achieved by enlargement of the heating area and maintenance of a median temperature of 43° C. This has been practised in adult $Po_2$ monitoring using a very wide heating ring (7).

The skin of a newborn infant is particularly sensitive to heat; we therefore constructed a ring with an adjustable heating area. Two additional circular heating elements were incorporated into the enlarged electrode ring (Fig. 1). The whole sensor consists of a standard electrode in normal use, with a heatable anode situated in the centre (interior heat) together with the new electrode ring fitted with two additional heating elements (middle and exterior heat). Fig. 2a and b shows the sensor viewed from the skin and in a cross section. It has a diameter of 22 mm, a height of 10 mm and is 8.5 g in weight, i.e. 2.24 g/cm³. The heating areas measure 16.5, 83.0 and 82.4 mm² from inside to outside. Fig. 3 shows a comparison between the test electrode and the normal electrode.

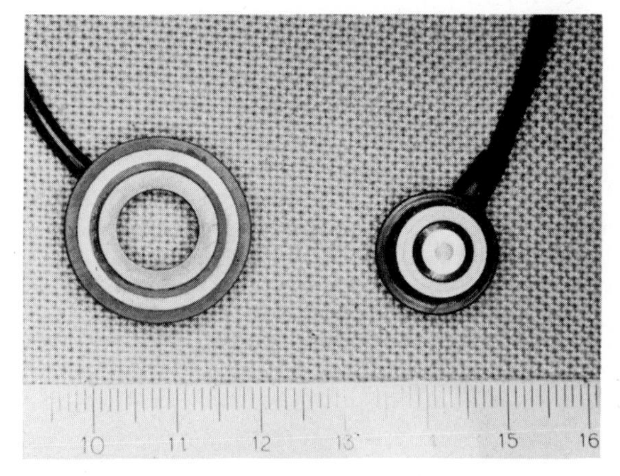


Fig. 1. The enlarged new electrode ring with the two different heating elements (left) and the standard $tcPo_2$ electrode (right) without any electrode ring.

Initial results of $Po_2$ measurements using this modified electrode are given for 60 newborn infants whose birth weights ranged from 690 g to 4300 g. For comparison, a standard $tcPo_2$ electrode was placed near the new electrode. Fifty of the infants had severe cardiopulmonary disorders (respiratory distress, sepsis, congenital heart defect) and ten infants had problems of adaptation, although their respiratory and circulatory condition at the time of testing was stable.

From more than 125 possibilities of regulating the temperature, 45° C was excluded to avoid unnecessary skin reactions during measurement. Lower temperatures of the additional heating elements compared to the standard electrode's temperature of 44° C took preference.

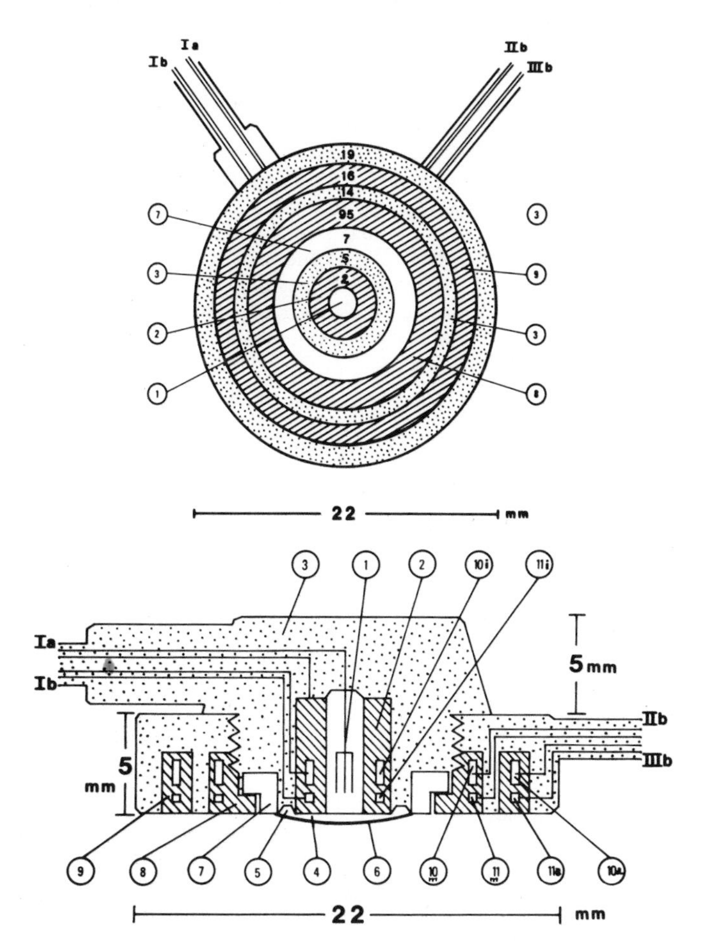


Fig. 2a + b. The test electrode viewed from the skin (a) and in a cross section (b). 1 = cathode, 2 = anode, 3 = araldit, 4 = electrolyte, 5 = electrolyte space, 6 = teflon membrane (25 µm), 7 = central heating ring, 8 = middle heating ring, 9 = peripheral heating ring, Ia, Ib, IIb = electric cables.

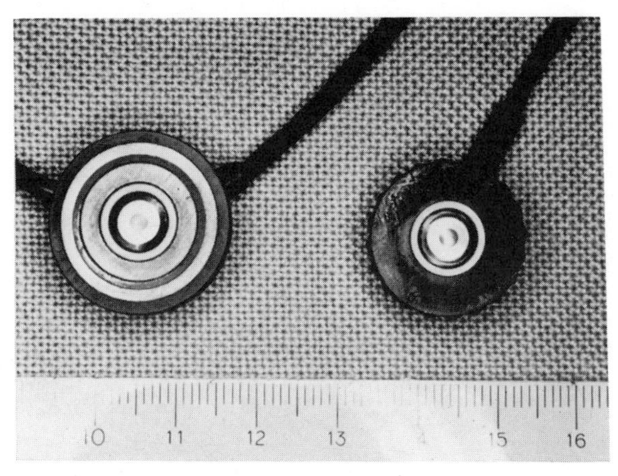

Fig. 3. Comparison between the test electrode and a standard electrode (left) (right)

## RESULTS

We found that with the central temperature kept at 37° C $tcPo_2$ never attained the standard $tcPo_2$ level when both middle and peripheral temperatures were 43° C or 44° C. This observation refers to both healthy and to sick neonates. The central temperature had to be kept at 43° C. When the middle and the peripheral temperatures were also kept at 43° C, the $Po_2$ values increased. The same level was reached when only the peripheral temperature was 43° C and the middle ring was unheated. The combined inner and outer heating of the test electrode at 43° C is significantly superior to the standard electrode heated also at a temperature of 43° C (Fig. 4), but the levels tend to be somewhat lower than those of the standard electrode heated to 44° C.

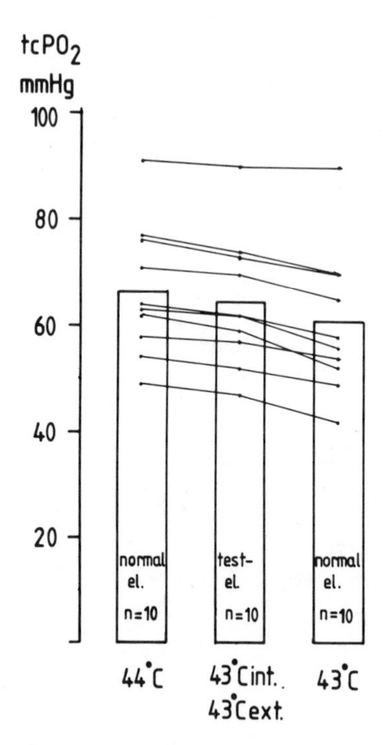


Fig. 4. $Po_2$ levels in 10 neonates with standard electrode kept at 44° C and at 43° C, respectively, and the test electrode with the central and the peripheral temperatures kept at 43° C.

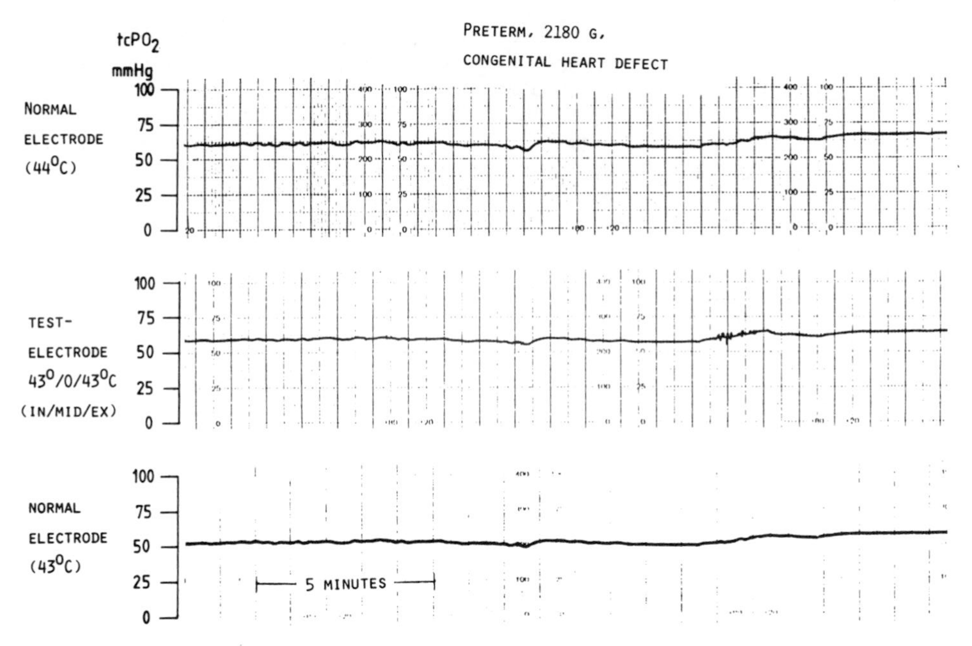


Fig. 5. Monitoring a preterm of 2180 g body weight, the $tcPo_2$ values at a temperature of 44° C are similar to those measured by use of the test electrode with inner and outer heating at 43° C.

Fig. 5 shows an original recording from a premature infant weighing 2180 g with stenosis of the aortic isthmus and arterial septum defect. The test electrode with central and peripheral temperatures of 43° C clearly gave a higher level than that registered by the standard electrode also set at 43° C.

## CONCLUSION

By enlarging the heating area it was possible to obtain $tcPo_2$ values at 43° C which were very close to those obtained by the use of a conventional sensor kept at 44° C. The latter are known to be of the same order of magnitude as the arterial levels in neonates. Application of this new sensor causes only minor skin lesions.

## REFERENCES

1. R. G. Boyle and W. Oh, Erythema following transcutaneous monitoring, Pediatrics 65:333 (1980).
2. C. S. M. Golden, Skin craters - a complication of transcutaneous oxygen monitoring, Pediatrics 67:514 (1982).
3. R. Huch, D. W. Lübbers and A. Huch, Reliability of transcutaneous monitoring of arterial $Po_2$ in newborn infants, Arch. Dis. Child. 49:213 (1974).
4. M. Kwan and J. Fatt, A non-invasive method of continuous arterial oxygen tension estimation from measured palpebral conjunctival oxygen tension, Anesthesiology 35:309 (1971).

5. H. Schachinger and D. Seiler, First experience with transjunctival $Po_2$ measurement, in: "Continuous transcutaneous blood gas monitoring", R. Huch and A. Huch (eds.), Marcel Dekker, Inc., New York and Basel, 387 (1983).
6. J. Schmauser, H. Schachinger, H.-D. Frank and D. Seiler, Vergleichende transcutane $Po_2$-Messungen am Thorax und an der Wangenschleimhaut, in: "Perinatale Medizin", Bd. 10, J. W. Dudenhausen und E. Saling (Hrsg.), Thieme Verlag, Stuttgart, 404 (1984).
7. J. M. Steinacker und R. E. Wodnik, Transcutaneous measurement of arterial $Po_2$ in adults: design of an improved electrode, in: "Continuous transcutaneous blood gas monitoring", R. Huch and A. Huch (eds.), Marcel Dekker, Inc., New York and Basel, 133 (1983).

MULTICHANNEL RECORDING AND ANALYSIS OF PHYSIOLOGICAL DATA

USING A PERSONAL COMPUTER

André R. van der Weil, T. Cornelis Jansen, Harry N. Lafeber and Willem P.F. Fetter

Department of Pediatrics, Subdivisions of Neonatology and Medical Electronics, Sophia Children's Hospital, Rotterdam, The Netherlands

SUMMARY

A flexible data acquisition system was designed for an Apple IIe computer. For implementation, we used a microcomputer capable of recording 4 channels simultaneously with a selectable sample rate; this information was saved on a floppy disc and the recordings could be evaluated later. This evaluation is accomplished by reproducing the analog signals on display and calculating a histogram after deleting artefacts in the signal.

INTRODUCTION

One problem with long term recording is the storage and evaluation of the acquired data. We faced this problem when we started a study with simultaneous registration of transcutaneous Po2,Pco2 and oxygen saturation in neonates and infants [1]. Our goal was to create a flexible recording and analysis system for more than one signal. It should be possible to analyse each individual signal but also to analyse their interrelationships.

METHODS

An Apple IIe is selected for our studies; this is an 8 bit microprocessor with 128 kbyte memory and a monochrome display. For storage, two Apple II floppy discdrives are used each with a capacity of 140 kbyte. Printing and plotting is handled by an Epson FX-80 matrixprinter. For analog to digital conversion a four channel converter Kronemuis APL-22 is included with a resolution of 8 bit and an adjustable input of 0 to 5 Volts full scale. For timing a clock, Kronemuis APL-05 is used. The program is written in Apple Pascal 1.2 and 6502 assembly. Overlays are used to fit the program in the limited memoryspace of the system. Furthermore the program is divided in two subprograms. The data acquisition subprogram is shown in figure 1.

data acquisition functions

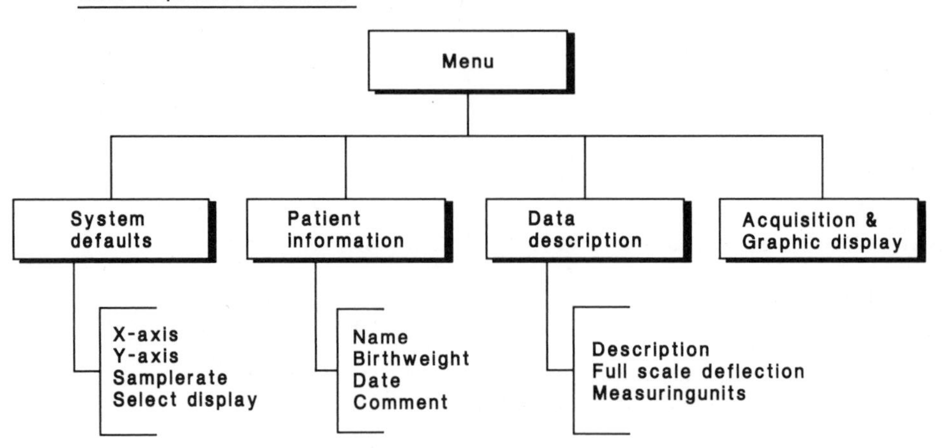


Figure 1. Data acquisition subprogram

The first function in this subprogram is used to set system parameters such as X- and Y- axis of the display. The X-axis can be set to 1, 4, 8, 12 or 24 hours of display. The Y-axis may range from 1 to 500. The sample rate is selectable between 1 and 60 second interval. The function "patient information" is used to store patient name, birthweight and date, followed by a single line mentioning the patient or measurement. Description of the data sources is defined by the name of the source, its full scale deflection on maximum output to the converter and the measuring units of the source. The values in the "system defaults" and "data description" functions are predefined for our application. The actual data collection and screen display is done by the fourth function which also stores the acquired data on a floppy disc. This storage is done in a "blockmode". Each block contains approximately 10 minutes of recording depending on the sample rate because a block contains 120 samples per channel. In this block there is also a time and date mark and a short data description. This adds to the integrity of the files and if necessary enables other software to read and interpret this information.
The other subprogram as shown in figure 2 has four functions for reviewing and evaluating the recorded signals. We found it possible to design the system to the users' needs such as the choice of an X- and Y-axis for displaying the data. The ranges for the axes are the same as in the data acquisition function. Calculating amplitude-time histograms is the other main option in this subprogram. In this function the user can remove artefacts by selecting the blocks of data which will be excluded from the histogram. This selection can be made from viewing the graphic display with its block numbers or from a printed list with compressed data from each block. After calculation one can plot the histogram with a set of selectable parameters such as increment, lower- and upper limits.

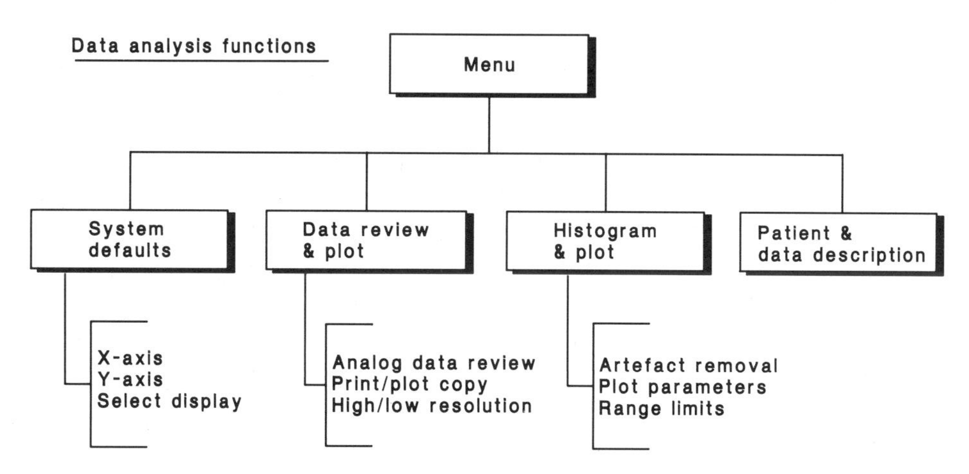


Figure 2. Data reviewing and analysis subprogram.

RESULTS

In the present application for the neonatal intensive care unit the sources are tcPo2 from our own device, tcSo2 and heart rate from a Nellcor N-101 Pulse Oximeter and tcPco2 from a Hellige TransCapnode system. The time window during acquisition is preset to 1 hour for monitoring purposes. In this application it is sufficient to use a sample interval of 5 seconds which gives about 48 hours recording time on one floppydisc.It is possible to change any of these parameters during the recording in order to adapt the system to the current situation.

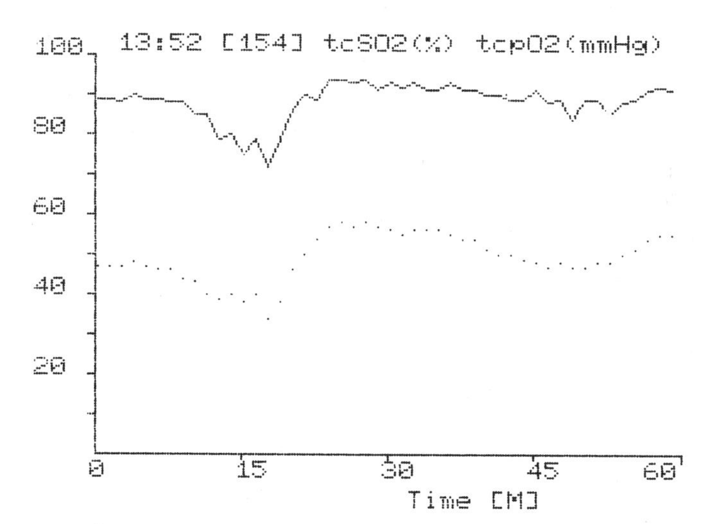


Figure 3. tcPo2 and tcSo2 in 1 hour display

On figure 3 a printed copy of an analog curve with a time window of one hour is shown. At the top of the display the start time, block number and sources are printed for the present plot. The two signals presented in this figure are transcutaneous Po2 with a dotted line and oxygen saturation with a solid line. The chosen amplitudes has been 100 mmHg for Po2 and 100 % for oxygen saturation. The close relationship between both parameters is clearly visible. Besides the time window one can also select the horizontal resolution of the display from one out of one, to 1 out of ten samples. This last setting provides for fast plotting of an averaged signal, while plotting each sample provides a more detailed result.

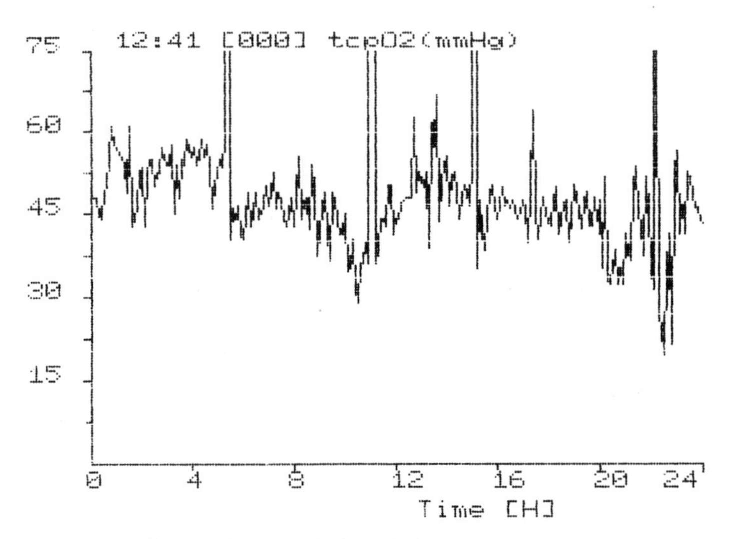


Figure 4. tcPo2 in 24 hour display

Figure 4 shows a recording with tcPo2 compressed with a time window of 24 hours. The solid line is the tcPo2. The calibrating and repositioning of the Po2 sensor approx. every 4 hours is visible here and serves to check upon the proper application of the electrode.
The results of another method of analyzing the data are shown in figure 5. A time-amplitude histogram over a given time period is calculated for one channel. The period of registration in which values were above or under a certain range can be detected and collected over the recording in order to give an estimate for hypoxemic and hyperoxemic periods. The printed copy of the result contains patient information, statistical data such as mean and standard deviation and the histogram itself, which is printed with a range and interval as specified by the user.

```
 Med.El.  Sophia Children's Hospital
---------------------------------------------------------------------------
Patient identification :100786HO
        Name           :---------
        Comment        :Bronchopulmonary dysplasia;on ventilator

Channel used for calculations :tcpO2

Start date mark :18-7   Last  date mark :20-7

Accumulated time period :  44:22 h  Time period within limits :  43:21 h
Mean Value              :     46 mmHg  Standard deviation      :      9 mmHg

Low limit               :     10 mmHg  Time period underrange  :    1.0 %
High limit              :    100 mmHg  Time period overrange   :    2.4 %
---------------------------------------------------------------------------
                1        10        20        30        40        50
 mmHg   %      |---------|---------|---------|---------|---------|

  100  0.0
   95  0.0
   90  0.0
   85  0.0
   80  0.0
   75  0.1
   70  0.1
   65  2.6 ===
   60 13.8 ==============
   55 15.3 ===============
   50 21.8 ======================
   45 21.2 =====================
   40 14.8 ===============
   35  7.1 =======
   30  1.5 ==
   25  0.7 =
   20  0.8 =
   15  0.3
   10  0.0
    5  0.0
```

Figure 5. Time-amplitude histogram.

DISCUSSION

We have developed a 4 channel data acquisition and analysis system which is easy to use and flexible for most current measurements in intensive care. This system has been used in the neonatal intensive care and in measurements involving older infants with bronchopulmonary dysplasia. In these applications its usefulness has been proved. Systems have been designed using larger computers or evaluating one signal in great detail [2]. However, the capabilities of the personal computer enable implementing multichannel data acquisition on small systems; these small systems can be preferable due to their flexibility and low costs.
At the moment we are preparing a transfer to the MS-DOS operating system to include the use of the IBM or compatible personal computers. Furthermore we plan to implement digital filtering and pattern recognition to improve our system.

REFERENCES


1. H.N. Lafeber, W.P.F.Fetter, A.R. van der Wiel, T.C. Jansen, Pulse oximetry and transcutaneous oxygen tension in hypoxemic neonates and infants with bronchopulmonary dysplasia. This volume.

2. F. Fallenstein, N.U. Bucher, R. Huch, A. Huch, The 'Dynamic Transcutaneous Po2 Histogram` or How to deal with immense Quantities of Monitoring Data, Pediatr. 75: 608-613 (1985)

# CONTINUOUS NON-INVASIVE BEAT-BY-BEAT BLOOD PRESSURE (B.P.) MEASUREMENT IN THE NEWBORN

P. Rolfe, P.P. Kanjilal, C. Murphy and P.J. Burton

Biomedical Engineering Centre
Churchill Hospital
Oxford, U.K.

## SUMMARY

Intermittent measurement of blood pressure (BP) is frequently performed in newborn babies under intensive care using an inflatable cuff encircling either the arm or leg. This paper describes work aimed at achieving continuous beat-by-beat measurement of the arterial pressure waveform by means of a finger cuff, the pressure of which is controlled by a feedback arrangement utilising a photoplethysmograph. Preliminary results are encouraging, but improvements to cuff design are still needed.

## INTRODUCTION

Arterial pressure measurement forms an important part of the clinical management of the pre-term baby. Furthermore, it is of great importance in the investigation of intra-cranial intra-ventricular haemorrhage and hypoxic-ischaemic brain injury, the major causes of mortality and morbidity in pre-term infants in the developed world. In many neonatal intensive care units it is customary to use an umbilical artery catheter, and this then allows continuous arterial pressure measurement to be performed with an hydraulically coupled pressure transducer. However, the possible hazards of this invasive procedure are well-recognised and an alternative non-invasive method would be preferred. Automated indirect blood pressure measuring instruments, based on inflatable limb-encircling cuffs, have become increasingly popular. Such instruments are based upon the straightforward procedure of arranging for the cuff pressure to be equal, sequentially, to the systolic and mean arterial pressures, with appropriate means for detecting when the cuff and arterial pressures are equal.

An important principle underlying many techniques for non-invasive BP measurement is that of 'vascular unloading', which was first described in 1876 by Marey[1]. If an artery is compressed by a pressure equal to mean arterial pressure (MAP) it will become 'unloaded', and the beat-by-beat volume pulsations of the artery will be at a maximum. This allows MAP to be determined non-invasively. An extension of this principle is to arrange for the compression force to be adjusted throughout the cardiac cycle in order to eliminate the arterial volume pulsations. This approach was proposed by Penaz[2] as a means of continuous non-invasive BP measurement, and forms the basis of the developments described here[3], and also pursued by others[4,5].

## METHODS

The Cuff. A rigid cyclindrical plastic tube, with a circular orifice at each end of appropriate dimensions to accept a finger, is used to construct the cuff. A thin flexible polymer tube is affixed concentrically within the outer rigid tube by means of attachment to the inner circumference of the orifices. The space between the inner and outer tubes may be pressurised, and the cuff pressure measured.

The Photoplethysmograph. An infra-red light-emitting diode and a phototransistor are attached to the flexible tube so as to detect the volume pulsations of the digit according to the classical photoplethysmographic principles.

Control of Cuff Pressure. A pneumatic/hydraulic connection is made between the cuff and two combined means of controlling cuff pressure: firstly, a stepper motor controlled syringe pump; secondly, an electromechanical 'shaker'. Cuff pressure is measured with a low volume-displacement transducer. The cuff and connecting tubing may be air or liquid filled.

Measurement of Mean Arterial Pressure. With a digit placed in the cuff, and cuff pressure a few mmHg above atmospheric pressure, the expected cardiac related pulsatile volume signal is seen from the photoplethysmograph. The syringe pump is then controlled to increase cuff pressure to above systolic pressure and the volume pulsations disappear. Cuff pressure can then be reduced progressively at a pre-determined rate related to heart rate. The volume pulsations re-appear when cuff pressure is equal to systolic pressure, and they reach a maximum magnitude when cuff pressure equals MAP.

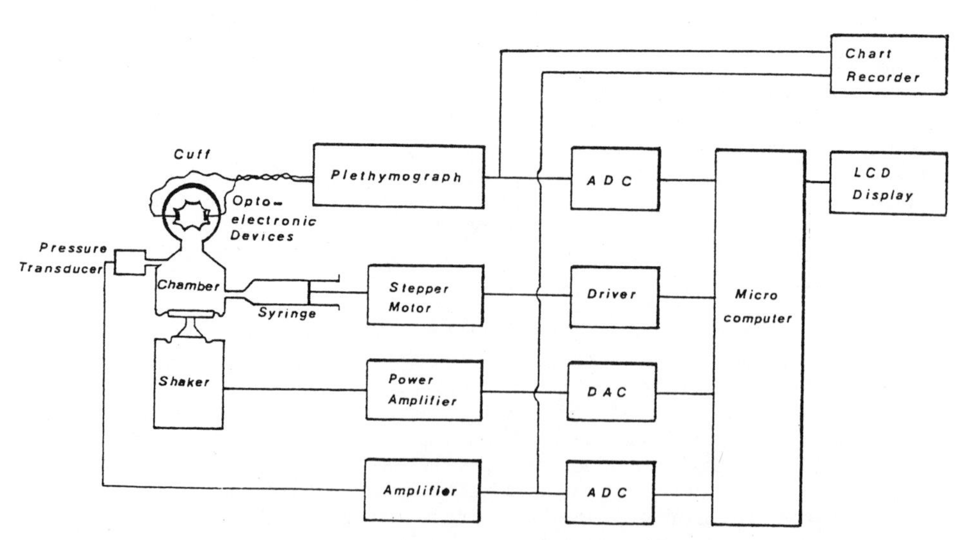


Figure 1: A schematic diagram of the BP measurement system.

Measurement of Arterial Pressure Waveform. Following determination of MAP the cuff pressure is set at that value by means of the syringe pump. A feedback control arrangement is then brought into operation, and the electro-mechanical shaker is driven to adjust cuff pressure very rapidly such that the digit volume pulsations tend towards a very low magnitude. Thus the cuff pressure tracks intra-arterial pressure, and the latter may therefore be estimated from the former.

## RESULTS

A variety of cuff sizes has been constructed to assess the technique in babies, children and adults. Initial tests concerned the comparison between the results obtained in the digit and in the upper arm with the new technique and with a conventional sphygmomanometer. Correlations for both MAP and systolic pressure have been found to be good, although in hypothermic subjects the peripheral pulsations may be very small and the digit pressure measurements do not then correlate well with upper arm measurements.

Comparisons have been made with direct arterial measurements, and values for MAP have agreed to within $\pm$ 3 mmHg in infants following cardiac surgery.

Recent clinical trials have also been carried out in adults undergoing intensive care, once again comparisons being made with direct arterial measurements. In several patients being studied the true pulse pressures were very large e.g. > 100 mmHg, and the closed-loop pressure controller was found to have insufficient dynamic range.

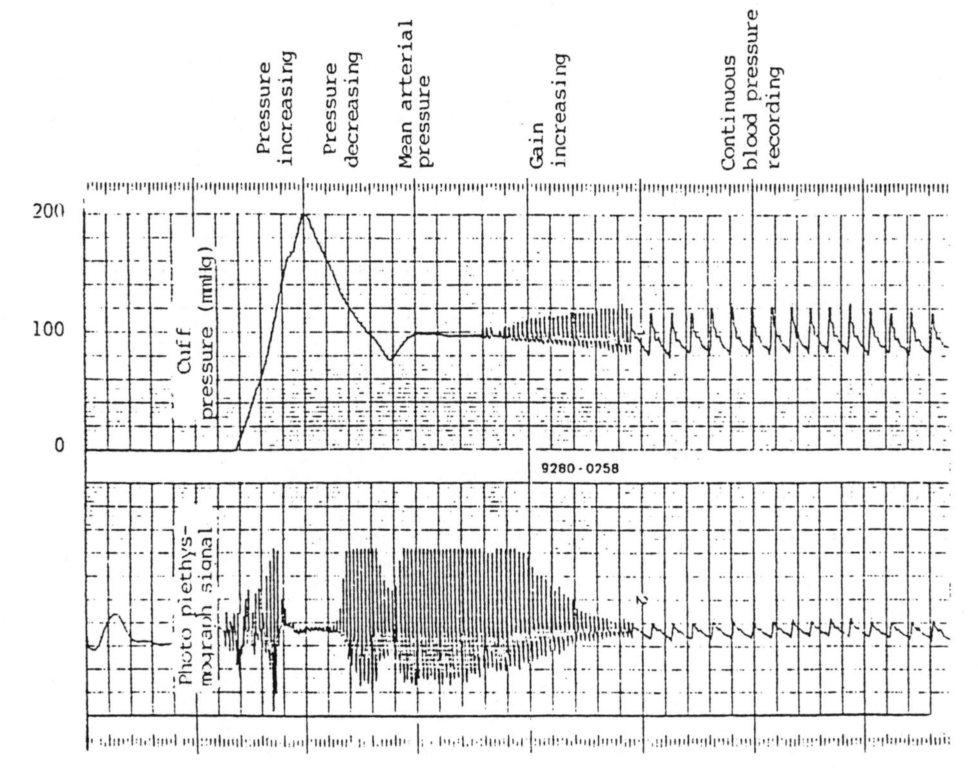


Figure 2: A continuous recording from the system, showing the transition from open-loop to closed-loop operation.

## DISCUSSION

The results to-date indicate that intermittent measurements of systolic and mean arterial pressure may be made with the non-invasive method in digits of babies, children and adults. Continuous non-invasive measurement has also proven to be feasible, although a comprehensive assessment has not yet been completed.

Development of an optimal performance closed-loop control system is complicated by the variable delays introduced by the finger and the essentially fixed delays of the cuff, connecting tubing and pressure drive arrangement. Operating the system at high closed-loop gain in order to minimise the measurement error can lead to instability. Ideally the controller transfer function needs to be adjusted automatically to compensate for variations in the mechanical properties of the digit and digital arteries.

During closed-loop operation perfusion of the digit is reduced. This means that long-term use may be inadvisable, although the prospect of alternating periods of measurement in two or more digits seems feasible. There are inevitably quantitative differences between digital artery pressure and brachial or aortic pressure due to pressure gradients and also possibly due to peripheral magnification. The precise relationships in various clinical conditions have yet to be elucidated. Furthermore, the impact of peripheral vasoconstriction may be very important and this requires detailed consideration.

## CONCLUSIONS

Preliminary clinical results indicate that it is possible to make intermittent and continuous non-invasive measurements of blood pressure via the digit. For continuous measurement in adults the dynamic range of the system described here must be increased. Cuff design requires further optimisation to improve the clinical applicability and the frequency response. The limitations of peripheral vasoconstriction must be fully assessed.

## ACKNOWLEDGEMENTS

British Heart Foundation, Oxfordshire Regional Health Authority, Wolfson Foundation.

## REFERENCES

1. E. J. Marey, "La methode graphique", E. Masson, ed., Paris (1876).
2. J. Penaz, Czech Patent 133205, Prague (1969).
3. C. Murphy et al, Med. Biol. Engng. Comput., 23(Pt I): 499-500 (1985).
4. Yamakoshi et al, IEEE Trans. Biomed. Eng. BME, 27(No 3): 150-155 (1980).
5. Wesseling et al, Funkt. Biol. Med., 1:249 (1982).

# THE RELATIVE ACCURACY OF THREE TRANSCUTANEOUS DUAL ELECTRODES AT 45 °C IN ADULTS

Colm Lanigan, Jose Ponté and John Moxham

Departments of Thoracic Medicine and Anaesthesia
King's College Hospital
London, U.K.

## INTRODUCTION

Conventional practice dictates that the results of any study comparing transcutaneous partial pressures of oxygen and carbon dioxide ($tcPO_2$ and $tcPCO_2$) with arterial blood gas values ($PaO_2$ and $PaCO_2$) should be presented in the form of a scattergram (Fig. 1). The relationship between the two is usually represented by a linear regression equation of the form

$$y\ (tcPO_2) = a + b\ x\ (PaO_2)$$

and a correlation coefficient, r, calculated (Godfrey, 1985). The probability that the transcutaneous and arterial blood gas values are related is given by a probability value, P (Table I). Does this statistical analysis give us the most useful information from transcutaneous monitoring for clinical purposes? We suggest that a recently proposed alternative method of analysis (Bland and Altman, 1986) may be more useful and we demonstrate its practical application in evaluating combined $tcPO_2$ and $tcPCO_2$ electrodes against endtidal oxygen and carbon dioxide ($ETPO_2$ and $ETPCO_2$) values in normal adults.

## MATERIAL

We tested three commercially available dual electrodes (Microgas Combisensor, Kontron Ltd = K; Novametrix 850 Commonsensor, Vickers Medical Ltd = N; and Radiometer E5270 electrode, V.A.Howe & Co Ltd = R). Each electrode was remembraned within 5 days, and given a 2 point dry gas calibration at 45°C before and after each study. A total of 12 non-smoking healthy volunteers (7 male, aged 23 to 48 years) were studied.

## METHODS

Electrodes were fixed to the skin with double-sided adhesive rings and operated at 45°C. Subjects were seated throughout the study and breathed through a mouthpiece from either room air or a 200 litre Douglas bag containing the previously analyzed gas mixture - either a hypoxic mixture (mean $O_2$ = 12.8%, range 12.3 to 13.4%, 7 subjects) or a hypercapnic mixture (mean $CO_2$ = 5.9%, range 4.2 to 6.7%, 7 subjects). Arterial gases were estimated from

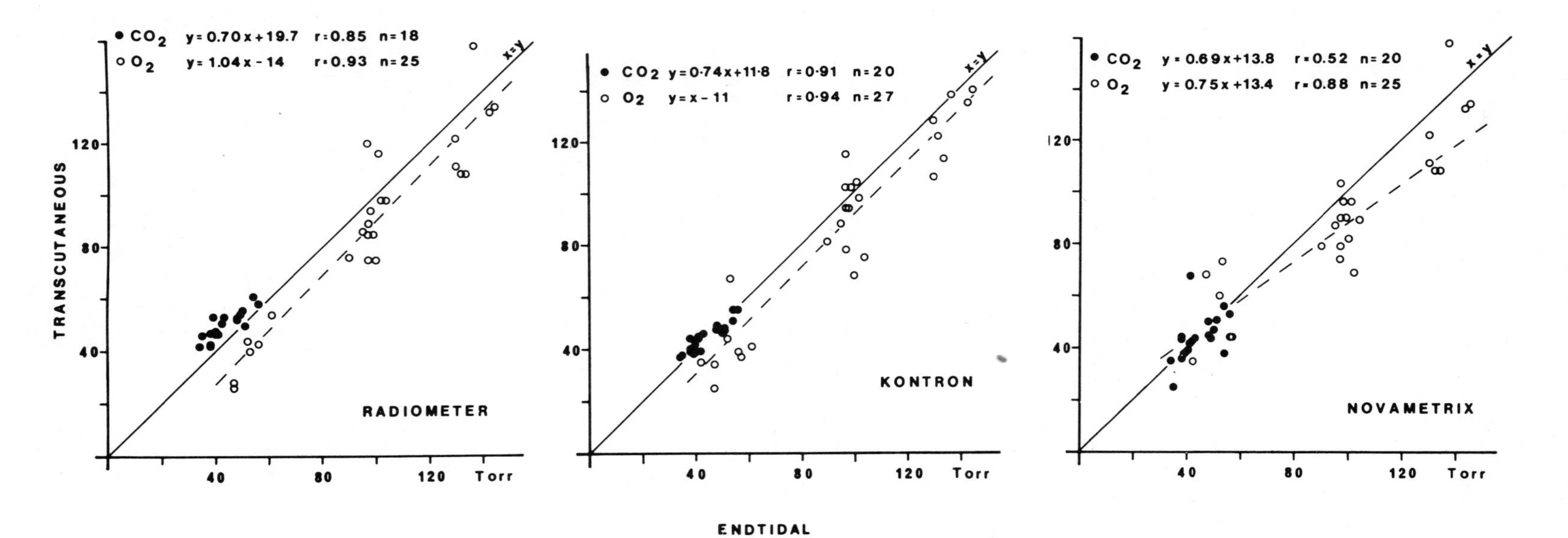


Fig. 1. The relationship between transcutaneous and endtidal gases. Line of identity (solid line) and regression line for oxygen (dashed line) shown for the three dual electrodes.

peak and trough endtidal gas concentrations, measured by a quadrupole mass spectrometer (Airspec 2000, drift less than 0.1% per hour). Values were corrected for barometric and water vapour pressure; values for oxygen were corrected for the arterio-alveolar gradient (taken as 5 torr breathing air and 0 torr during hypoxia). Readings were allowed to stabilize for 30 minutes before recordings were made. Steady state values were recorded in air for endtidal gases, $tcPO_2$ and $tcPCO_2$ and averaged over 2 minutes before exposing the subject to a step change in inspired gas concentrations until a new steady state was reached, before finally being returned to air.

## ANALYSIS

$TcPO_2$ was plotted against $ETPO_2$; then regression equations, r, and P values were calculated for each electrode (Fig. 1), (Godfrey, 1985). The difference for each electrode between $tcPO_2$ and $tcPCO_2$ were measured and plotted against the average $PO_2$ value: $(tcPO_2 + ETPO_2) / 2$ (Bland and Altman, 1986). Average differences, and the 95% confidence intervals for those differences were then calculated and plotted (Gardner and Altman, 1986); see Fig. 2. $TcPCO_2$ and $ETPCO_2$ values were treated in the same way (Fig. 1 and 2B).

## RESULTS

$TcPO_2$ values were highly correlated with $ETPO_2$ for all three electrodes ($P < 0.001$, Table 1 and Fig. 1). Although $tcPCO_2$ was less highly correlated with $ETPCO_2$ compared to the equivalent relationship for oxygen, $tcPCO_2$ differed from $ETPCO_2$ by a smaller and less variable amount than $tcPO_2$ from $ETPO_2$ (Table 1 and Fig. 2).

## DISCUSSION

Although transcutaneous blood gas values are not equivalent to arterial blood gases, they are widely used as non-invasive trend indicators of arterial blood gases. Most studies report a statistical relationship between the two which is highly significant ($P < 0.001$). However this high level of probability that the two are related is hardly suprising and is of little use in interpreting the isolated transcutaneous blood gas value. Variance ($r^2$) tells us the amount of variability in one parameter that can be explained by change in the other e.g. when $r = 0.85$ (Fig. 1), only 72% of the variability in $ETPCO_2$ can be accounted for by change in $tcPCO_2$. Variance therefore only serves as a general guide to overall accuracy. The standard error of the estimate (Sy, x) provides an index of the scatter of values about the regression line. It is calculated by obtaining the standard deviation of the differences between the actual values for y and those predicted from the regression equation. Its disadvantage lies in the fact that unless the slope of the regression equation is 1, the standard error of the estimate for y calculated from x will differ from the standard error of the same estimate for x calculated from y (Godfrey, 1985).

A more useful clinical index is the average difference (bias) between the transcutaneous and arterial blood gas value (Bland, and Altman, 1986). It is easily calculated and understood. The precision of this measurement i.e. how closely the transcutaneous or arterial blood gas value will reflect each other on repeated sampling - is given by the 95% confidence intervals (the average differences plus and minus 1.96 times the standard deviation of the differences): for example -

Table 1. The Relationship and Differences between Transcutaneous and Endtidal Gas Values in Adults.

| | n | a | b | r | P | d | d+2SD | d-2SD |
|---|---|---|---|---|---|---|---|---|
| *KO2 | 27 | -11 | 1.00 | 0.94 | 0.001 | 9.6 | 34.3 | -15.1 |
| RO2 | 25 | -14 | 1.04 | 0.93 | 0.001 | 10.0 | 35.2 | -15.2 |
| NO2 | 25 | 13 | 0.75 | 0.88 | 0.001 | 14.3 | 41.0 | -18.2 |
| KCO2 | 20 | 12 | 0.74 | 0.91 | 0.001 | -0.5 | 5.2 | -6.1 |
| RCO2 | 18 | 20 | 0.70 | 0.85 | 0.001 | -6.6 | 0.4 | -13.5 |
| NCO2 | 20 | 14 | 0.69 | 0.52 | 0.02 | 0.0 | 15.7 | -15.7 |

* Kontron, Radiometer and Novametrix dual electrodes; n = number of paired samples analyzed; a, b = intercept and slope of the linear regression equation $y = a + b x$ where y = transcutaneous oxygen or carbon dioxide and x = endtidal oxygen or carbon dioxide; r = correlation coefficient, P = probability value; d = average difference between transcutaneous and endtidal values; d+2SD and d-2SD = 95% confidence intervals for the average difference.

(a) the Kontron electrode provided $tcPCO_2$ values which, 19 times out of 20, were within +5 and -6 torr of the $ETPCO_2$ value.

(b) the precision of the Radiometer electrode in predicting $ETPCO_2$ from $tcPCO_2$ was very similar to the Kontron electrode, but it over-estimated $ETPCO_2$ by an average of 6 torr more than the Kontron electrode (Fig. 2B).

Differences between the transcutaneous and arterial blood gases are easily plotted against the average estimated arterial blood gas value (Fig. 2), and any tendency towards increasing inaccuracy with higher or lower blood gas values can be quickly seen. If such a tendency is seen, the data can be normalized by plotting the difference from unity of the ratio of transcutaneous to arterial blood gas values. We conclude that calculation of the average difference between transcutaneous and arterial blood gas values provides an index which is more easily understood and much more helpful in clinical practice than correlation coefficients.

SUMMARY

Measurements made by transcutaneous electrodes can be compared with arterial blood gases in several different ways. The relationship between them is commonly expressed by a linear regression equation, and a correlation coefficient, r, calculated. However calculation of the bias and the precision of transcutaneous electrode readings is more helpful in clinical practice. The differences between the two methods are explained by comparing the relationship of three transcutaneous dual electrodes to endtidal gases in adults.

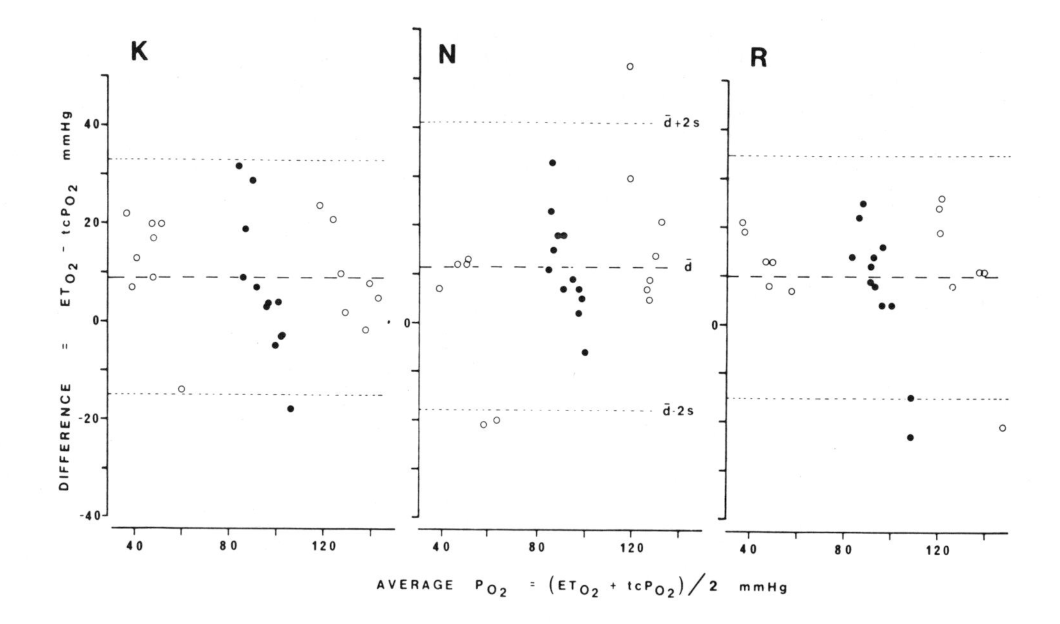


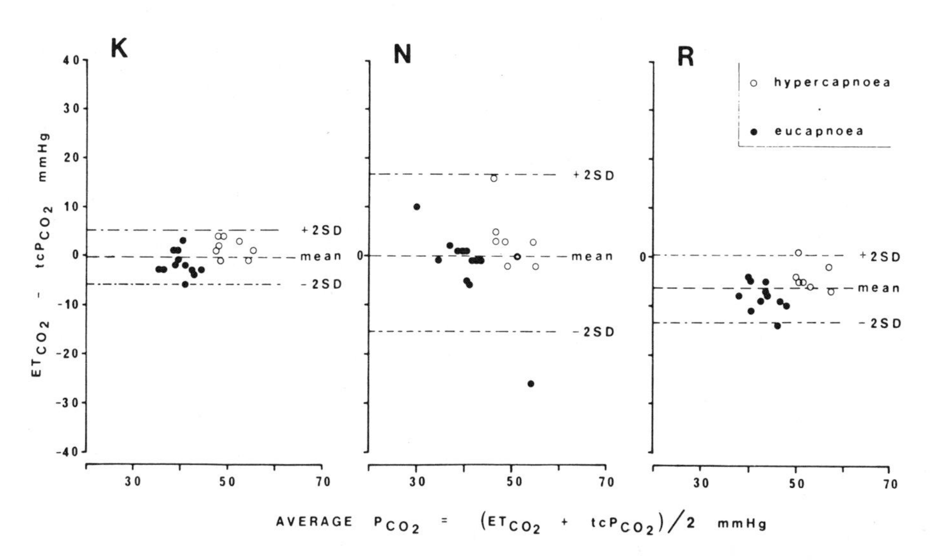


Fig. 2. Differences between Transcutaneous and Endtidal Values. The average differences between steady state transcutaneous oxygen (panel A) and carbon dioxide (panel B) values, and endtidal values for Kontron (K), Novametrix (N) and Radiometer (R) dual electrodes at 45°C in 12 normal adults breathing air (closed circles), and hypoxic or hypercapnic mixtures (open circles).

REFERENCES

Bland, J.M., and Altman, D.G., Statistical methods for assessing agreement between two methods of clinical measurement, Lancet, 1: 307 (1986).

Gardner, M.J., and D.G. Altman, Confidence intervals rather than P values: estimation rather than hypothesis testing, Br. Med. J., 1986, 292:746.

Godfrey, K., (1985), Simple linear regression in medical research, N Engl J Med, 313: 1629.

CONTRIBUTORS

W.A. van Asselt, Department of Pediatrics, University Hospital, 59, Oostersingel, NL-9713 EZ Groningen

P. Bäckert, Neonatologie, Departement für Frauenheilkunde, Frauenklinikstr. 10, CH-8091 Zürich

S. Bambang Oetomo, Department of Pediatrics, University Hospital, 59, Oostersingel, NL-9713 EZ Groningen

E. Bancalari, Division of Neonatology, University of Miami, P.O. Box 016960, Miami, FL 33101

S. Basler, Institut für Biomedizinische Technik der Universität Zürich, Moussonstrasse 18, CH-8044 Zürich

K.H.P. Bentele, Universitäts-Kinderklinik, Martinstr. 52, D-2000 Hamburg 20

A. Bollinger, Departement für Innere Medizin, Angiologie, Universitätsspital, CH-8091 Zürich

E. Bossi, Medizinische Universitäts-Kinderklinik und Poliklinik, CH-3010 Bern

M. Brülisauer, Departement für Innere Medizin, Angiologie, Universitätsspital, CH-8091 Zürich

F. Caligara, Europäisches Institut für Transurane, Postfach 2266, D-7500 Karlsruhe

L. Caspary, Abteilung Angiologie, Medizinische Hochschule Hannover, Postfach 610180, D-3000 Hannover 61

U. Ewald, Department of Pediatrics, University Hospital, S-75 185 Uppsala

F. Fallenstein, Perinatalphysiologisches Labor, Departement für Frauenheilkunde, Klinik und Poliklinik für Geburtshilfe, Frauenklinikstr. 10, CH-8091 Zürich

S. Fanconi, Intensive Care Unit, University Children's Hospital, CH-8032 Zurich

B. Friis-Hansen, The Neonatal Department, Rigshospitalet, Blegdamsvej 9, DK-2100 Kopenhagen

P.M. Gaylarde, Department of Dermatology, Royal Free Hospital & School of Medicine, GB-London NW3 2QG

W.B. Geven, Department of Pediatrics, St. Radboud Hospital, Geert Groteplein Zuid 20

I.H. Gøthgen, Department of Anesthesia, Gentofte Hospital and Rigshospitalet, DK-2100 Copenhagen

B.J. Gray, King's College London, Thoracic Medicine, Denmark Hill, GB-London SE5 8RX

W.W. Hay Jr., University of Colorado Health Sciences Center, Department of Pediatrics B-195, 4200 E. 9th Avenue, Denver, CO 80262

A. Hodgson, Department of Pediatrics, University of Vermont, College of Medicine, Burlington, VT 05405

U. Hölscher, Medizintechnik, Drägerwerk AG, Postfach 1339, D-2400 Lübeck 1

A. Huch, Departement für Frauenheilkunde, Klinik und Poliklinik für Geburtshilfe, Frauenklinikstrasse 10, CH-8091 Zürich

R. Huch, Departement für Frauenheilkunde, Klinik und Poliklinik für Geburtshilfe, Frauenklinikstrasse 10, CH-8091 Zürich

D.C.S. Hutchison, Department of Thoracic Medicine, Kings's College School of Medicine, London SE5

A. Jensen, Universitäts-Frauenklinik, Geburtshilfe und Gynäkologie, Klinikstrasse 32, D-6300 Giessen

H.P. Keller, Chirurgische Universitätsklinik, D-8700 Würzburg

E. Konecny, Forschung und Entwicklung, Drägerwerk AG, Postfach 1339, D-2400 Lübeck 1

H.N. Lafeber, University Hospital Rotterdam, Department of Neonatology, P.O. Box 70029, NL-3000 LL Rotterdam

C. Lanigan, Department of Anaesthetics, King's College Hospital Denmark Hill, GB-London SE5, 8RX

R. Lemke, Städtische Krankenanstalten, Medizinische Klinik, Beurhausstr. 40, D-4600 Dortmund 1

C.C. Liu, Department of Obstetrics & Gynecology, Cleveland Metropolitan General Hospital, 3395 Scranton Road Cleveland, OH 44109

D.W. Lübbers, Max-Planck-Institut, Systemphysiologie, Rheinlanddamm 201a, D-4600 Dortmund

J.F. Lucey, Department of Pediatrics, University of Vermont, College of Medicine, Burlington, VT 05405

U. v. Mandach, Perinatalphysiologisches Labor, Departement für Frauenheilkunde, Klinik und Poliklinik für Geburtshilfe, Frauenklinikstr. 10, CH-8091 Zürich

R.J. Martin, Department of Pediatrics, University Hospital of Cleveland, Childrens' Hospital, 2101 Adelbert Road, Cleveland OH 44106

J. Messer, Hopital de Hautepierre, Service de Pédiatrie II, Avenue Molière, F-Strassbourg Cedex

M.R. Neuman, Department of Obstetrics & Gynecology, Cleveland Metropolitan General Hospital, 3395 Scranton Road, Cleveland, Ohio 44109

C. Nickelsen, Department of Obstetrics & Gynecology, Rigshospitalet, University of Copenhagen, Blegdamsvej 9, DK-2100 Copenhagen

A. Ott, Hautklinik, Klinikum Charlottenburg (FB 3), Augustenburger Platz 1, D-1000 Berlin 65

J.L. Peabody, Department of Pediatrics, Division of Neonatology, Loma Linda University, Loma Linda, CA 92350

P. Rolfe, Department of Pediatrics, John Radcliffe Hospital, University of Oxford, GB-Headington, Oxford, OX3 9DU

G. Rooth, Oefre Slottsgatan 14 C, S-75235 Uppsala

H. Saner, Internistische Angiologie, Medizinische Universitätsklinik, Inselspital, CH-3100 Bern

H. Schachinger, Freie Universität Berlin, Kinderklinik und Poliklinik Heubnerweg 6, D-1000 Berlin 19

M.E. Schläfke, Abteilung für Angewandte Physiologie, Medizinische Fakultät der Ruhr-Universität Bochum, D-4630 Bochum 1

H. Seifert, Departement für Innere Medizin, Angiologie, Universitätsspital, CH-8091 Zürich

J.W. Severinghaus, Anesthesia Research Center, 1386 HSE, University of California, Medical Center, San Francisco, CA 94143

J.M. Steinacker, Universität Ulm, Sportmedizinische Untersuchungsstelle, Postfach 4066, D-7900 Ulm

A. Talbot-Pedersen, Department of Biomedical Engineering, Case Western Reserve University, Cleveland, OH 44106

J.E. Tooke, University of London, Department of Physiology, Fulham Palace Road, GB-London W6 8RF

P. Tuchschmid, Kinderspital, Neonatologie, Steinwiesstr. 75, CH-8032 Zurich

N. Weindorf, Universitäts-Hautklinik, St.-Josef-Hospital, Gudrunstrasse 56, D-4630 Bochum 1

A.R. van der Wiel, Sophia Childrens' Hospital, Department of Neonatology, Gordelweg 160, NL-3038 GE Rotterdam

I. Yamanouchi, Children's Hospital Center, Okayama National Hospital, Minamigata 2-13-1, Okayama 700,

# Index

Acid-base, 123
  balance, 128
Adjustable heating area, 299
Adult subjects, 41, 55, 61, 67, 71, 75, 79, 83, 188, 191, 215, 220, 227, 232, 235, 244, 256, 259, 263, 270, 286
Aerobic-anaerobic threshold, 61
Aerosol, 71
Airway caliber, 71
Allergic vasculitis, 81
Alternating heating, 285
Alveolar arterial oxygen difference, 42, 61, 72
American Society for Testing and Materials, 278
Anesthesia, 3
Apnea, 89, 90
Artefact elimination, 181
Arterial carbon dioxide tension - _see_ Arterial $Pco_2$
Arterial ischemia, 227
Arterial occlusion, 27, 239, 270
Arterial occlusive disease, 227, 235
Arterial oxygen saturation - _see_ Oxygen saturation
Arterial oxygen tension - _see_ Arterial $Po_2$
Arterial $Pco_2$, 51, 72, 95, 101, 106, 115, 129
Arterial pH, 106, 129
Arterial $Po_2$, 55, 67, 71, 75, 95, 102, 110, 115, 129, 146, 155, 160, 165, 172, 181, 259, 263, 269
Arterial $So_2$ - _see_ Oxygen saturation
Arterial $tcPo_2$ index, 259
Arterio-alveolar gradient, 317
Artificial ventilation, 96, 105, 165, 191
Assisted ventilation, 36
Athletes, 61
Autonomic respiratory control, 89
Autonomous nervous system, 85

Backscatter oximetry, 137
Base deficit, 30
Beer-Lambert law, 136, 196
Blood flow
  adrenal, 129
  arterial, 263
  brainstem, 127
  carcass, 130
  cerebellum, 129
  cerebrum, 129
  cutaneous, 263
  intestines, 127, 130
  kidney, 127
  muscles, 130
  myocardial, 129
  skin - _see_ Skin blood flow
  spleen, 127
  tissue, 223
  uterine, 128
Blood pressure, 128, 138, 146, 259
  arterial waveform, 211, 313
  continuous non-invasive beat-by-beat, 311
  mean arterial, 312
Bronchoconstrictors, 71
Bronchopulmonary dysplasia, 46, 181, 309
Buffering of tissues, 30

Calibration (_see_ in vivo calibration and _in vitro_ calibration)
Calibration curves, 14
Capillary blood pressure, 209
Capillary blood volume, 138
Capillary diameters, 219
Capillary dome, 9
Capillary flow rate, 209
Capillary loop, 9, 209, 221, 269
Capillary resistance, 210
Carboxyhemoglobin, 3, 177, 195, 201
Cardiopulmonary disease, 51
Cardiopulmonary illness, 101
Catecholamines, 128

Catheter
  arterial, 36, 51, 68, 76, 115, 146, 152, 165, 172, 178
  radial, 51, 159, 172
  tibial, 51
Chemosensitive afferents, 99
Chronic hypoxemia, 181
Chronic preparation, 127
Cicatricial changes, 105, 111
Circulatory hyperbola, 11, 263
Claudication, 236
Cold exposure, 215
Compression therapy, 244
Computer recording, 181, 305
Congenital cyanotic heart disease, 181
Contact dermatitis, 79
CO-oximeter, 177, 201
Correction factors, 36, 51, 103, 124
Countercurrent diffusion, 241
Countercurrent $O_2$ flow, 254
Critical care, 3
Cutaneous $Pco_2$, 23
Cutaneous $Po_2$, 23
Cyanotic cardiac malformation, 172

Deoxyhemoglobin, 135, 187, 201
Diabetes, 25, 83, 213, 244
Disorders of adaption, 101
Doppler blood flow measurement, 223, 260, 263
Dynamic computer model, 269

ECG, 96
ECG electrode, 291
EEG, 96
EOG, 96
Ear oximetry, 4, 72, 138
Electrode differences, 19, 259 285, 315
Electrode drift, 41, 48, 76, 124, 289
Electrodes, combined
  alternating $tcPo_2$ and $tcPo_2$, 285
  ECG and $tcPo_2$, 103, 291
  pH and $Po_2$, 295
  $tcPo_2$ and heating ring, 299
  $tcPo_2$ and Laser Doppler, 231
  $tcPo_2$ and $tcPco_2$, 41, 45, 101, 115, 278, 292, 315
Empiric algorithms, 4
End expiratory - see End-tidal
End-tidal $Pco_2$, 41, 72, 95, 115, 315
End-tidal $Po_2$, 41, 62, 72, 315
Epidermis, 9, 269
Epinephrine, 129
Ergometer, 68
Erythrocyte column, 219
Excessive hyperemia, 11
Exchange transfusion, 152
Exercise, 27, 61, 67, 244, 256
Extinction coefficients, 152, 195

False alarms, 163, 167
Faults
  frequency, 279
Fetal acidosis, 123
Fetal hemoglobin - see Hemoglobin
Fetal sheep, 127
Fick's law, 242
Flow stop, 215
Fourier transformation, 223
Funduscopy, 105

Gold electrodes, 295

Halogen light, 219
Heart rate, 103, 146, 291, 308
Heat loss, 249
Heat production, 249
Hemoglobin
  Adult, 151, 195, 202
  Fetal, 145, 151, 160, 165, 178, 195, 201
Hemoglobin concentration, 12
Hemoglobin oxygen saturation - see Oxygen saturation
Histamine acid phosphate, 71
Hyaline membrane disease, 171
Hyperbaric conditions, 282
Hypercapnia, 103, 116
Hypercapnic gas mixture, 315
Hypercarbia - see Hypercapnia
Hyperoxemia, 79, 116, 145, 165, 171, 174, 179, 181, 191
Hyperoxia, 6, 105, 150, 158, 175, 267
Hypertension, 209
Hypocapnia, 103, 116
Hypocarbia - see Hypocapnia
Hypoglossal nerve, 96
Hypoxemia, 73, 89, 95, 115, 145, 159, 171, 174, 181, 192
Hypoxia, 73, 158
  tissue, 231
Hypoxic gas mixture, 315

Indirect calorimetry, 249
Indocyanine Green, 219
Indomethacin, 25
Induction plethysmography
  abdominal, 96
  thoracic, 96
Infants of smoking mothers, 25, 201
Infrared fluorescence videomicroscopy, 219,
Initial slope index, 259
Intensive care
  adult,55, 75

neonatal, 45, 101, 159, 308
pediatric, 159
Intermittent claudication, 228
Internal scattering, 4
International Electrotechnical Commission, 277
International Organization for Standardization, 277
Intrapartum fetal pulse oximeter, 142,
In vitro calibration, 75
In vivo calibration, 67, 72, 75
In vivo response time, 191

Ketoacidosis, 27

Labor, 123
Lactate, 29, 61, 128
Laser Doppler, 227, 231, 235
Laser transmission of fetal head, 143
Leg dependency, 237
Leg ulcers, 244
Leukotriene D4, 71

Mass spectrometer, 41, 72, 317
Maternal hyperventilation, 125
Mathematical model, 9, 55, 195, 253, 263, 269
Mechanical ventilation, 109
Membrane diffusion, 242
Membrane properties, 14, 259
Metabolic acidosis, 25
Metabolism
aerobic, 61
anaerobic, 61
Methemoglobin, 3, 177, 195, 201
Microangiopathy, 25, 83, 209
Microcirculation, 9, 125, 215, 253
Microcomputer control system, 286
Microinjection, 209
Micromanipulator, 209
Microsurgery, 188
Microvascular flux, 227
Microvascular tissue transfer, 188
Motion artefacts, 140, 145
Multichannel recording, 305

Nailfold capillaries, 209, 215, 220
Newborn infants, 10, 36, 45, 51, 115, 121, 145, 151, 159, 165, 171, 177, 181, 191, 249, 311
Nicotinic acid, 83
Norepinephrine, 129

Ondine's Curse Syndrome, 95
Ophthalmoscopy, 110
Optical density, 137
Orthostatis, 228
Overhydration, 106

Oximetry, 3, 135, 195, 201 (see also Pulse Oximetry)
Oxygen conductivity, 270
Oxygen consumption, 9, 43, 253, 263, 270
Oxygen content, 11, 55
Oxygen delivery, 10
Oxygen diffusion, 9
Oxygen diffusion coefficient, 43, 254
Oxygen dissociation curve, 9, 72, 151, 255
Oxygen permeability, 43, 269
Oxygen saturation, 3, 55, 96, 129, 135, 145, 151, 159, 165, 171, 177, 181, 187, 195, 201, 305
Oxygen sensitivity, 298
Oxygen solubility, 254
Oxygen therapy, 105, 109, 181
Oxygen transport, 9, 151, 269
Oxygen treatment, 171
Oxyhemoglobin, 177, 187, 195, 201

Palladium-palladium oxide pH-electrode, 296
Perfusion pressure, 263
Periodic breathing, 89
Peripheral blood flow, 249
Peripheral circulation, 259
Peripheral resistance, 209
Pharmacological stimulation test, 85
Phenoxybenzamine, 245
Photodermatosis, 80
Photolithography, 295
Photometric reflection oximetry, 187
Photoplethysmographic pulse, 139
Photoplethysmography, 311
Photo-diodes, 5
Physiological data-personal computer storage, 305
Plasma oncotic pressure, 210
Postischemic hyperemia, 24
Postocclusive reactive hyperemia, 231
Predictive value, 171
Pregnancy, 123
Premature infants, 6, 45, 89, 105, 109, 145, 151, 165, 171, 181, 249
Pre-heating, 35
Pt wire diameter, 259
Pulmonary emphysema, 67
Pulsatile light, 3
Pulse oximetry, 3, 57, 72, 96, 135, 145, 151, 159, 165, 171, 177, 181, 191, 195, 201, 308
principle, 5, 135
Pulse rate, 139

Quality control, 277

Raynaud's syndrome, 213, 215, 244
Reactive hyperemia, 25, 83, 226, 228, 231, 235
Receiver-operating characteristic curves, 166
Reliability, 277
Replanted extremities, 187
Respiratory $Co_2$ response, 96
Respiratory distress syndrome, 121, 182
Respiratory efforts
  abdominal, 90
  thoracic, 90
Respiratory rate, 211
Response time, 5, 242, 288
Reticular formation, 99
Retinopathy of prematurity, 105, 109, 145, 165
Reversible visual disorder, 215
Rhodium-rhodium oxide electrode, 295
Rotating interference filters, 187

Safety control, 277
Second degree burn, 35, 145, 299
Second stage of labor, 126
Sensitivity, 160, 167, 171
Sensors - (see also Electrodes)
  microelectronic, 295
Servonulling technique, 209
Skin anatomy, 10, 269
Skin blood flow, 9, 23, 73, 128, 152, 215, 223, 227, 231, 235, 241, 249, 253, 259, 263, 269
Skin hyperemia, 286
Skin leasion, 286
Skin pigmentation, 145
Skin thickness, 101, 242
Skin respiration, 241
Sleep hypoxemia 95
Sleep states, 90
Sleep syndromes, 95
Smoking mothers, 25, 201
Specificity, 160, 167
Spectrophotometry, 135
Spectroscopic multicomponent analysis, 198
Standardization of electrodes, 261
Stripping, 83, 243, 272
Subepidermal venous plexus, 9
Suction application of electrode, 123
Sudden Infant Death Syndrome, 95
Sulfhemoglobin, 201
Surfactant therapy, 121

Sympathetic tone, 212

Temperature coefficient, 35
Temperature control, 286
Temperature correction, 51
Tissue transfer, 187
Tolazoline, 116
Transcutaneous carbon dioxide tension, see Transcutaneous $Pco_2$
Transcutaneous dual electrodes - see Electrodes, combined
Transcutaneous index, 263
Transcutaneous measurements
  errors, 20
Transcutaneous oxygen tension - see Transcutaneous $Po_2$
Transcutaneous oxygen tension at 37°C, 23, 79, 83, 231, 235, 241, 253, 302
Transcutaneous $Pao_2$-indicator of lactate, 61
Transcutaneous $Pco_2$, 9, 23, 35, 41, 45, 51, 61, 71, 89, 96, 101, 115, 121, 123, 253, 291, 305, 315
Transcutaneous $Po_2$, 9, 23, 35, 41, 45, 55, 61, 67, 71, 75, 79, 83, 89, 95, 101, 105, 109, 115, 121, 145, 152, 159, 165, 171, 181, 191, 231, 235, 241, 253, 259, 263, 269, 285, 291, 299, 305
Transcutaneous technique
  theoretical analysis, 9
Transmission oximetry, 135
Treadmill, 62
TV-microscopy, 215

Umbilical vascular resistance, 129
Unspecific respiratory stimulus, 95
UV-Erythema, 79

Vasoactive drugs, 209
Vasoconstrictor tonus, 85
Vasodilator response, 245
Vasodilators, 73, 244
Vasomotor phenomena, 25, 232
Vasospastic syndrome, 215
Venous occlusion, 235
Ventilation/perfusion mismatching, 71
Ventral medullary surface, 96
Videodensitometric analysis, 220
Videomicroscopy, 219
Vital capacity, 72